Chemo-Biological Systems for CO_2 Utilization

Chemo-Biological Systems for CO_2 Utilization

Edited by

Ashok Kumar and Swati Sharma

CRC Press is an imprint of the
Taylor & Francis Group, an **informa** business

First edition published 2021
by CRC Press
6000 Broken Sound Parkway NW, Suite 300, Boca Raton, FL 33487-2742

and by CRC Press
2 Park Square, Milton Park, Abingdon, Oxon, OX14 4RN

Library of Congress Cataloging-in-Publication Data
Names: Kumar, Ashok, 1949– editor. | Sharma, Swati, editor.
Title: Chemo-biological systems for CO_2 utilization / edited by Ashok Kumar and Swati Sharma.
Description: First edition. | Boca Raton, FL : CRC Press, 2020. | Includes index.
Identifiers: LCCN 2020012058 (print) | LCCN 2020012059 (ebook) | ISBN 9780367321932 (hardback) | ISBN 9780429317187 (ebook)
Subjects: LCSH: Carbon dioxide—Industrial applications. | Carbon dioxide mitigation.
Classification: LCC TP244.C1 C44 2020 (print) | LCC TP244.C1 (ebook) | DDC 665.8/9—dc23
LC record available at https://lccn.loc.gov/2020012058
LC ebook record available at https://lccn.loc.gov/2020012059

ISBN: 978-0-367-32193-2 (hbk)
ISBN: 978-0-429-31718-7 (ebk)

Typeset in Times
by codeMantra

Contents

Preface

This book will describe the various advanced tools and techniques developed in the past decade for carbon dioxide capture and its utilization. In the ecosystem, CO_2 is primarily absorbed by plants, oceans, algae, and soil contents. But in the 21st century, increased industrialization and urbanization result in a tremendous production of CO_2, which is considered as one of the challenging factors of global warming and a dreadful pollutant that affects the human health. Researchers attempted to develop various techniques using chemical, microbiological, enzymatic, and biomolecular systems in order to absorb the increased concentration of CO_2 in the environment and produce some value-added compounds using CO_2. In this context, various biopolymers, nanomaterials, bioinspired surfaces, polysaccharides, organic solvents, chemicals, enzymes, and metal–organic frameworks (MOFs) were developed and utilized for the sequestration of CO_2 into various carbonates. The biomolecular system has its own importance in the conversion of the atmospheric CO_2 into carbonates. Carbonic anhydrase is one of the fastest enzymes and most studied for the conversion of CO_2. Microflora on the Earth present various classes of methanogens, thermophilic bacteria, phosphate-solubilizing bacteria, and acetogens, which play important roles in CO_2 absorption. Amine-based chemical compounds, porphyrins, ionic liquids, eutectic solvents, ceramics, biochar, and organic solvents have also great potential to convert CO_2 into other fine chemicals or useful products. The major advantages such as stability and efficiency of the systems will be discussed in the various sections of this book.

Nowadays, climate change is considered as one of the major issues, and the massive generation of greenhouse gases from automobiles, industries, and carbonaceous fuels have tremendously promoted the alteration in temperature from its normal cycle. This book will emphasize on the energy generation in the form of biofuels, bioelectricity, or biogas from CO_2 using chemicals; nanomaterials; and microbial, enzymatic, and chemo-enzymatic-integrated systems. This book has been divided into four sections. The chapters 1–4 described the importance and utilization of CO_2 in the living system, and various fundamental methods, policies, and techniques involved in CO_2 conversion. The chapters 5–8 focused on the adsorption and fixation of CO_2 using various ionic liquids, organic solvents, amine-based solutions, electrocatalytic reduction, nanomaterials, porphyrins, ceramics, MOFs, and activated carbons (in particular, biological materials). This section draws the insight of various chemical engineers and researchers working across the globe for CO_2 conversion. The chapters 9 and 10 give the emphasis on the production of value-added products using CO_2 and will mainly focus on the production of biomethanol, industrial carbonates, lime, liquid and gaseous fuels, industrially useful precursors, etc. The chapters 11–13 discuss the potential of the microbial system, enzymes involved in the sequestration of CO_2, and CO_2 utilization. The chemo-enzymatic system developed for the utilization of CO_2 will be discussed in the respective chapters of this book. This reliable information from various active researchers and groups will be helpful to find out the alternative methods for clean energy and mitigating the climate change. This book will help the researchers and industrialists to better understand the correlation between microbial, biological, and chemical products and their roles in the conversion of CO_2 into useable energy and related products.

Editors

Dr. Ashok Kumar is working as an assistant professor in the Department of Biotechnology and Bioinformatics, Jaypee University of Information Technology, Waknaghat, India. He holds an extensive 'research and reaching' experience of more than 10 years in the field of microbial biotechnology, with research expertise focusing on various issues pertaining to 'nanobiocatalysis, biomass, and bioenergy', and 'climate change'. He holds International work experiences in South Korea, India, Malaysia, and Republic of China. He worked as a post-doctoral fellow in the State Key Laboratory of Agricultural Microbiology, Huazhong Agricultural University, Wuhan, China. He also worked as a brain pool researcher at Konkuk University, Seoul, South Korea. Dr. Ashok has a keen interest in microbial enzymes, biocatalysis, CO_2 conversion, climate change issues, nanobiotechnology, waste management, biomass degradation, biofuel synthesis, and bioremediation. His work has been published in various internationally reputed journals, namely *Chemical Engineering Journal*, *Bioresource Technology*, *Scientific Reports*, *Energy*, *International Journal of Biological Macromolecules*, *Science of Total Environment*, and *Journal of Cleaner Production*. He has published 60 research articles, 25 book chapters, and 5 books. He is also a member of the editorial board and reviewer committee of the various journal of international repute.

Dr. Swati Sharma is working as an assistant professor at the University Institute of Biotechnology, Chandigarh University, Mohali, India. She has completed her Ph.D. at the Universiti Malaysia Pahang, Malaysia. She worked as a visiting researcher in the college of life and environmental sciences at Konkuk University, Seoul, South Korea. She has completed her master's degree (M.Sc.) from Dr. Yashwant Singh Parmar University of Horticulture and Forestry, Nauni, India. She has also worked as a program co-coordinator at the Himalayan Action Research Centre, Dehradun, India, and senior research fellow at India Agricultural Research Institute, Pusa, New Delhi, in 2013–2014.

She has published her research papers in reputed international journals. Presently, her research is in the field of bioplastics, hydrogels, nano-fibres and nano-particles, biodegradable polymers and polymers with antioxidant and anti-cancerous activities, and sponges. Dr. Swati has published 20 research papers in various internationally reputed journals, four books (with Springer and Taylor & Francis), and a couple of book chapters.

Contributors

Hamidah Abdullah
Faculty of Chemical and Process Engineering Technology, College of Engineering Technology
Universiti Malaysia Pahang
Gambang, Malaysia

Sumaiya Zainal Abidin
Faculty of Chemical and Process Engineering Technology, College of Engineering Technology
Universiti Malaysia Pahang
Gambang, Malaysia
and
Centre of Excellence for Advanced Research in Fluid Flow (CARIFF)
Universiti Malaysia Pahang
Gambang, Malaysia

Komal Agrawal
Bioprocess and Bioenergy Laboratory, Department of Microbiology
Central University of Rajasthan
Kishangarh, India

Yumlembam Rupert Anand
College of Agricultural Engineering and Post-Harvest Technology
Central Agricultural University
Gangtok, India

Zenab T. Baig
School of Environment
Tsinghua University
Beijing, China
and
Department of Environmental Sciences
University of Haripur
Haripur, Pakistan

Priya Banerjee
Environmental Studies, DDE
Rabindra Bharati University
Kolkata, India

Anirban Biswas
Department of Chemical Engineering
Jadavpur University
Kolkata, India

Clemente Capasso
Department of Biology, Agriculture and Food Sciences
Institute of Biosciences and Bioresources - CNR
Napoli, Italy

Chi Cheng Chong
Faculty of Chemical and Process Engineering Technology, College of Engineering Technology
Universiti Malaysia Pahang
Gambang, Malaysia

Chin Sim Yee
Faculty of Chemical and Process Engineering Technology, College of Engineering Technology
Universiti Malaysia Pahang
Gambang, Malaysia
and
Centre of Excellence for Advanced Research in Fluid Flow (CARIFF)
Universiti Malaysia Pahang
Gambang, Malaysia

Papita Das
Department of Chemical Engineering
Jadavpur University
Kolkata, India

Avirup Datta
Department of Environmental Science
University of Calcutta
Kolkata, India

Hanglem Sonibala Devi
School of Agriculture, School of Horticulture, Department of Horticulture and Department of Agricultural Microbiology
Pandit Deen Dayal Upadhyay Institute of Agricultural Sciences
Bishnupur, India

Anand Giri
Department of Environmental Sciences
Central University of Himachal Pradesh
Kangra, India

Rahul Kumar Goswami
Bioprocess and Bioenergy Laboratory, Department of Microbiology
Central University of Rajasthan
Kishangarh, India

S. Gurumurthy
Division of Basic Science
Indian Institute of Pulses Research, ICAR
Kanpur, India

Mahima Khandelwal
Department of Materials Science and Engineering
Korea University
Seoul, Republic of Korea

Ashok Kumar
Department of Biotechnology and Bioinformatics
Jaypee University of Information Technology
Waknaghat, India

Suvendu Manna
Department of Health Safety and Environment, School of Engineering
University of Petroleum and Energy Studies
Dehradun, India

Sanjeet Mehariya
Bioprocess and Bioenergy Laboratory, Department of Microbiology
Central University of Rajasthan
Kishangarh, India
and
ENEA, Italian National Agency for New Technologies, Energy and Sustainable Economic Development, Department of Sustainability - CR Portici
Portici, Italy
and
Department of Engineering
University of Campania "L.Vanvitelli"
Aversa, Italy

Tracila Meinam
School of Agriculture, School of Horticulture, Department of Horticulture and Department of Agricultural Microbiology
Pandit Deen Dayal Upadhyay Institute of Agricultural Sciences
Bishnupur, India

Koijam Melanglen
School of Agriculture, School of Horticulture, Department of Horticulture and Department of Agricultural Microbiology
Pandit Deen Dayal Upadhyay Institute of Agricultural Sciences
Bishnupur, India

Antonio Molino
ENEA, Italian National Agency for New Technologies, Energy and Sustainable Economic Development, Department of Sustainability - CR Portici
Portici, Italy

Aniruddha Mukhopadhyay
Department of Environmental Science
University of Calcutta
Kolkata, India

Dino Musmarra
Department of Engineering
University of Campania "L.Vanvitelli"
Aversa, Italy

Khumukcham Nongalleima
Branch Laboratory Imphal (BLIM)
CSIR-North East Institute of Science and Technology
Jorhat, India

Osarieme Uyi Osazuwa
Faculty of Chemical and Process Engineering Technology, College of Engineering Technology
Universiti Malaysia Pahang
Gambang, Malaysia

Deepak Pant
School of Chemical Sciences
Central University of Haryana
Mahendragarh, India

Maryam Takht Ravanchi
Catalysis Research Group, Petrochemical Research and Technology Company
National Petrochemical Company
Tehran, Iran

Uttariya Roy
Department of Chemical Engineering
Jadavpur University
Kolkata, India

Saeed Sahebdelfar
Catalysis Research Group, Petrochemical Research and Technology Company
National Petrochemical Company
Tehran, Iran

Jinus S. Senjam
College of Horticulture, FEEDS Group of Institutions
Kangpokpi, India

Herma Dina Setiabudi
Faculty of Chemical and Process Engineering Technology, College of Engineering Technology
Universiti Malaysia Pahang
Gambang, Malaysia
and
Centre of Excellence for Advanced Research in Fluid Flow (CARIFF)
Universiti Malaysia Pahang
Gambang, Malaysia

Abhishek Sharma
Department of Biotechnology
Himachal Pradesh University
Shimla, India

Swati Sharma
University Institute of Biotechnology
Chandigarh University
Mohali, India

Tanvi Sharma
Department of Biotechnology and Bioinformatics
Jaypee University of Information Technology
Waknaghat, India

Veerbala Sharma
Department of Environmental Sciences
Central University of Himachal Pradesh
Kangra, India

Tan Ji Siang
School of Chemical and Energy Engineering, Faculty of Engineering
Universiti Teknologi Malaysia
Johor Bahru, Malaysia

Thiyam Jefferson Singh
Department of Silviculture and Agroforestry, College of Horticulture and Forestry
Central Agricultural University
Imphal, India

Thoudam Santosh Singh
School of Agriculture, School of Horticulture, Department of Horticulture and Department of Agricultural Microbiology
Pandit Deen Dayal Upadhyay Institute of Agricultural Sciences
Bishnupur, India

Mansooreh Soleimani
Department of Chemical Engineering
Amirkabir University of Technology (Tehran Polytechnic)
Tehran, Iran

Abdul F. Soomro
School of Environment
Tsinghua University
Beijing, China

Claudiu Supuran
Department of NUROFARBA
University of Florence
Florence, Italy

Shabnam Thakur
Department of Environmental Sciences, Central University of Himachal Pradesh
Kangra, India

Ramya Thangamani
Department of Environmental Sciences
Bharathiar University
Coimbatore, India

Dhanaraj Singh Thokchom
Ethno-Medicinal Research Centre, Foundation for Environment and Economic Development Services
Kangpokpi, India

Kangjam Tilotama
Ethno-Medicinal Research Centre, Foundation for Environment and Economic Development Services
Kangpokpi, India

Sunita Varjani
Gujarat Pollution Control Board
Gandhinagar, India

Pradeep Verma
Bioprocess and Bioenergy Laboratory, Department of Microbiology
Central University of Rajasthan
Kishangarh, India

Lakshmanaperumal Vidhya
Department of Chemical Engineering
Sethu Institute of Technology
Pulloor, India

Dai-Viet N. Vo
Center of Excellence for Green Energy and Environmental Nanomaterials (CE@GrEEN)
Nguyen Tat Thanh University
Ho Chi Minh City, Vietnam

Arun K. Vuppaladadiyam
School of Civil and Transportation Engineering
Shenzhen University
Shenzhen, Republic of China

Varsha S. S. Vuppaladadiyam
Department of Civil Engineering, AVN Institute of Science and Technology
Hyderabad, India

1 Recent Developments in CO_2-Capture and Conversion Technologies

Tanvi Sharma, Abhishek Sharma, Swati Sharma, Anand Giri, Ashok Kumar, and Deepak Pant

CONTENTS

1.1 INTRODUCTION

Carbon dioxide (CO_2) is one of the greenhouse gases causing a threat to the environment at an increased level. The development of technologies for converting CO_2 into value-added products can slow down global warming and reduce the energy crisis too. Thus, the excess amount of CO_2 can be reduced by converting it into industrial products, which are useful for the industries and mankind. Therefore, CO_2 can be considered as a raw substrate for the synthesis of various commercially important products. CO_2 is emitted from different sources such as cement production units; automobiles; and burning of fossil fuels such as oils, coal, and natural gases (Ashley et al. 2012; Billig et al. 2019). An effective method to reduce the CO_2 level is the primary necessity of time. To date, various methods, including enzymatic, chemical, photochemical, and electrochemical systems, have been exploited for CO_2 capture on laboratory-scale and large-scale applications (Figure 1.1) (Wang et al. 2017).

In the literature, various examples have already been given for CO_2 capture using living organisms and synthetic materials. Using plants and algae to reduce CO_2 is a sustainable and green approach to decrease global carbon footprint. Many plants, such as *Pinus radiata* and *Malus domestica*, and algal species, such as *Chlorella vulgaris* and *Scenedesmus quadricauda*, have been widely studied for their remarkable CO_2-capture capacities (Wu et al. 2012; Pavlik et al. 2017). Among the living organisms, bacteria also play a vital role in alleviating the CO_2 level by using their enzymes such as carbonic anhydrase (CA), formate dehydrogenase, and decarboxylase (Chen 2019). Particularly, CO_2 capture using plants, algae, and microbes may be proven as an eco-friendly approach. Nowadays, amine-based solvents and ionic liquids (ILs) have also been used to reduce the CO_2 from industries such as steel manufacturing and fossil fuel power plants. Yet, these solvents have some disadvantages such as high energy demand and low CO_2-capture efficiency (Vega et al. 2018). To overcome these problems, different nanoporous materials, such as metal–organic frameworks (MOFs), nanoparticles, nanotubes, nanosized zeolites, and activated carbon, are seen as a promising alternative with high CO_2-capture efficiency.

Conversion of CO_2 into an industrial product is required for greenhouse gas mitigation. Nowadays, extensive work is being carried out by researchers for the conversion of CO_2 into hydrocarbons such as methane, formic acid, and methanol. It is still a tedious task to convert CO_2 efficiently, economically, and selectively into products of commercial value (Chen and Mu 2019). In this chapter, we emphasize the development of techniques and advances in CO_2 capture and its conversion into valuable fuels and chemicals.

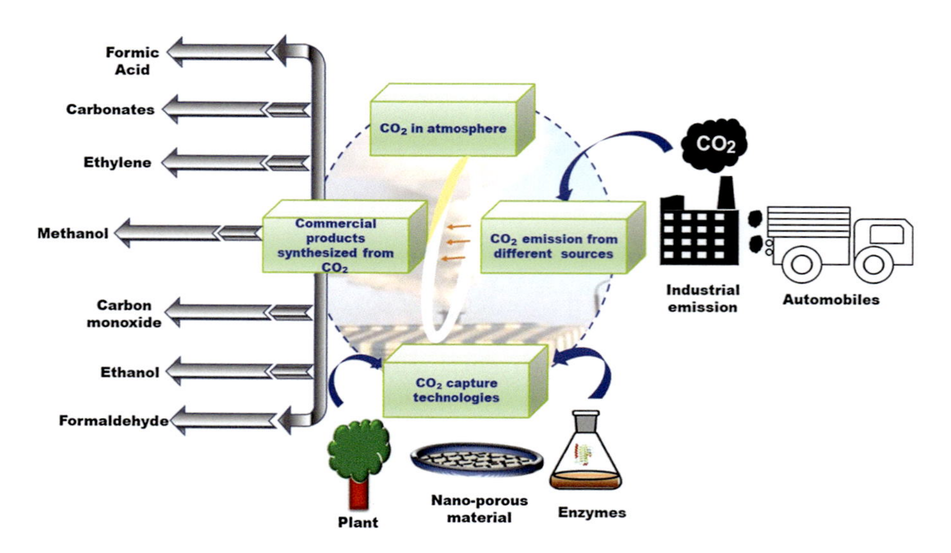

FIGURE 1.1 Schematic representation of various systems used for CO_2 capture and conversion.

1.2 CO_2 CAPTURE IN NATURE

Plant, algae, and microbial enzymes capture the CO_2 by carbon-concentrating mechanism and play a significant role in maintaining the CO_2 concentration in the atmosphere (Table 1.1). The conversion of atmospheric CO_2 by the plants and algae occurs by photosynthesis, and in this process, enzyme ribulose bisphosphate carboxylase/oxygenase (RuBisCo) plays a vital role. The CO_2 released from various industries can be used as a source of carbon for the growth of plants and algae (Mistry et al. 2019). But the RuBisCo has a low affinity for CO_2 over O_2, which showed a low CO_2 fixation rate. Many photosynthetic and autotropic bacteria have CA that helps in CO_2 fixation and its conversion. The major benefit of using microorganisms to convert CO_2 is that they have a natural capability to take up CO_2 through their metabolic pathways (Rittmann 2008). CO_2 sequestration using microorganisms not only provides a green approach to alleviate global warming, but also produces numerous value-added products.

1.2.1 CO_2 Capture by Plants

Different terrestrial ecosystems such as agricultural land, forest, and orchards play a vital role in CO_2 capture. For example, *P. radiata* sequesters about 300–500 tonnes of carbon dioxide per hectare. It was also reported that the CO_2 sequestration capacity of the Chinese forest was 41 tonnes of carbon per hectare (Wu et al. 2012; Mistry et al. 2019). The disadvantage of using trees like *M. domestica* for carbon management is that the CO_2 sequestration capability reduces with the age of the tree (Wu et al. 2012).

TABLE 1.1
CO_2 Capture in Nature Using Plants, Algae, and Microbial Enzymes

S. No	Natural System	Examples	Advantages	Disadvantages	Reference
1.	Plants	*P. radiate, Pinus sylvestris, Cynodon dactylon*	Orchards and agricultural crops can be used to fix CO_2, and they also provide food.	CO_2 fixation capability reduces with the age of the tree.	Luttge (2004)
2.	Algae	*C. vulgaris, S. quadricauda, Chlamydomonas reinhardtii, Nannochloris* sp.	Biofuel production such as biodiesel, biohydrogen, and bioethanol	Require light during growth and poor utility in industrial reaction.	Batista et al. (2015)
3.	Microbial enzymes	CA, formaldehyde dehydrogenase	Economic viability; the growth rate of microbes is faster than that of plants.	Low stability	Long et al. (2017); Kumar et al. (2019)

Various plants such as crassulacean acid metabolism (CAM) plants, C3 plants, and C4 plants use different carbon fixation pathways. In C3 plants, the product of photosynthesis is phosphoglyceric acid. Common examples of C3 plants are *Oryza sativa*, *Triticum* sp, *Hordeum vulgare*, *Pinus sylvestris*, and *Arachis hypogaea*. To decrease the photorespiration, CAM and C4 plants have photosynthetic adaptation, and the rate of CO_2 conversion is higher in these plants as compared to C3 plants (Grodzinski, Jiao, and Leonardos 1998). The C4 plants are the most productive plants, and some examples of these plants are *Sorghum bicolor*, *Zea mays*, *Bouteloua gracilis*, and *Saccharum officinarum*. CAM plants are mostly found in a dry environment, and to reduce photorespiration, these plants use the CAM pathway. Some examples of CAM plants are *Hoya carnosa*, *Ananas comosus*, *Tillandsia usneoides*, and *Crassula argentea* (Mistry et al. 2019; Nimmo 2000; Sage, Sage, and Kocacinar 2012).

Increasing the atmospheric CO_2 level results in enhanced photosynthesis, but after a certain level, the excessive concentration causes a decrease in the nutritional level of the plant (Raven and Karley 2006; Nogia et al. 2016). Thus, to overcome the negative effects of global warming, the researchers are trying to improve naturally occurring photosynthetic reactions by transferring genes from efficient photosynthetic systems, such as C4 plants, microalgae, and cyanobacteria, to inefficient photosynthetic systems, like C3 plants.

1.2.2 CO_2 Capture by Algae

Algae are one of the most widely researched organisms exploited for CO_2 fixation. The major advantages of algae, including biodegradability, non-toxicity, high CO_2 fixation efficiency, and production of value-added products, make it an ideal candidate for large-scale applications. The CO_2 emitted from various power plants can be used as a carbon source for algal production. Algae use the CO_2 as a carbon source and fix it into organic compounds using the Calvin–Benson pathway (Ullah et al. 2015). These organic compounds can be converted into numerous value-added cellular components such as proteins, lipids, vitamins, and carbohydrates that can be used as feedstock for animal feed, biofuels, and functional foods. Algal species such as *C. vulgaris*, *S. quadricauda*, *Nannochloropsis* sp., *Chlamydomonas reinhardtii*, and *Nannochloris* sp. have been studied to sequester CO_2 (Pavlik et al. 2017). The biofixation of CO_2 and biomass production depend on the characteristics of microalgal species, pH, nutrients, temperature, and CO_2 concentration. The microalgae like *Chlorella* sp., are capable of growing in pH of 5.5–6.0, at the temperature of 30°C, and at 40% CO_2 (v/v). The CO_2 concentration above 5% (v/v) is lethal for algal growth. But the flue gas from the coal-fired plants often includes 10%–15% CO_2 that was reported to use for the cultivation of algae. In flue gas-assisted algal growth, the CO_2 that escapes from microalgal solution due to the high rate of gas flow and low solubility becomes non-toxic to algal species (Zhao and Su 2014; Chen et al. 2014). To improve CO_2 fixation efficiency, screening and domestication of algal species will be major promising strategies.

The most proficient microalgal species such as *C. vulgaris*, *Scenedesmus obliquus*, *Nannochloropsis oculata*, and *Botryococcus braunii* are used for biofuel production due to their high lipid contents (Kamyab, Chelliapan, Nadda, et al. 2019;

Kamyab et al. 2018). Microalgae like *S. obliquus* is able to produce biohydrogen that is considered as the recyclable and non-pollutant. Meanwhile, *Dunaliella* and *Spirulina* have also been reported as microalgal species widely used for feed production or as functional foods due to their high nutritional value (Song et al. 2019; Kim et al. 2017; Kamyab, Chelliapan, Lee, et al. 2019; Kamyab et al. 2017).

1.2.3 CO_2 Capture Using Microbial Enzymes

The enzymatic conversion of CO_2 can be applied not only for efficient CO_2 utilization but also for capturing waste CO_2 from power plants' flue gas. Generally, enzyme immobilization or absorption technologies have been employed for capturing and long-term storage of CO_2. CA converts CO_2 into bicarbonate ions, which can be further converted to calcium carbonates in the presence of calcium ions. $CaCO_3$ is the product formed during CO_2 conversion that can be separated easily and used as a raw material for many industrial applications such as ceramics, cement, sugar refining, steel, iron, and glass production units (Sharma, Sharma, Kamyab, et al. 2019). In search of potent CA, various strategies such as the isolation of bacteria from different habitats, the use of protein engineering tools, and immobilization of enzyme on various matrixes to achieve high CO_2 conversion become essential (Sharma, Sharma, et al. 2018; Sharma, Sharma, Sharma, et al. 2019; Kumar et al. 2020; Kumar, Wu, et al. 2018; Thakur et al. 2018). Various bacterial genera having CA, such as *Serratia* sp., *Pseudomonas* sp., *Bacillus* sp., *Vibrio* sp., and *Lactobacillus* sp., have been studied for the conversion of CO_2 into $CaCO_3$ (Kumar, Sundaram, et al. 2018; Kumar et al. 2019).

Besides CA, several other types of decarboxylases have been explored for the conversion of CO_2 into eco-friendly chemicals. For example, 4-hydroxybenzoate decarboxylase purified from *Enterobacter cloacae*, *Chlamydophila*, *Pneumoniae* sp., and *Clostridium hydroxybenzoicum* catalyses the reversible carboxylation of phenol with CO_2 to yield 4-hydroxybenzoate (Shi et al. 2015). Another promising route for the reduction of CO_2 into methanol using a multienzymatic system such as formaldehyde dehydrogenase, formate dehydrogenase, and alcohol dehydrogenase has been gaining attention. Conversion of CO_2 into methanol is very beneficial as it is cheaper to produce, is less inflammable, and is advantageous in many industries (Aresta et al. 2014). Methanol is gaining popularity as an alternative to petroleum-based fuels and is beneficial for a safer and cleaner environment. Another product formed from the enzymatic conversion of CO_2 is formic acid, which can be used in textile finishing, paper production, animal feed additive, and chemical intermediates (Alvarez-Guerra, Quintanilla, and Irabien 2012; Rees and Compton 2011). Formate dehydrogenase from *Pseudomonas oxalaticus* is used in the reduction of CO_2 into formate using oxidized methyl viologen as an electron relay (Lu et al. 2006; Long et al. 2017). In this regard, the enzymatic conversion of CO_2 into various products using a free or immobilized biocatalyst seems to be a promising approach.

1.3 CO_2 CAPTURE BY SYNTHETIC MATERIALS

Various chemical solvents (e.g. diglycolamine (DGA), monoethanolamine (MEA), diethylenetriamine) and solid absorbents (nanoparticles, graphene oxide, MOFs) have been used to capture CO_2. The CO_2-capture efficiency of these solvents and solid absorbent materials is high. However, the implementation of CO_2-capture plant at industrial levels such as in iron and steel manufacturing, fossil fuel power plants, and cement production requires novel solvent and solid absorbent formulations with highly equipped machinery (Sharma, Sharma, Kamyab, et al. 2019). The nanoporous material and chemical solvents will be a key solution to enhance CO_2 capture.

1.3.1 CO_2 Capture by Nanoporous Materials

Nanotechnology can provide a viable material for the global reduction of CO_2. Nanoporous materials such as metal and metal oxide nanoparticles, nanotubes, nanosized zeolites, nanosheets, and MOFs have been used as adsorbents for CO_2 capture (Hedin, Chen, and Laaksonen 2010; Sharma, Verma, et al. 2018; Ghodake et al. 2019). Nanoporous materials are one of the best options for capturing CO_2 due to their large surface area, easy surface functionalization, and low cost (Yu et al. 2019). Due to the acidic nature of the CO_2 molecule, the large surface area of nanoporous materials is essential for CO_2-capture applications. Copper oxide nanoparticles show enhanced CO_2 absorption capacity as CO_2 molecules have the electron acceptor property and these nanoparticles have the electron donor property, which results in an efficient CO_2 capture (Kim, Cho, and Park 2010).

Nowadays, MOFs are gaining attention due to their three-dimensional structure, large pore volume, large surface area, and good potential for CO_2 capture. Copper porphyrin-based MOFs are emerging tools as they mimic photosynthesis and offer a method for designing a photocatalyst for CO_2 reduction and its capture (Liu et al. 2013). Various nanostructures have been studied for CO_2 capture; among them, nanotubes, nanofibers, and nanosheets have CO_2 adsorption capacities between 0.26 and 4.15 mmol g L^{-1} (Rodriguez Acevedo, Cortes, and Franco 2019; Wang et al. 2014). Recent advances in nanomaterials like chemically or physically introducing a nanoparticle into a base solvent can obtain liquid nano-absorbents having high CO_2-capture efficiency, improved solvent stability, and decreased solvent vapour pressure (Agarwal, Qi, and Archer 2010). These nanoporous structures have enhanced CO_2 absorption capacity as compared to conventional absorbents.

1.3.2 CO_2 Capture by Graphene Oxides

Recently, graphene, chemically modified graphene, and graphene oxide have been widely used in gas separation, desalination, organic separation, and water filtration processes. Due to their outstanding chemical stability, thermal stability, and mechanical strength, graphene-based materials have attracted the attention of researchers. As such, graphene-based materials are recognized for their potential in capturing CO_2 from the combustion of fossil fuels (Ali, Razzaq, and In 2019; Chowdhury and Balasubramanian 2016). It was found that graphene hydrogel prepared via the hydrothermal treatment of graphene oxide, having high surface area, three-dimensional structure, and large pore volume, showed high CO_2 adsorption capacity (Sui and Han 2015). In another study, it was reported that polyethylenimine loading on 3D graphene led to an increase in CO_2-capture capacity (Liu et al. 2015). Moreover, the nanoporous graphene (NPG) membrane has the ability to separate CO_2 from CO_2/CH_4, CO_2/O_2, and CO_2/N_2. The performance of the NPG membrane used for the separation of gaseous impurities is governed by its monoatomic thickness and high permeability for gas molecules (Lee and Aluru 2013). Although great progress has been achieved on the benefits of graphene-based membranes for CO_2 capture, there are still some challenges that need to be addressed, such as resistance against impurities (SO_2 and NO_2), temperature, pressure, and graphene production cost.

1.3.3 CO_2 by Chemicals

Among various technologies employed for CO_2 capture, chemical absorption using aqueous amine solutions such as DGA, methyl diethanolamine (MDEA), MEA, and pentamethyldiethylenetriamine (PMDETA) is one of the most widely used methods (Dutcher, Fan, and Russell 2015; Mazari et al. 2015). During CO_2 capture by an amine, one CO_2 molecule reacts with two amine molecules to form a stable carbamate ion and protonated amine, limiting the loading capacity to 0.5 mol CO_2/mol amine (Stowe and Hwang 2017). MEA is one of the highly efficient amines employed for CO_2 absorption with 90% efficiency. These amine solutions possess some drawbacks such as high energy consumption in regeneration, high equipment corrosion rate, and low CO_2-capture efficiency. So, the researchers are focusing on novel solvent blends because they possess high thermal stability and are resistant to corrosion and solvent degradation; moreover, these solvent blends require low energy for solvent regeneration (Lee et al. 2012). These novel solvent blends consist of fast kinetic solvents like MEA and other slow kinetic solvents like MDEA (Tong et al. 2013). These blends improve individual CO_2 absorption capacity.

Nowadays, IL has engrossed the attention of researchers due to its unique properties such as non-volatility, high thermal stability, non-toxicity, high polarity, and low energy requirement for energy regeneration (Yu, Huang, and Tan 2012; Cullinane and Rochelle 2004). IL selectively absorbs acidic gases, and these ILs are organic salts with low vapour pressure and increased boiling point. Although IL possesses higher selectivity and CO_2 solubility for CO_2-capture application, some researchers stated that amino-functionalized IL has more CO_2 absorption capacity. Recently, amino-functionalized ILs such as triethylenetetramine lysine and diethylenetriamine lysine were used, and CO_2 uptake capacity of this system was found to be 2.59 and 2.13 mol CO_2/mol, respectively (Jing et al. 2018). But the viscosity and production cost of these ILs are high, which may be one of the concerns in their practical applications.

1.4 SYNTHESIS OF INDUSTRIAL PRODUCTS BY CO_2 CONVERSION

Transformation of CO_2 into useful chemicals/fuels is a promising strategy as such a system not only provides value-added products by consuming CO_2, but also alleviates global warming. Additionally, the CO_2 conversion into industrial products also possesses the potential to fulfil the energy demand in a sustainable manner. CO_2 is a notorious greenhouse gas and a cheaper source of carbon too (Ali, Razzaq, and In 2019; Wang, Yu, and Huang 2018). The development of a new system for converting

CO_2 into hydrocarbon and alcohol is an attractive subject. Various methods are employed for CO_2 conversion, including photochemical, enzymatic, chemical, and electrochemical methods. These methods will vary in the economic value of the product, CO_2 fixation time, and volume of CO_2 utilized (Long et al. 2017). Therefore, CO_2 concentration in future technologies and processes needs to be reduced in a controlled manner that allows the conversion of CO_2 into industrial products.

1.4.1 Methanol

Biomethanol is an alternative fuel to petrochemical industries, and it can be produced by the reduction of CO_2. Firstly, the electrochemical reduction of CO_2 into methanol using formate dehydrogenase, methanol dehydrogenase, and pyrroloquinoline quinone (PQQ) as a cofactor was reported. This approach provides a facile route for the production of methanol from CO_2 under mild conditions (Shi et al. 2015). The catalytic hydrogenation of CO_2 and H_2 is the basis of the syngas process. Various homogeneous and heterogeneous catalytic systems are used for catalytic hydrogenation of CO_2 into methanol. For over 40 years, $Zn/Cu/Al_2O_3$ catalyst system is used for the industrial synthesis of methanol (Yang et al. 2017). Carbon Recycling Inc (CRI) plant in Iceland produces low-carbon-intensity methanol that is utilized for gasoline blending and biodiesel production (Pontzen et al. 2011; Sharma, Shadiya, et al. 2019). CRI is the first company to capture CO_2 from flue gas and to be involved in the large-scale conversion of CO_2 into methanol.

1.4.2 Carbon Monoxide

Carbon monoxide (CO) is considered as waste gas, yet it is a vital feedstock for the synthesis of many fuels and chemicals. Mostly, the conversion of CO_2 to CO occurs in the presence of a catalyst, which results in a decrease in reaction rate and an increase in reaction velocity. Recently, the rhenium tricarbonyl catalyst attracted the attention of researchers for the conversion of CO_2 to CO. Carbon monoxide dehydrogenase from *Moorella thermoacetica* was the first reported biological catalyst for the electrochemical reduction of CO_2, and it exhibits no overpotential. However, the direct electrochemical reduction of CO_2 requires 1–2-V overpotential. Carbon monoxide dehydrogenase is another important metalloenzyme that contains nickel and iron in its inactive site, which catalyses the reversible oxidation of CO to CO_2. Several metal-centred catalysts such as nickel, iron, cobalt, and ruthenium catalyse the conversion of CO_2 into a value-added product (Agarwal et al. 2012; Shin et al. 2003). Further insights in this field are required to develop an effective catalytic system in terms of selectivity and overpotential.

1.4.3 Methane (CH_4)

Methane is one of the most efficient means to store electric energy. The most common methods for methane synthesis are Sabatier reaction and Fischer–Tropsch process, but both of these methods are expensive. For a catalytic transformation of CO_2 to methane, $Rh/\gamma\text{-}Al_2O_3$ catalyst is one of the promising compounds, as this works at low pressure and temperature as compared to other conventional methods that require higher temperatures (Beuls et al. 2012). It was reported that nitrogenase containing molybdenum iron protein also catalyses the reduction of CO_2 to methane (Yang et al. 2012). Nowadays, research has been going for implementing TiO_2 nanotubes for CO_2 conversion; this technique is eco-friendly as it depends on solar energy as an input.

1.4.4 Formic Acid

Formic acid can be used as animal feed additive, silage preservation, fuel for low-temperature fuel cells, and textile finishing, and in paper and pulp industry. The various homogeneous and heterogeneous systems have been studied for the reduction of CO_2 into formic acid (Alvarez-Guerra, Quintanilla, and Irabien 2012). In a homogeneous system, Formate dehydrogenase (FDH) is used for the hydration of CO_2 into formic acid. The catalyst uses NADPH as an electron donor and is embedded in alginate–silica hybrid gel nanostructures; this process occurs at low temperature and neutral pH. In addition to the homogeneous system, several heterogeneous systems have also been developed in the past decades (Lu et al. 2006). The catalytic hydrogenation of CO_2 into formic acid using $Ru(II)Cl(OAc)(PMe_3)_4$ as a catalyst in the presence of alcohol and base shows the high turnover frequency for formic acid production (Munshi et al. 2002). Various other metal complexes such as Ni, Rh, Pd, and Cu have been studied for formic acid formation with excellent yields. Climostat Ltd. (Cheshire, UK) has filed a patent application for the conversion of CO_2 and methane into formic acid using enzymatic catalysis (Fothergill et al., 2014).

1.5 CONCLUSION

Worldwide, CO_2 capture seems to be one of the serious issues. There is a critical need to capture CO_2 from the atmosphere, in order to prevent climate change. Although various efforts have been made to capture CO_2, there are still some challenges that need to be addressed. Various impurities in the flue gases may affect the activity and stability of chemical solvents and absorbent materials. Conversion of captured CO_2 into industrial products at large scale is still needed to be explored.

ACKNOWLEDGEMENT

The financial support from the Jaypee University of Information Technology, Waknaghat, Solan, is thankfully acknowledged to undertake this study. Further, the authors have no conflict of interest either among themselves or with the parent institution.

REFERENCES

Agarwal, Jay, Etsuko Fujita, Henry F. Schaefer, and James T. Muckerman. 2012. "Mechanisms for CO production from CO_2 using reduced rhenium tricarbonyl catalysts." *Journal of the American Chemical Society* 134 (11):5180–6. doi: 10.1021/ja2105834.

Agarwal, Praveen, Haibo Qi, and Lynden A. Archer. 2010. "The ages in a self-suspended nanoparticle liquid." *Nano Letters* 10 (1):111–5. doi: 10.1021/nl9029847.

Ali, Shahzad, Abdul Razzaq, and Su-Il In. 2019. "Development of graphene based photocatalysts for CO_2 reduction to C1 chemicals: A brief overview." *Catalysis Today* 335:39–54. doi: 10.1016/j.cattod.2018.12.003.

Alvarez-Guerra, Manuel, Sheila Quintanilla, and Angel Irabien. 2012. "Conversion of carbon dioxide into formate using a continuous electrochemical reduction process in a lead cathode." *Chemical Engineering Journal* 207–208:278–84. doi: 10.1016/j.cej.2012.06.099.

Aresta, Michele, Angela Dibenedetto, TomaszBaran, Antonella Angelini, Przemyslaw Labuz, and Wojciech. Macyk. 2014. "An integrated photocatalytic/enzymatic system for the reduction of CO_2 to methanol in bioglycerol-water." *Beilstein Journal of Organic Chemistry* 10:2556–65. doi: 10.3762/bjoc.10.267.

Ashley, Michael, Charles Magiera, Punnamchandar Ramidi, Gary Blackburn, Timothy G Scott, Rajeev Gupta, Kerry Wilson, Anindya Ghosh, and Abhijit Biswas. 2012. "Nanomaterials and processes for carbon capture and conversion into useful by-products for a sustainable energy future." *Greenhouse Gases: Science and Technology* 2 (6):419–44. doi: 10.1002/ghg.1317.

Batista, Ana, Paula Lucas Ambrosano, Sofia Graca, Catarina Sousa, Paula Marques, Belina Ribeiro, Botrel Elbris, Castro Neto Pedro, and Lusia Gouveia. 2015. "Combining urban wastewater treatment with biohydrogen production--an integrated microalgae-based approach." *Bioresource Technology* 184:230–5. doi: 10.1016/j.biortech.2014.10.064.

Beuls, Antoine, Colas Swalus, Marc Jacquemin, George Heyen, Alejandro Karelovic, and Patricio Ruiz. 2012. "Methanation of CO_2: Further insight into the mechanism over Rh/γ-Al_2O_3 catalyst." *Applied Catalysis B: Environmental* 113–114:2–10. doi: https://doi.org/10.1016/j.apcatb.2011.02.033.

Billig, Eric, Maximilian Decker,Walther Benzinger, F. Ketelsen, Peter Pfeifer,. Ralf Peters, Detlef Stolten, and Daiela Thrän. 2019. "Non-fossil CO_2 recycling—The technical potential for the present and future utilization for fuels in Germany." *Journal of CO_2 Utilization* 30:130–41 doi: 10.1016/j.jcou.2019.01.012.

Chen, Fanbing. 2019. "Cloning, expression and characterization of two beta carbonic anhydrases from a newly isolated CO_2 fixer, Serratia marcescens Wy064." *Indian Journal of Microbiology* 59 (1):64–72. doi: 10.1007/s12088-018-0773-6.

Chen, Wei-Hsin, Ming-Yueh Huang, Jo-Shu Chang, and Chun-YenChen.2014. "Thermal decomposition dynamics and severity of microalgae residues in torrefaction." *Bioresource Technology* 169:258–64 doi: 10.1016/j.biortech.2014.06.086.

Chen, Yu, and Tiancheng Mu. 2019. "Conversion of CO_2 to value-added products mediated by ionic liquids." *Green Chemistry* 21 (10):2544–74. doi: 10.1039/C9GC00827F.

Chowdhury, Shamik, and Rajasekhar Balasubramanian. 2016. "Holey graphene frameworks for highly selective post-combustion carbon capture." *Scientific Reports* 6 (1):21537. doi: 10.1038/srep21537.

Cullinane, J. Tim, and Gary Thomas Rochelle. 2004. "Carbon dioxide absorption with aqueous potassium carbonate promoted by piperazine." *Chemical Engineering Science* 59 (17):3619–30. doi: https://doi.org/10.1016/j.ces.2004.03.029.

Dutcher, Bryce, Maohong Fan, and Armistead G. Russell. 2015. "Amine-based CO_2 capture technology development from the beginning of 2013—A review." *ACS Applied Materials & Interfaces* 7 (4):2137–48. doi: 10.1021/am507465f.

Fothergill MichaelDavid., Timothy David Gibson., Gerard Garcia Sobany. 2014. Method of enzyme conversion using an immobilized composition consisting of at least two enzymes and cofactor. *UK Patent, US 6440711,B.*

Ghodake, Gajanan, Surendra Shinde, Rijuta Ganesh Saratale, Avinash Kadam, Ganesh Dattatraya Saratale, Rahul Patel, Ashok Kumar, Sunil Kumar, and Dae-Young Kim. 2019. "Whey peptide-encapsulated silver nanoparticles as a colorimetric and spectrophotometric probe for palladium(II)." *Microchimica Acta* 186 (12):763. doi: 10.1007/s00604-019-3877-8.

Grodzinski, Bernard., Jirong Jiao, and Evangelos D. Leonardos. 1998. "Estimating photosynthesis and concurrent export rates in C3 and C4 species at ambient and elevated CO21,2." *Plant Physiology* 117 (1):207–15. doi: 10.1104/pp.117.1.207.

Hedin, Niklas, LiJun Chen, and Aatto Laaksonen. 2010. "Sorbents for CO_2 capture from flue gas—aspects from materials and theoretical chemistry." *Nanoscale* 2 (10):1819–41. doi: 10.1039/C0NR00042F.

Jing, Guohua, Yuhao Qian, Xiaobin Zhou, Bihong Lv, and Zuoming Zhou. 2018. "Designing and screening of multi-amino-functionalized ionic liquid solution for CO_2 capture by quantum chemical simulation." *ACS Sustainable Chemistry & Engineering* 6 (1):1182–91. doi: 10.1021/acssuschemeng.7b03467.

Kamyab, Hesam, Shreeshivadasan Chelliapan, Chew Tin Lee, Shahabaldin Rezania, Amirreza Talaiekhozani, Tayebeh Khademi, and Ashok Nadda. 2019. "Microalgae cultivation using various sources of organic substrate for high lipid content." In *International Conference on Urban Drainage Modelling*, 893–8.

Kamyab, Hesam, Shreeshivadasan Chelliapan, Mohd Fadhil Md Din, Shahabaldin Rezania, Tayebeh Khademi, and Ashok Nadda. 2018. "Palm oil mill effluent as an environmental pollutant." In *Palm Oil* edited by Vidhuranga Waisundara, 13–28. London, UK: Intech Open Limited.

Kamyab, Hesam, Shreeshivadasan Chelliapan, Mohd Fadhil Md Din, Reza Shahbazian-Yassar, Shahabaldin Rezania, Tayebeh Khademi, and Ashok Nadda. 2017. "Evaluation of Lemna minor and Chlamydomonas to treat palm oil mill effluent and fertilizer production." *Journal of Water Process Engineering* 17:229–36 doi: 10.1016/j.jwpe.2017.04.007.

Kamyab, Hesam, Shreeshivadasan Chelliapan, Ashok Nadda, Shahabaldin Rezania, Amirreza Talaiekhozani, Tayebeh Khademi, Parveen Fatemeh Rupani, and Swati Sharma. 2019. "Microalgal biotechnology application towards environmental sustainability." In *Application of Microalgae in Wastewater Treatment* edited by Sanjay Kumar Gupta and Bux Faizal 445–65. Singapore: Springer Singapore.

Kim, Byung- Joo, Ki-Sook Cho, and Soo-Jin Park. 2010. "Copper oxide-decorated porous carbons for carbon dioxide adsorption behaviors." *Journal of Colloid and Interface Science* 342 (2):575–8. doi: 10.1016/j.jcis.2009.10.045.

Kim, G. Y., J. Heo, H. S. Kim, and J. I. Han. 2017. "Bicarbonate-based cultivation of Dunaliella salina for enhancing carbon utilization efficiency." *Bioresource Technology* 237:72–7 doi: 10.1016/j.biortech.2017.04.009.

Kumar, A., Gaobing Wu, Zuo Wu, Narender Kumar, and Ziduo Liu. 2018. "Improved catalytic properties of a serine hydroxymethyl transferase from Idiomarina loihiensis by site directed mutagenesis." *International Journal of Biological Macromolecules* 117:1216–23 doi: 10.1016/j.ijbiomac.2018.05.003.

Kumar, Ashok, Renata Gudiukaite, Alisa Gricajeva, Mikas Sadauskas, Vilius Malunavicius, Hesam Kamyab, Swati Sharma, Tanvi Sharma, and Deepak Pant. 2020. "Microbial lipolytic enzymes – promising energy-efficient biocatalysts in bioremediation." *Energy* 192:116674. doi: https://doi.org/10.1016/j.energy.2019.116674.

Kumar, Ashok, Tanvi Sharma, Sikandar I. Mulla, Hesam Kamyab, Deepak Pant, and Swati Sharma. 2019. "Let's protect our earth: Environmental challenges and implications." In *Microbes and Enzymes in Soil Health and Bioremediation*, edited by Ashok Kumar and Swati Sharma, 1–10. Singapore: Springer Singapore.

Kumar, Manish, Smita Sundaram, Edgard Gnansounou, Christian Larroche, and Indu Shekhar Thakur. 2018. "Carbon dioxide capture, storage and production of biofuel and biomaterials by bacteria: A review." *Bioresource Technology* 247:1059–68 doi: https://doi.org/10.1016/j.biortech.2017.09.050.

Lee, Joonho, andNarayana R. Aluru. 2013. "Water-solubility-driven separation of gases using graphene membrane." *Journal of Membrane Science* 428:546–53 doi: 10.1016/j.memsci.2012.11.006.

Lee, Sang-Sup., Seong-Man Mun, Won-Joon Choi, Byoung-Moo Min, Sang-Won Cho, and Kwang-Joong Oh. 2012. "Absorption characteristics of new solvent based on a blend of AMP and 1,8-diamino-p-menthane for CO_2 absorption." *Journal of Environmental Sciences (China)* 24 (5):897–902. doi: 10.1016/s1001-0742(11)60788-2.

Liu, Fa-Qian, Wei Li, Jie Zhao, Wei-Hua Li, Dong-Mei Chen, Li-Shui Sun, Lei Wang, and Rong-Xun Li. 2015. "Covalent grafting of polyethyleneimine on hydroxylated three-dimensional graphene for superior CO_2 capture." *Journal of Materials Chemistry A* 3 (23):12252–8. doi: 10.1039/C5TA01536G.

Liu, Yuanyuan, Yanmei Yang, Qilong Sun, Zeyan Wang, Baibiao Huang, Ying Dai, Xiaoyan Qin, and Xiaoyang Zhang. 2013. "Chemical adsorption enhanced CO_2 capture and photoreduction over a copper porphyrin based metal organic framework." *ACS Applied Materials & Interfaces* 5 (15):7654–8. doi: 10.1021/am4019675.

Long, Nguyen, Jintae Lee, Kee-Kahb Koo, Patricia Luis, and Moonyong Lee. 2017. "Recent progress and novel applications in enzymatic conversion of carbon dioxide." *Energies* 10:473, 1–19. doi: 10.3390/en10040473.

Lu, Yang, Zhong-Yi Jiang, Song-Wei Xu, and Hong Wu. 2006. "Efficient conversion of CO_2 to formic acid by formate dehydrogenase immobilized in a novel alginate–silica hybrid gel." *Catalysis Today* 115 (1):263–8. doi: https://doi.org/10.1016/j.cattod.2006.02.056.

Luttge, U. 2004. "Ecophysiology of crassulacean acid metabolism (CAM)." *Annals of Botany* 93 (6):629–52. doi: 10.1093/aob/mch087.

Mazari, Shaukat A., Brahim Si Ali, Badrul M. Jan, Idris Mohamed Saeed, and S. Nizamuddin. 2015. "An overview of solvent management and emissions of amine-based CO_2 capture technology." *International Journal of Greenhouse Gas Control* 34:129–40 doi: https://doi.org/10.1016/j.ijggc.2014.12.017.

Mistry, Avnish Nitin, Upendar Ganta, Jitamanyu Chakrabarty, and Susmita Dutta. 2019. "A review on biological systems for CO_2 sequestration: Organisms and their pathways." *Environmental Progress & Sustainable Energy* 38 (1):127–36. doi: 10.1002/ep.12946.

Munshi, Pradip, A. Denise Main, John C. Linehan, Chih-Cheng Tai, and Philip G. Jessop. 2002. "Hydrogenation of carbon dioxide catalyzed by ruthenium trimethylphosphine complexes: The accelerating effect of certain alcohols and amines." *Journal of the American Chemical Society* 124 (27):7963–71. doi: 10.1021/ja0167856.

Nimmo, H. G. 2000. "The regulation of phosphoenolpyruvate carboxylase in CAM plants." *Trends in Plant Science* 5 (2):75–80. doi: 10.1016/s1360-1385(99)01543-5.

Nogia, Panchsheela, Gurpreet Kaur Sidhu, Rajesh Mehrotra, and Sandhya Mehrotra. 2016. "Capturing atmospheric carbon: biological and nonbiological methods." *International Journal of Low-Carbon Technologies* 11 (2):266–74. doi: 10.1093/ijlct/ctt077.

Pavlik, David, Yingkui Zhong, Carly Daiek, Wei Liao, Robert Morgan, William Clary, and Yan Liu. 2017. "Microalgae cultivation for carbon dioxide sequestration and protein production using a high-efficiency photobioreactor system." *Algal Research* 25:413–20 doi: 10.1016/j.algal.2017.06.003.

Pontzen, Florian, Waldemar Liebner, Veronika Gronemann, Martin Rothaemel, and Bernd Ahlers. 2011. "CO_2-based methanol and DME – Efficient technologies for industrial scale production." *Catalysis Today* 171:242–50 doi: 10.1016/j.cattod.2011.04.049.

Raven, John A., and Alison J. Karley. 2006. "Carbon sequestration: Photosynthesis and subsequent processes." *Current Biology* 16 (5):R165–7. doi: 10.1016/j.cub.2006.02.041.

Rees, Neil, and Richard Compton. 2011. "Sustainable energy: A review of formic acid electrochemical fuel cells." *Journal of Solid State Electrochemistry* 15:2095–100. doi: 10.1007/s10008-011-1398-4.

Rittmann, Bruce. 2008. "Opportunities for renewable bioenergy using microorganisms." *Biotechnology and Bioengineering* 100 (2):203–12. doi: 10.1002/bit.21875.

Rodriguez Acevedo, Elizabeth., Farid B Cortes, and Camilo A Franco. 2019. "An enhanced carbon capture and storage process (e-CCS) applied to shallow reservoirs using nanofluids based on nitrogen-rich carbon nanospheres." *Materials* 12 (13):2088. doi: 10.3390/ma12132088.

Sage, Rowan F., Tammy L. Sage, and Ferit Kocacinar. 2012. "Photorespiration and the evolution of C4 photosynthesis." *Annual Review of Plant Biology* 63 (1):19–47. doi: 10.1146/annurev-arplant-042811-105511.

Sharma, Abhishek, Shadiya, Tanvi Sharma, Rakesh Kumar, Khemraj Meena, and Shamsher Singh Kanwar. 2019. "Biodiesel and the potential role of microbial lipases in its production." In *Microbial Technology for the Welfare of Society*, edited by Pankaj Kumar Arora, 83–99. Singapore: Springer Singapore.

Sharma, Abhishek, Tanvi Sharma, Khem Raj Meena, Ashok Kumar, and Shamsher Singh Kanwar. 2018. "High throughput synthesis of ethyl pyruvate by employing superparamagnetic iron nanoparticles-bound esterase." *Process Biochemistry* 71:109–17 doi: 10.1016/j.procbio.2018.05.004.

Sharma, Abhishek, Taruna Sharma, Tanvi Sharma, Shweta Sharma, and Shamsher Singh Kanwar. 2019. "Role of microbial hydrolases in bioremediation." In *Microbes and Enzymes in Soil Health and Bioremediation*, edited by Ashok Kumar and Swati Sharma, 149–64. Singapore: Springer Singapore.

Sharma, Swati, Ambika Verma, Ashok Nadda, and Hesam Kamyab. 2018. "Magnetic nanocomposites and their industrial applications." *Nano Hybrids and Composites* 20:149–72 doi: 10.4028/www.scientific.net/NHC.20.149.

Sharma, Tanvi, Swati Sharma, Hesam Kamyab, and Ashok Kumar. 2019. "Energizing the CO_2 utilization by chemo-enzymatic approaches and potentiality of carbonic anhydrases: A review." *Journal of Cleaner Production* 247:(2020) 119138. doi:10.1016/j.jclepro.2019.119138.

Shi, Jiafu., Yanjun Jiang. Zhongyi Jiang, Xueyan Wang., Shaohua Zhang, Pingping Han, and Chen Yang. 2015. "Enzymatic conversion of carbon dioxide." *Chemical Society Reviews* 44 (17):5981–6000. doi: 10.1039/c5cs00182j.

Shin, Woonsup, Sang Hee Lee, Jun Won Shin, Sang Phil Lee, and Yousung Kim. 2003. "Highly selective electrocatalytic conversion of CO_2 to CO at −0.57 V (NHE) by carbon monoxide dehydrogenase from Moorella thermoacetica." *Journal of the American Chemical Society* 125 (48):14688–9. doi: 10.1021/ja037370i.

Song, Chunfeng, Qingling Liu, Yun Qi, Guanyi Chen, Yingjin Song, Yasuki Kansha, and Yutaka Kitamura. 2019. "Absorption-microalgae hybrid CO_2 capture and biotransformation strategy—A review." *International Journal of Greenhouse Gas Control* 88:109–17 doi: 10.1016/j.ijggc.2019.06.002.

Stowe, Haley M., and Gyeong S. Hwang. 2017. "Fundamental understanding of CO_2 capture and regeneration in aqueous amines from first-principles studies: Recent progress and remaining challenges." *Industrial & Engineering Chemistry Research* 56 (24):6887–99. doi: 10.1021/acs.iecr.7b00213.

Sui, Zhu-Yin, and Bao-Hang Han. 2015. "Effect of surface chemistry and textural properties on carbon dioxide uptake in hydrothermally reduced graphene oxide." *Carbon* 82:590–8 doi: 10.1016/j.carbon.2014.11.014.

Thakur, Neha, Ajay Kumar, Abhishek Sharma, Tek Chand Bhalla, and Dinesh Kumar. 2018. "Purification and characterization of alkaline, thermostable and organic solvent stable protease from a mutant of *Bacillus* sp." *Biocatalysis and Agricultural Biotechnology* 16:217–24 doi: 10.1016/j.bcab.2018.08.005.

Tong, Danlu, Geoffrey C. Maitland, Martin J. P. Trusler, and Paul S. Fennell. 2013. "Solubility of carbon dioxide in aqueous blends of 2-amino-2-methyl-1-propanol and piperazine." *Chemical Engineering Science* 101:851–64 doi: 10.1016/j.ces.2013.05.034.

Ullah, Kifayat, Mushtaq Ahmad, Sofia, Vinod Sharma, Pengmei Lu, Adam Harvey, Muhammad Zafar, and Shazia Sultana. 2015. "Assessing the potential of algal biomass opportunities for bioenergy industry: A review." *Fuel* 143:414–23 doi: 10.1016/j.fuel.2014.10.064.

Vega, Fernando, Mercedes Cano, Sara Camino, Luz M. Gallego Fernández, Esmeralda Portillo, and Benito Navarrete. 2018. "Solvents for carbon dioxide capture." In *Carbon Dioxide Chemistry, Capture and Oil Recovery*, edited by Iyad Karame, Janah Shayaand Hassan Srour, 141–62. London, UK: Intech Open Limited.

Wang, Junya, Liang Huang, Ruoyan Yang, Zhang Zhang, Jingwen Wu, Yanshan Gao, Qiang Wang, Dermot O'Hare, and Ziyi Zhong. 2014. "Recent advances in solid sorbents for CO_2 capture and new development trends." *Energy & Environmental Science* 7 (11):3478–518. doi: 10.1039/C4EE01647E.

Wang, Minghui, Junjie Zhao, Xiaoxue Wang, Andong Liu, and Karen K. Gleason. 2017. "Recent progress on submicron gas-selective polymeric membranes." *Journal of Materials Chemistry A* 5 (19):8860–86. doi: 10.1039/C7TA01862B.

Wang, Yuan, Yulv Yu, and Jin Huang. 2018. "Catalytic conversion of CO_2 to value-added products under mild conditions." *ChemCatChem* 10(21): 4849–53. doi: 10.1002/cctc.201801346.

Wu, Ting, Yi Wang, Changjiang Yu, Rawee Chiarawipa, Xinzhong Zhang, Zhenhai Han, and Lianhai Wu. 2012. "Carbon sequestration by fruit trees - Chinese apple orchards as an example." *PLOS One* 7 (6):e38883. doi: 10.1371/journal.pone.0038883.

Yang, Haiyan, Chen Zhang, Peng Gao, Hui Wang, Xiaopeng Li, Liangshu Zhong, Wei Wei, and Yuhan Sun. 2017. "A review of the catalytic hydrogenation of carbon dioxide into value-added hydrocarbons." *Catalysis Science & Technology* 7 (20):4580–98. doi: 10.1039/C7CY01403A.

Yang, Zhi- Yong., Moure Vivian-Rotuno, Dean Dennis R, and Seefeldt Lance. 2012. "Carbon dioxide reduction to methane and coupling with acetylene to form propylene catalyzed by remodeled nitrogenase." *Proceedings of the National Academy of Sciences of the United States of America* 109 (48):19644–8. doi: 10.1073/pnas.1213159109.

Yu, Cheng-Hsiu, Chih-Hung Huang, and Chung-Sung Tan. 2012. "A review of CO_2 capture by absorption and adsorption." *Aerosol and Air Quality Research* 12 (5):745–69. doi: 10.4209/aaqr.2012.05.0132.

Yu, Wei, Tao Wang, Ah-Hyung Alissa Park, and Mengxiang Fang. 2019. "Review of liquid nano-absorbents for enhanced CO_2 capture." *Nanoscale* 11 (37):17137–56. doi: 10.1039/C9NR05089B.

Zhao, Bingtao, and Yaxin Su. 2014. "Process effect of microalgal-carbon dioxide fixation and biomass production: A review." *Renewable and Sustainable Energy Reviews* 31:121–32. doi: 10.1016/j.rser.2013.11.054.

2 Heterogeneous Catalytic Hydrogenation of CO_2 to Basic Chemicals and Fuels

Saeed Sahebdelfar and Maryam Takht Ravanchi

CONTENTS

2.1 INTRODUCTION

Due to economic development, energy consumption, including fossil fuel, is increasing all around the world. The use of fossil fuels for energy production in houses, power plants, and vehicles is the main source of anthropogenic CO_2 emissions. In 2017, CO_2 concentration in the atmosphere reached 405 ppm (Liu et al., 2019). It is predicted that by the end of this century, CO_2 concentration in the atmosphere will continue to rise to 570 ppm. Unfortunately, it is not possible to propose a truly low-carbon technology to the public market in the short term. The best approach will be CO_2 utilization (Araújo et al., 2014).

Due to the worldwide concern about global warming and climate change, CO_2 emissions into the atmosphere must be controlled. It was reported that by burning 1 ton of carbon in fossil fuels, more than 3.5 tons of CO_2 was obtained. International Energy Agency (IEA) reported that for limiting temperature increase within 2°C till 2050, the annual CO_2 emissions should not exceed 15 gigatonnes. Consequently, a simultaneous increase in energy efficiency and use of renewable sources has the best effect.

For CO_2 utilization, there are some barriers, the main important of which are capture and storage costs, required energy for CO_2 conversion, market scale, and socio-economic driving force. The selected process must be an environmentally benign one; an industrially useful chemical must be produced; and renewable energy must be involved to conserve carbon sources (Huang and Tan, 2014).

The concept of carbon capture and storage (CCS) alone is no longer useful, and carbon capture, utilization, and storage (CCUS) must be applied, which is a combination of CCS and CCU (carbon capture and utilization).

Although CO_2 is produced with high purity in certain processes such as ammonia synthesis, ethylene oxide production, and fermentation, it is mostly present in the flue gases in low concentrations along with impurities. This renders the separation more costly.

There are several technologies for utilizing the captured CO_2. Among those technologies, enhanced oil recovery (EOR), enhanced coal bed methane (ECBM), mineralization, fuels and chemicals, biofuels from algae, and enhanced geothermal systems have the highest CO_2 consumption potential (Norhasyima and Mahlia, 2018).

In EOR, CO_2 is injected to the reservoir to pressurize the rock formation in order to release the trapped oil (or gas). It is a growing mature technology and has been applied. The challenges are a large number of parameters involved, CO_2 transportation to the site, and fluctuating oil price. CO_2-ECBM is a technology similar to EOR. The outcome of both technologies is increased fossil fuel production, which in turn results in CO_2 emissions.

In mineralization, CO_2 reacts with natural alkaline earth silicates, in the same way as natural weathering of the rocks. It is thermodynamically feasible, but a very slow process. Thus, high pressures and temperatures are necessary, which render the *in situ* process costly. It can be accelerated in *ex situ* mode by grinding the particles and using elevated pressures. Mineralization of carbon to insoluble carbonates (e.g. $CaCO_3$ and $MgCO_3$) at low CO_2 concentrations can be facilitated by using amines like monoethanolamine (MEA), as illustrated in slurry reactor system tests using amine looping strategy (Liu and Gadikota, 2018).

Injection of carbon dioxide into deep saline aquifers is a promising technology. Compared with other geological storage methods (EOR and ECBM), it has the largest capacity for the storage of CO_2 (Leung et al., 2014). Despite its great potential, there is little information about this method.

Chemical fixation of CO_2, especially in bulk chemicals and fuels, is a promising solution to CO_2 emission by recycling carbon as a supplement for natural photosynthesis. At the same time, it reduces the dependence on petroleum as a carbon source irrespective of its contribution in reducing CO_2 emissions. The main challenge here

TABLE 2.1
Major CO_2 Utilization Technologies

Technology	Advantages (or Opportunities)	Disadvantages (or Challenges)
Geological storage (including EOR, ECBM)	Mature technology	Risk of leakage Large number of parameters Transportation cost Location dependent
Mineralization	Chemical free Not sensitive to CO_2 impurities	Unfavourable kinetics Large number of parameters Severe operating conditions Energy intensive and costly
Chemical utilization	Reduced dependence on fossil fuels as feedstock for fuels and chemicals Sustainable carbon cycle	Severe operating conditions Low catalyst lifetimes Low selectivity

is the need for renewable energy (hydrogen) and developing catalysts with a high yield of the desired product and suitable stability.

CO_2 utilization technologies have been reviewed elsewhere (Norhasyima and Mahlia, 2018). Table 2.1 compares the major utilization technologies.

Currently, non-renewable fossil-based resources (such as oil, natural gas, and coal) are the main feedstock of fuels and base chemicals. Upon burning fossil fuels, CO_2 is the end product. It can be a waste product in the chemical industry as well. In the near future, by the depletion of fossil fuels, CO_2 will be the main carbon source. Consequently, CO_2 hydrogenation and CO_2 conversion with other H sources to fuels and other value-added chemicals are of high importance, in order to close the carbon cycle (Figure 2.1).

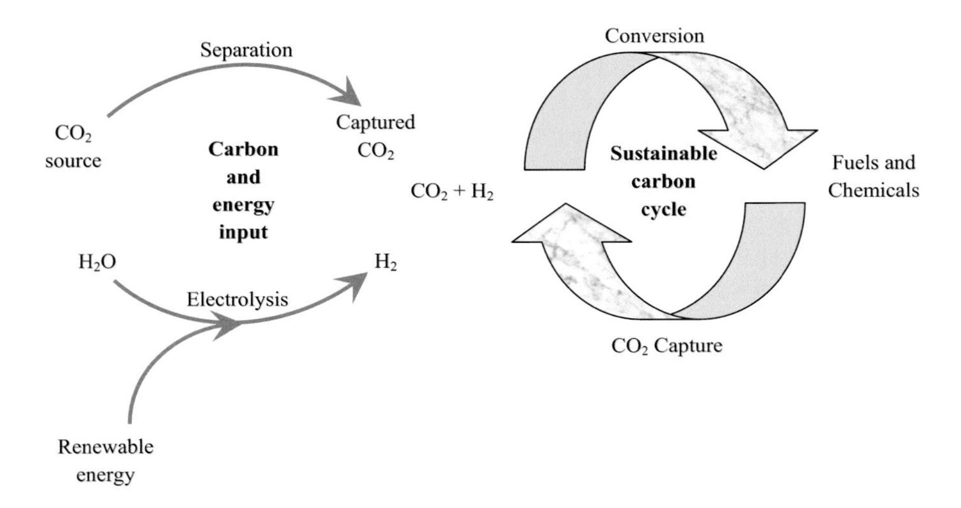

FIGURE 2.1 Sustainable carbon cycle.

The desirable future is when renewable fuels obtained from CO_2 are delivered by pipelines to various industries, homes, and buildings. Carbon dioxide-based refineries that are in service by novel catalysts being powered by renewable forms of solar, electrical, and thermal energy produce renewable fuels, i.e. a sustainable carbon-neutral carbon cycle (Jia et al., 2017).

2.2 CHEMISTRY OF CARBON DIOXIDE

The chemistry of carbon dioxide received much attention due to environmental considerations and using a cheap source of carbon. CO_2 can be used as a carbon source and building block in C_1 chemistry for producing a variety of chemicals. Many books, monographs, and review papers have been devoted to the conversion of CO_2 to the value-added products (Halmann, 1993; Ma et al., 2009; Mikkelsen et al., 2010; Aresta, 2010; Tahriri Zangeneh et al., 2011; Styring et al., 2015; Li et al., 2015).

Carbon dioxide is the highest oxidized form of carbon, which is relatively inert, and its reactions are energetically unfavourable. Thus, catalysis plays an important role in the conversion of carbon dioxide by lowering the high activation energy. The reduction of CO_2 also requires high-energy electron donors like the high-energy reducing agent H_2.

Carbon dioxide has a strong affinity towards nucleophiles and electron-donating reagents due to electron deficiency of the carbonyl carbon (Omae, 2012). The chemical activity of CO_2 is determined by polarization of carbon–oxygen bonds, and its chemistry is dominated by its reaction with nucleophiles that react with the carbon atom. The nucleophile may be a neutral species (an amine), an electron-rich π bond (phenolate), or carbon–metal σ bond (Grignard reagent) (North, 2015). The coordination of CO_2 to metal complexes is also important as it alters the electron distribution and molecular geometry of CO_2, which is the basis for the induction of reactions with metals and catalysis by metals. At least 13 coordination geometries in CO_2–metal complexes have been identified (North, 2015). The coordination of CO_2 to complexes of transition metals is a promising route as it promotes C–C bond formation and metal-catalysed reactions.

Reactive substrates such as oxygen-containing compounds (e.g. alcohols and epoxides), nitrogen-containing compounds (e.g. amines), unsaturated hydrocarbons (alkynes, alkenes, and aromatics), and hydrogen have been investigated for the utilization of carbon dioxide for the production of organic chemicals (Figure 2.2). Otto et al. (2015) listed 23 CO_2 reactions for the production of bulk chemicals with much more for fine chemicals.

Figure 2.2 reveals that the reactions of CO_2 can be divided into two classes: (i) reactions in which the entire molecule is involved, e.g. fixation onto an organic substrate, and (ii) reactions that convert CO_2 into another C_1 molecule or C_n molecules (Aresta and Dibenedetto, 2003). In the first group, relevant to fine chemicals, the substrate denotes (most of) the required energy of reaction and the reaction occurs at lower temperatures (–30°C–130°C). In the latter, exemplified by hydrogenation and dry reforming reactions, hydrogen or heat is used as an energy source and the reactions occur at high temperatures (>250°C). These reactions are relevant to the production of bulk and energy chemicals (Havran et al., 2011).

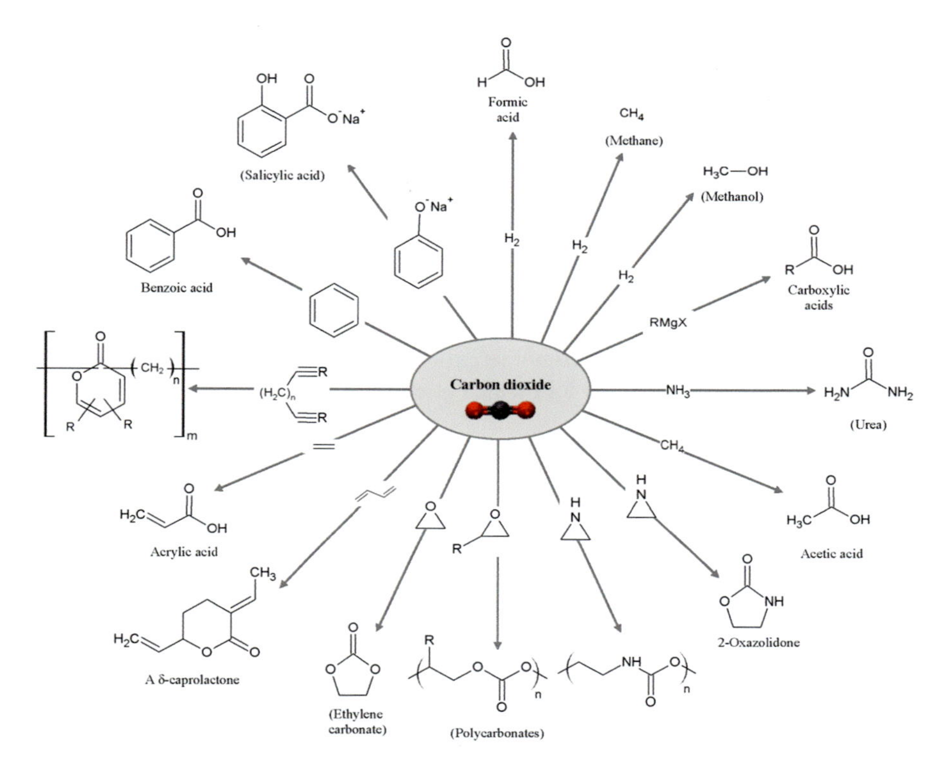

FIGURE 2.2 Representative heterogeneously catalysed reactions of CO_2 (names in parentheses refer to commercialized CO_2-based processes).

Heterogeneous catalytic reactions of carbon dioxide are still much more limited than homogeneous processes, but include potentially important reactions such as hydrogenation reactions and dry reforming of methane. Heterogeneous catalysts are more suited for the production of bulk chemicals due to the minimum difficulty in their separation from products although they typically need more severe operating conditions.

Carbon dioxide is already being used for the synthesis of a few chemicals, including urea, salicylic acid, cyclic carbonates, and organic polycarbonates, but the overall CO_2 consumption is not significant compared to global CO_2 emission (<1%). Many processes have been employed to utilize CO_2, including dry (or CO_2) reforming of methane to produce synthesis gas (a mixture of CO and H_2); hydrogenation to produce methanol, dimethyl ether (DME), and methane; and the synthesis of dimethyl carbonate (a carbonylation agent as phosgene substitute for aromatic polycarbonates using supported copper catalysts) and cyclic carbonates (polymer precursors, via cycloaddition of CO_2 to epoxide using zeolite catalysts). Compared to cyclic carbonates, the commercial production of polycarbonates is a much more recent development and most processes are still at pre-production stage (North, 2015).

Carbon dioxide (or dry) reforming of methane is an important heterogeneously catalysed CO_2 conversion reaction because it consumes two greenhouse gases

(GHGs) to produce CO-rich synthesis gas, which is useful for producing valuable oxygenates, which is given in the following equation:

$$CO_2 + CH_4 \Leftrightarrow 2CO + 2H_2 \qquad \Delta H^0_{298} = 260.5 \text{ kJ mol}^{-1} \tag{2.1}$$

The reaction is highly endothermic and is operated at high temperatures (>900°C). It is catalysed by noble metals and Ni. Nickel-based catalysts have been most extensively studied due to high activity and reasonable price although the noble metals are less susceptible to deactivation by coke formation (Jafarbegloo et al., 2016).

To obtain a syngas composition appropriate for the synthesis of methanol and hydrocarbons, bi-reforming (a combination of steam and dry reforming) has been proposed, which is given in the following equation:

$$CO_2 + 2H_2O + 3CH_4 \Leftrightarrow 4CO + 8H_2 \qquad \Delta H^0_{298} = 657.5 \text{ kJ mol}^{-1} \tag{2.2}$$

The resulting synthesis gas is termed 'metgas' because of its stoichiometric H_2/CO ratio (2 mol mol^{-1}), which corresponds to that of methanol synthesis (Olah et al., 2018a).

Carbon dioxide can be used as a mild oxidant, as an alternative to oxygen or air, to avoid over-oxidation of the reactants. In these reactions, CO_2 does not act as a carbon source and the reactions are typically endothermic. Examples include oxidative dehydrogenation of styrene and propane by CO_2 in which CO_2 is reduced to CO (Tahriri Zangeneh et al., 2016). The catalysts are metal oxides such as Cr_2O_3 and Ga_2O_3.

2.3 CO_2 HYDROGENATION REACTIONS

Carbon dioxide can be transformed into a variety of important chemicals via catalytic hydrogenation according to the catalyst type and process conditions (Figure 2.3). Several researchers published thorough reviews with the subject of catalytic hydrogenation of carbon dioxide (Wang et al., 2011; Gnanamani et al., 2014; Saeidi et al., 2014; Jalama, 2017; Saeidi et al., 2017; Guo et al., 2018; Roy et al., 2018, Liu et al., 2019).

Photocatalytic reduction of CO_2 with H_2O to simple molecules (CO, methanol, methane, and ethanol) over semiconductors such as TiO_2, ZnO, and CdS is a green conversion route, but the efficiency is still low. Metal–organic frameworks (MOFs), which are porous coordination polymers composed of organic–inorganic components, have been recognized as promising photocatalysts in several reactions, including CO_2 reduction. The incorporation of porphyrin (a group of heterocyclic macrocycle organic compounds) linkers into MOFs enhances CO_2 reduction performance. Cu^{2+} in a porphyrin-based MOF enhances CO_2-to-methanol conversion under visible light irradiation. The CO_2 conversion with Cu^{2+} is increased by a factor of seven times compared with the sample without Cu^{2+} (Liu et al., 2013).

2.3.1 Thermodynamic Considerations

Although carbon dioxide is the most oxidized form of carbon and energetically highly stable, many hydrogenation reactions, e.g. to produce methane and methanol,

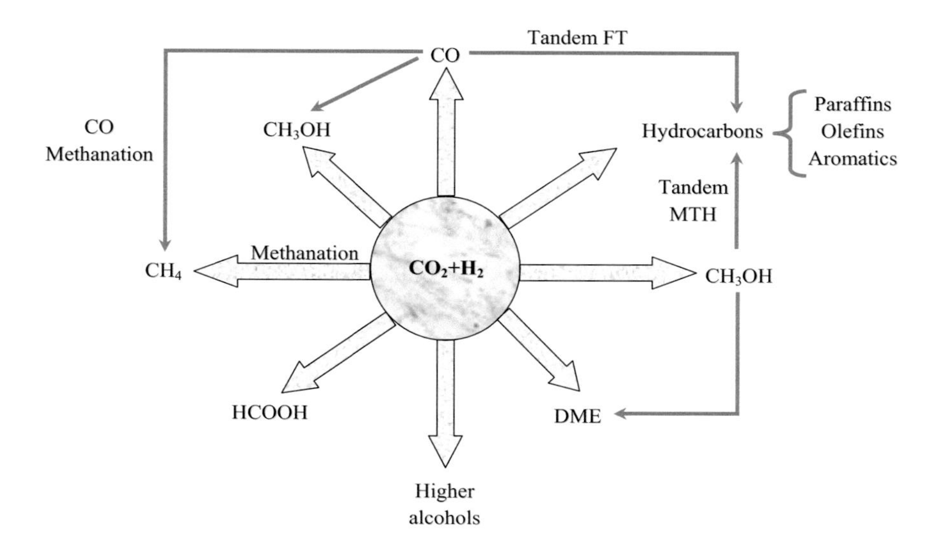

FIGURE 2.3 Reaction paths for catalytic hydrogenation of CO_2 (⟹: direct routes; ⟶: tandem routes).

are energetically favourable (Heyn, 2003). Hydrogen acts as a co-reactant with higher Gibbs free energy, making the reaction thermodynamically more favourable (Jiang et al., 2019).

Figure 2.4 shows the gas-phase Gibbs free energy change of reaction for different CO_2 hydrogenation products per mole of CO_2. With the exception of CO via endothermic reverse water–gas shift (RWGS), the reactions become more favourable at lower temperatures. This implies that highly active (and selective) catalyst should be developed to promote the specific reaction selectively at a low operating temperature.

Lower paraffinic compounds (especially methane) are thermodynamically most favoured products at lower temperatures. The molecular weight has a negative impact on the favourability of formation. In contrast, methanol is a thermodynamically least favoured hydrogenation product among lower alcohols. Formic acid is the unfavourable product under low temperature conditions. It should be reminded that the equilibrium conversion also depends on H_2/CO_2 ratio in the feed and operating pressure.

Heterogeneously catalysed hydrogenation reactions are mostly equilibrium-limited, and the product yields are rarely 100% (Heyn, 2003). Furthermore, the simultaneous occurrence of several equilibria complicates the system. A technical catalyst should be able to selectively promote a single (or a family of) reaction(s). Therefore, catalyst plays a critical role in the development of viable processes for chemical utilization of CO_2 at a commercial scale.

Because of the predominant equilibrium limitation, thermodynamic equilibrium analysis of the desired reaction (with the accompanying reactions, if any) is quite useful for catalyst and process development. Several thermodynamic analyses have been reported for individual hydrogenation products such as methane (Sahebdelfar and Takht Ravanchi, 2015), methanol, DME (Shen et al., 2000; Ateka et al., 2017a),

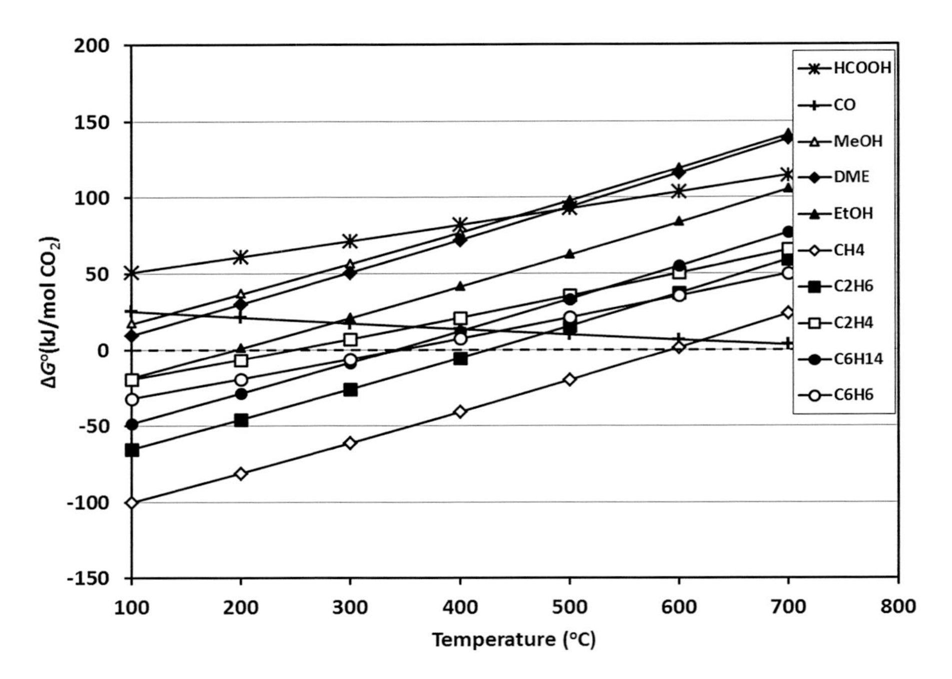

FIGURE 2.4 Gibbs free energy change of representative CO_2 hydrogenation products as a function of temperature.

hydrocarbons (Torrente-Murciano et al., 2014), and different products (Ahmad and Upadhyayula, 2019).

Swapnesh et al. (2014) performed a comparative study on the thermodynamics of CO_2 utilization reactions. They also compared their thermodynamic simulation results with experimental data. Coke formation was found to be insignificant in CO_2 methanation process. Jia et al. (2016) considered a higher diversity of products, including CO, alcohols, aldehydes, carboxylic acids, and C_1–C_4 hydrocarbons, and identified the favourable product for each group. Methanation was found to be the most favourable reaction with 100% equilibrium methane yield at moderate conditions (<600°C), while 100% CO selectivity was obtained at higher temperatures (>750°C). They, however, did not consider coke formation in their analysis.

Because of the presence of carbon-containing compounds in feed and products, coke formation remains an issue in CO_2 hydrogenation over heterogeneous catalyst. Although coke amount may be slight, its effect on reaction can be significant due to catalyst deactivation by fouling (covering active sites or blocking the pores). Solid carbon or 'coke' can be formed by the reduction of CO (by CO or H_2) or the decomposition of other carbon-containing products. From the thermodynamic point of view, coke formation potential will be decreased by increasing temperature and decreasing pressure.

Although thermodynamics determines the limit for reaction(s), the yield of a specific product and operating conditions are largely governed by catalyst type used, i.e.

reaction kinetics. The following sections cover these aspects for different desirable hydrogenation products.

2.3.2 CO_2 to Formic Acid and Derivatives

Formic acid is considered as an energy/hydrogen storage material because it can readily decompose to CO-free hydrogen and CO_2 which can easily be separated. No hydrogen is lost as water for its synthesis.

The direct synthesis of formic acid by the hydrogenation of CO_2 is thermodynamically highly unfavourable, which is given in the following equation:

$$CO_2\,(g) + H_2\,(g) \Leftrightarrow HCOOH\ (l) \qquad \Delta G^0_{298} = +\ 32.9\ \text{kJ mol}^{-1} \tag{2.3}$$

However, the presence of a solvent alters the thermodynamics of the reaction and the reaction becomes slightly exergonic when operated in the aqueous phase, which is expressed in the following equation (Alvarez et al., 2017):

$$H_2\,(aq) + CO_2\,(aq) \Leftrightarrow HCOOH(aq) \qquad \Delta G^0_{298} = -4\ \text{kJ mol}^{-1} \tag{2.4}$$

So far, most of the catalysts tested for the system have been homogeneous metal complexes. The best performances have been observed for complexes of rows 2 and 3 of groups 8–10 and 12 metals such as Ru and Rh, usually with halides and hydrides as anionic ligands and phosphenes as neutral ligands (Hao et al., 2011). Despite their reasonable yields, homogeneous catalysts are not desirable for large-scale applications due to difficulties in their separation from the products.

Preti et al. (2011) studied the reverse reaction of decomposition of formate salts to H_2 and CO_2 over transition metals of groups 8–11 such as Fe, Co, Ni, Cu, Ru, Rh, Pd, Ag, Ir, Pt, and Ru at 40°C for screening the active metals. Only Au black showed gas evolution, implying that it is also capable of catalysing the desirable hydrogenation of CO_2. Later works illustrated that Au in triethylamine (Et_3N) exhibits good performance although the aggregation of metal particles causes the deactivation of the catalyst. Therefore, stabilization over support is necessary to increase the stability.

Au, Pd, Rh, Ru, and Ni, as well as their combinations, have been tested as bulk or supported catalysts for the synthesis of formic acid from CO_2. Pd and Au are the most verified and their performance can be improved by using support (Alvarez et al., 2017).

For Pd-based catalysts, the hydrophobic carbon supports were given the best results, whereas for Au (and Ru)-based catalysts, the hydrophilic supports such as Al_2O_3 and TiO_2 resulted in best catalytic performances (Alvarez et al., 2017). Au(1 wt.%)/Al_2O_3 has shown higher activity and stability compared to the unsupported catalyst, with turn-over number (TON) = 850 at 40°C and 18 MPa in the presence of Et_3N.

In the decomposition process of formic acid solutions, the order of activity was Pd-Au/C>Pd-Ag/C>Pd-Cu/C>Pd/C, in which the highert activity of bimetallic catalysts could be attributed to the high resistance of Ag and Au to CO poisoning (Singh et al., 2016). More research is necessary for elucidating the structure–property relationships of the catalysts.

As noted above, to overcome the thermodynamic limitations, the reaction is performed in a basic solution. The amine type used is also effective in product yield. Among different amines tested over Au/TiO_2 (being commercially available as AUROlite), Et_3N shows the highest $HCOO^-$ yields. The reason is not clear. Perhaps the optimum zwitterionic interaction between CO_2 and Et_3N plays a role. Furthermore, the formate/Et_3N adduct has ionic liquid properties and changes the dipole of the solution (Gunasekar et al., 2019). The produced HCOOH can be separated by amine exchange by adding a high boiling point amine (e.g. tri-n-hexylamine) followed by distillation.

Another class of catalysts is heterogenized molecular catalysts. Homogeneous catalysts show superior performance compared to their heterogeneous counterparts by providing more active sites that molecularly dispersed. To eliminate the problems associated with the separation of the catalyst from the product, they are immobilized on a support. However, the immobilization of homogeneous catalysts results in declining their catalytic performance by several orders of magnitude (Gunasekar et al., 2016).

2.3.3 Reverse Water–Gas Shift Reaction

Hydrogen can reduce CO_2 to CO at elevated temperatures through reverse water-gas shift (RWGS) reaction, which is expressed in the following equation:

$$CO_2 + H_2 \Leftrightarrow CO + H_2O \qquad \Delta H^0_{298} = 41.1 \text{ kJ mol}^{-1} \qquad (2.5)$$

The reaction is mildly endothermic and is favoured at higher temperatures (>600°C). It provides a route for converting CO_2 to oxygenates and hydrocarbons via CO by syngas chemistry. RWGS occurs in many processes where CO_2 and H_2 occur together. Many researchers consider RWGS as a key step in CO_2 hydrogenation reactions. When RWGS and CO hydrogenation are conducted separately, the resulting water from the former should be avoided because it can deactivate the subsequent step (e.g. methanol or Fischer–Tropsch (FT) synthesis) catalysts. The high operating temperature of RWGS complicates the cascade reaction. Regarding conversion and energy consumption, an intermediate temperature is preferred.

The catalysts employed commercially in water–gas shift (WGS) can be potentially applied in RWGS as well. As a consequence, an obvious choice is to try the existing WGS catalysts to the reverse reaction. However, the higher operating temperatures of the latter can damage the catalyst. The catalysts investigated for RWGS con be classified as supported metal catalysts, metal oxides, and transition metal carbides (TMCs).

The conventional Cu-based $Cu/ZnO/Al_2O_3$ low temperature shift conversion (LTSC) catalyst has the advantage of no methane formation activity (as a competing reaction) in RWGS, but its thermal stability is low due to the sintering of Cu. Fe and K_2O promoters can improve the thermal stability of Cu catalysts. Two mechanisms have been proposed over Cu catalysts, namely the redox and the formate decomposition mechanisms (Porosoff et al., 2016). The former can be simplified as the following reaction sequence:

$$CO_2 + 2Cu^0 \Leftrightarrow Cu_2O + CO \tag{2.6}$$

$$Cu_2O + H_2 \Leftrightarrow 2Cu^0 + H_2O \tag{2.7}$$

The formate formation mechanism suggests that CO can be formed by the decomposition of the formate intermediate (HCOO*) that resulted from the hydrogenation of CO_2 followed by its decomposition to CO and surface OH* via C=O bond cleavage.

Ni-based catalysts like Ni/CeO_2-ZrO_2 have been tested for RWGS. A drawback of Ni as a RWGS active component is its activity in methanation. It has been shown that the selectivity of RWGS increases with reducing Ni loading (less than 3 wt.%) or decreasing NiO particle size (<5 nm). In contrast, higher loadings and larger particle sizes (>30 nm) promote the methanation reaction (Lu and Kawamoto, 2014).

The commercial Fe_2O_3/Cr_2O_3 catalyst for high temperature shift conversion (HTSC) also shows poor performance in RWGS because of the reduction of iron oxide to Fe metal and extensive coke formation.

Several metal oxides have been introduced as alternatives to metal-based catalysts. Oxide catalysts are less susceptible to sintering and sulphur poisoning compared to metal-based catalysts. ZnO/Al_2O_3 and $ZnAlO_2$ are effective at higher temperatures (400°C–700°C); however, they are unstable due to the volatilization of zinc oxide. Ga_2O_3 and In_2O_3 are other active oxide catalysts with the activity order $In_2O_3 > Ga_2O_3$ (Sun et al., 2014). CeO_2 promotes both catalysts by favouring CO_2 adsorption and by the generation of oxygen vacancies, which can enhance the dissociative adsorption of H_2 (Zhao et al., 2012; Wang et al., 2016). Mixed oxides like barium zirconate perovskite are another class of oxide catalysts being applied at high temperatures with reasonable performance (Kim et al., 2014).

TMCs are a new class of catalysts with hybridized electron configuration similar to noble metals. They have been used at higher temperatures where other reactions like methanol formation become thermodynamically unfavourable. Mo_2C is more active than other carbide catalysts, which can be attributed to facile oxygen transfer on its surface. Mo_2C exhibits dual properties, namely dissociation of H_2 and rupture of C=O bonds, and it acts as a reducible oxide. The catalyst breaks C=O bonds and forms surface CO* and O*. CO* can be desorbed as CO, whereas the O* as oxycarbide (Mo_2C-O) is released by H_2 (Su et al., 2017). The use of a support can further improve the catalytic performance.

Mo_2C can be used as a support or co-catalyst in hydrogenation reactions. It is also active in the formation of methane and methanol. The drawbacks of this type of catalyst are the high production cost and low surface area.

2.3.4 Methanol and Dimethyl Ether Syntheses

Methanol and DME are important chemical intermediates and also potential fuels for combustion engines and fuel cell applications. The synthesis reaction shown in Eq. (2.8) is performed at relatively low temperatures (typically 220°C–270°C). Higher pressure conditions increase the equilibrium yield of methanol.

$$CO_2 + 3H_2 \Leftrightarrow CH_3OH + H_2O \qquad \Delta H^0_{298} = -49.4 \text{ kJ mol}^{-1} \tag{2.8}$$

The number of research articles on methanol synthesis by the catalytic hydrogenation of CO_2 has been greatly increased in the last two decades (from about 80 to 500 in 1997 to 2017 according to science Citation Index Expanded (SCI-Expanded) search by Dang et al. (2019)). The interested researcher is referred to the valuable review papers published in these fields (Jadhav Prakash et al., 2014; Ganesh, 2014; Tursunov et al., 2017; Catizzone et al., 2018; Dang et al., 2019).

The hydrogenation of CO_2 over heterogeneous metal catalysts leads directly to methanol and methane. Spectroscopic studies revealed formates, formaldehyde, and formyl as intermediates over the surface (Olah et al., 2018b). The hydrogenation of CO_2 is interrelated with that of CO via WGS reaction, and their distinction is arbitrary. Isotope labelling studies illustrated that CO_2 is the primary source of methanol in CO hydrogenation, which is expressed in the following equation (Olah et al., 2018b):

$$CO + 2H_2 \Leftrightarrow CH_3OH \qquad \Delta H^0_{298} = -90.6 \text{ kJ mol}^{-1} \tag{2.9}$$

The conventional multicomponent Cu/Zn/Al/Cr catalyst used for the hydrogenation of CO to methanol shows poor performance in CO_2-to-methanol conversion because of the negative impact of the produced water on strongly hydrophilic alumina, which results in rapid deactivation of the catalyst by Cu sintering. Therefore, novel catalysts should be developed. The metal of choice is Cu with suitable promoters. The most useful form is Cu/ZnO-based with metal oxide promoters. The $CuO/ZnO/ZrO_2$ catalyst system shows higher activity and stability compared to Al-containing catalysts due to its higher hydrogen storage capacity and higher thermal stability (Wang et al., 2019).

X-ray photoelectron spectroscopic (XPS) studies of $Cu/ZnO/Al_2O_3$ catalyst showed that part of Cu is oxidized to Cu^{2+} after the long-term operation. Deactivation of the catalyst is a consequence of Cu oxidation and ZnO agglomeration during the hydrogenation of CO_2 (Liang et al., 2019). Growth of Cu particles is not observed after 72 h, which is attributed to low reaction temperature (200°C). The water formed during reaction accelerates the agglomeration of Cu particles.

Noble metals such as Pt and Pd are also effective in the synthesis of methanol from CO_2. Pd-based catalysts show high activity and selectivity (Yang et al., 2017). Bimetallic Pd-Cu/SiO_2 catalysts prepared by the co-impregnation method exhibit strong synergic effects within Pd/(Pd+Cu) atomic ratios of 0.25–0.34 (Jiang et al., 2019). Temperature-programmed desorption (TPD) analyses showed that Pd-Cu alloys shift adsorption towards weakly bonded H_2 and CO_2, which is related to CH_3OH promotion.

Indium oxide is also an effective catalyst with high selectivity in the hydrogenation of CO_2 to methanol (Sun et al., 2015). Oxygen vacancies formed by the reaction of H_2 and In_2O_3 are CO_2 adsorption and activation sites (Ye et al., 2012). In_2O_3/ZrO_2

catalyst prepared by the impregnation method shows high activity and selectivity (~100%) over a long period of time (1000h). The results are accounted for by the continuous creation and loss of oxygen vacancies and the electronic interactions with zirconia carrier-promoting formation of oxygen vacancies on In_2O_3 surface (Martin et al., 2016).

Transition metal carbides such as Mo_2C and Fe_3C show high CO_2 conversions and good methanol selectivities, but they need improvements.

DME can be obtained by the dehydration of methanol, which is expressed as follows:

$$2CH_3OH \Leftrightarrow CH_3OCH_3 + H_2O \qquad \Delta H^0_{298} = -23.6 \text{ kJ mol}^{-1} \quad (2.10)$$

DME can also be synthesized by CO_2 hydrogenation in a single-step process over a bifunctional catalyst or via a two-step process. By a single-step route, the CO_2 conversion can be increased through the removal of methanol product, which shifts equilibrium to more products. The overall reaction is a combination of Eqs. (2.8) and (2.10), which is expressed in the following equation:

$$2CO_2 + 6H_2 \Leftrightarrow CH_3OCH_3 + 3H_2O \qquad \Delta H^0_{298} = -122.2 \text{ kJ mol}^{-1} \quad (2.11)$$

The operating temperature for the aforementioned reaction (250°C–275°C) is similar to that for the hydrogenation of CO_2 to methanol.

The bifunctional catalysts for the single-step DME synthesis contain active sites for methanol synthesis, mostly Cu-based, but also Pd, and and a solid acid such as γ-Al_2O_3, HZSM-5, HY, and SAPOs (Zhou et al. 2016). γ-Al_2O_3 is an effective catalyst for methanol dehydration to DME due to its moderate acidity, high selectivity, and low hydrocarbon (which can be transformed into coke) formation. However, alumina is hydrophilic in nature, and the water produced in RWGS competes with methanol for active sites, thus limiting its application in direct CO_2-to-DME conversion (Ateka et al. 2017b). As a consequence, other solid acids have been incorporated into the catalyst formulations.

The bifunctional catalyst can be prepared by physical or mechanical mixing of the respective catalysts, but this approach reduces the intimate contact of their active sites. Precipitation–deposition preparation method is not effective as the metal precursors can be adsorbed on zeolite active sites due to their cation exchange capability. Solid-state synthesis has been proposed as an alternative method. Hydrated nitrates of metal precursors are mixed with oxalic acid as cation ligand and ground into a muddy mixture which is mixed with the zeolite, followed by drying and calcination (300°C–600°C). Over CuO-ZnO-ZrO_2/HZSM-5 bifunctional catalysts prepared by this method, CO_2 conversion and DME selectivity decrease with calcination temperature due to the reduction in Cu surface area by sintering. With 67.7% CO_2 conversion, the DME yield is 15% (at 250°C, 30MPa, gas hourly space velocity (GHSV)=3600mL $g^{-1}h^{-1}$) over the catalyst calcined at 300°C, which is about 50% larger than DME yield over the corresponding catalysts prepared by physical mixing (Li et al., 2016).

2.3.5 Synthesis of Higher Alcohols

Ethanol and higher alcohols are more desirable than methanol as liquid fuels and fuel additives because of their easier handling, safer transport, higher energy density, and better compatibility with gasoline.

Equation 2.12 shows that the stoichiometric H_2/CO_2 ratio is independent of the carbon number of alcohol product. The overall reaction can be considered as a combination of ethanol (or higher alcohol) synthesis from syngas ($CO+H_2$ mixture) and RWGS. Therefore, it is expected that the catalysts which are active in both reactions will also be active in the synthesis of higher alcohols.

$$nCO_2 + 3nH_2 \Leftrightarrow C_nH_{2n+1}OH + (2n-1)H_2O \tag{2.12}$$

The hydrogenation of CO_2 to higher alcohols has been less extensively studied than that for methanol. The catalysts used for ethanol synthesis are similar to those used for methanol synthesis, but they need centres for promoting C–C bond formation. This is a challenge in catalyst development as the catalyst should contain active sites activating CO molecules both associatively and dissociatively. Based on this concept, the earliest attempts were devoted to composite catalysts comprising Fe-based and Cu-based components for FT synthesis and oxygenate formation, respectively (Inui, 1996). Physically mixed catalysts were also applied (Inui et al., 1999). The catalysts used for the synthesis of higher alcohols can be classified as modified Cu catalysts (Li et al., 2013a), promoted MoS_2 catalysts (Nieskens et al., 2011; Liu et al., 2017a), and Rh-based catalysts (Kusama et al., 1994).

Effective conversion of CO_2 to EtOH remains a challenge because methanol or C_1–C_4 mixed alcohols are the main products hindering the commercial implementation. Large-scale direct hydrogenation of CO_2 to C_{2+} alcohols has not accomplished yet.

2.3.6 Methanation

Methanation has been used in industries for many decades for the removal of residual CO_x, especially in ammonia plants. Currently, the hydrogenation of carbon dioxide to methane (synthetic natural gas, SNG) via Sabatier reaction, expressed in Eq. (2.13), has no practical justification although research on catalyst and reaction condition optimization is very active because of its potential applications. Recently, researchers published review papers on this topic (Ghaib et al., 2016; Su et al., 2016; Frontera et al., 2017; Navarro et al., 2018).

$$CO_2 + 4H_2 \Leftrightarrow CH_4 + 2H_2O \qquad \Delta H^0_{298} = -165.1 \ \text{kJ mol}^{-1} \tag{2.13}$$

The reaction is thermodynamically highly favourable; however, it is kinetically limited as it involves the reduction of fully oxidized carbon. Therefore, an effective catalyst is required. Various metals, including Ru, Rh, Co, Ni, and Fe supported on metal oxides (e.g. Al_2O_3, SiO_2, TiO_2, CeO_2, CeO_2-ZrO_2, and zeolites), have been examined in the temperature range of 275°C–400°C. For high metal loadings (as in Ni-based catalysts), co-precipitation or precipitation–deposition techniques are preferable. The order of activity is Ru>Rh>Ni>Fe>Co, and that of selectivity is Pd>Pt>Ir>Ni>Rh>Co (Younas et al., 2016). Noble metals show the best performance, but their application is limited due to their high cost. Ru/TiO_2 can promote the reaction even at room temperature.

The metal oxide support is important in the dispersion of the active metal and affects the specific activity of the metal, which in turn influences the catalytic performance. Metal-support interactions are important in methanation catalysts. Over Ni, Pd, and Ru catalysts, metal nanoparticle size, as well as the partial pressure of hydrogen, controls the CO/CH_4 ratio in the product. Smaller particles promote CO formation (Wu et al., 2015).

Nickel-based catalysts are most intensively studied because of their high activity and low cost. However, they show low sulphur tolerance and are susceptible to sintering and coke formation. The loading of Ni affects the metal–support interaction and metal dispersion (or particle size). The support influences the performance of 10 wt.% Ni-loaded catalysts prepared by the impregnation method in the order $Ni/CeO_2 > Ni/\alpha\text{-}Al_2O_3 > Ni/TiO_2 > Ni/MgO$ (Tada et al., 2012). The equilibrium CO_2 conversion for the CeO_2-supported catalyst (90% conversion with 100% methane selectivity) is achieved at 350°C (GHSV = 10,000 h^{-1}), which is 100°C lower than that for other catalysts. The superior performance of CeO_2 as a support can be attributed to the coverage of CeO_2 surface with species resulted from CO_2 and partial reduction of CeO_2. Promoters include MgO, La_2O_3, and CeO_2. Ceria has been used as an electronic and structural promoter in Ni-based catalysts.

To improve the performance, bimetallic catalysts have also been developed. The introduction of a second metal (in low quantities) can change the electronic and geometric structures of the base metal. Methanation of carbon dioxide over $Ni\text{–}M/ZrO_2$ (M = Fe, Co, Cu) catalysts prepared by the co-impregnation method shows that the incorporation of iron (3 wt.%) enhances the catalyst activity by improving the reduction and dispersion of Ni and also by the partial reduction of zirconia (Ren et al., 2015).

The proposed reaction mechanisms are classified into two main categories from experimental and theoretical calculation studies. In one reaction pathway, carbon dioxide is converted to CO as an intermediate followed by the methanation of CO. The other mechanism involves the formation of formate species as the main intermediate during the direct hydrogenation of CO_2 (Muroyama et al., 2016).

2.3.7 Conversion to Higher Hydrocarbons

Higher (C_{2+}) hydrocarbons are more valuable than methane because of their higher reactivity as chemical feedstocks and easier handling and higher energy density as fuels. Lower olefins (C_2–C_4) and aromatics are, in particular, the important building blocks for producing a vast variety of commodity chemicals.

Recently, Yang et al. (2017) and Li et al. (2018) published valuable review articles on the subject of CO_2 conversion to the value-added hydrocarbons. The hydrogenation of carbon dioxide to hydrocarbons can broadly be divided into two routes, namely methanol-mediated routes and non-methanol-mediated or direct routes. In the former, CO_2 is first hydrogenated to methanol as an intermediate, which is then converted to hydrocarbons over the acidic sites of a hybrid catalyst. In the second route, the reaction passes through RWGS-FT synthesis. The product composition varies depending upon the reaction route and the catalyst composition.

The composite catalyst (e.g. Cu-ZnO-Al_2O_3/ZSM-5) used in the methanol-mediated route, as shown in Eq. (2.14), contains active sites (mostly Cu based) for methanol synthesis and an acidic component (mostly a zeotype) for the dehydration of the methanol formed. The zeolite component determines the hydrocarbon composition through shape selectivity. SAPO-34 gives high light olefin selectivities but it is much prone to coke formation than the less selective ZSM-5 zeolite.

$$nCH_3OH \rightarrow (-CH_2-)_n + nH_2O \qquad n \geq 2 \tag{2.14}$$

Under the high temperatures required for the conversion of methanol to aromatics (400°C–550°C), metal oxides exhibit better performance compared to the metal-based catalyst in the synthesis of methanol from carbon dioxide. Similarly, Zn-ZrO/SAPO (Li et al., 2017) and In_2O_3-ZrO_2/SAPO-34 (Dang et al., 2018) bifunctional catalysts have been employed for lower olefin synthesis from CO_2. As might be noted, the main difference between aromatic and olefin-selective catalysts is their acidic component promoting the methanol-to-hydrocarbon (MTH) conversion.

Aromatics have been produced over tandem catalysts of this type such as ZnO/ZrO_2-ZSM-5 (Zhang et al., 2019a) and Zn-CrO_x–zeolite composites (Zhang et al., 2019b). The aromatic selectivity showed a volcano-like plot with temperature with a peak at about 320°C. The Zn-CrO_x–Zn-ZSM-5 catalyst ($SiO_2/Al_2O_3 = 140$) exhibited 19.9% CO_2 conversion, 29.8% hydrocarbon selectivity, and 81.1% aromatic selectivity in C_{5+} products ($T = 320$°C, $P = 5$ MPa, feed flow = 2000 mL $g^{-1}h^{-1}$).

The conventional FT synthesis produces oxygenates and hydrocarbons via the hydrogenation of carbon monoxide. The well-known catalysts are Fe, Co, and Ru. The product is characterized by Anderson–Schulz–Flory (ASF) distribution with CH_4 and C_2 hydrocarbons showing deviations. Because CO is the chain growth agent in FT, CO_2 produces lighter products ($<C_{5+}$). Therefore, the chain growth probability, α, for CO_2-FT is typically much smaller than that for conventional FT. According to ASF distribution, the maximum C_5–C_{15} selectivity of 60 wt.% is achieved for $\alpha = 0.8$ (Rodemerck et al., 2013).

Many of the studies on CO_2 hydrogenation have been dealt with the conversion of CO/CO_2 mixtures or CO_2 over commercial FT catalysts. However, the yields were low and small amounts of C_{2+} were produced. Therefore, for the successful application in CO_2 conversion, improvements in catalyst formulations are necessary.

RWGS activity of the catalyst is important when FT route for CO_2 hydrogenation is under consideration. RWGS produces the CO required for conventional FT synthesis. Among FT catalysts, only iron-based catalysts show an appreciable RWGS activity.

Mössbauer studies showed that in CO_2-FT over iron-based catalysts, χFe_5C_2, FeC_3, and Fe_3O_4 are present (Riedel et al., 1999). In conventional iron-based catalysts, the active-phase Fe_5C_2 (Hägg carbide) forms *in situ* from the syngas feed. It has been said that RWGS occurs over oxide phase (Fe_3O_4) and tandem conversion to hydrocarbons and chain growth occurs during carbide phase. However, Infra-red

(IR) spectroscopic studies showed that both reactions have common intermediates (Chakrabarti et al., 2015).

The performance can be improved by a second metal (Cu, Mn, La, ...), alkali metal (Na, K, Rb), and/or metal oxide (α-Al_2O_3, TiO_2) as a promoter in catalyst formulation. Nevertheless, the C_{5+} yield is still very low for CO_2-FT. Potassium has widely been used as a promoter in iron-based catalysts by acting as an electronic promoter (Wang et al., 2011). It facilitates Hägg carbide formation and increases α. Incorporation of K increases CO_2 adsorption but decreases H_2 adsorption. The optimum amount of K is K/Fe = 0.5 mol/mol. Excess K results in the growth of carbide sites and pore blockage by coke (Sai Prasad et al., 2008).

Catalyst deactivation is a consequence of phase transfer. FeC_3 is inactive and is formed as a result of the carburization of Fe_5C_2 (Li et al., 2018). The water produced by CO_2-FT deactivates the catalyst through the oxidizing active phase and promoting its sintering; therefore, the rate of hydrogenation of CO_2 is typically slower than that of CO. The main challenges are to reduce rapid deactivation rate by coke formation, phase change, and sintering.

Unlike iron, cobalt is active as metal. Cobalt and ruthenium show little RWGS activity; therefore, methanation is the main reaction. The activity and selectivity are size dependent. Thus, for cobalt particles smaller than 10–11 nm, the activity and C_{5+} selectivity decrease with the particle size, whereas CH_4 selectivity shows an opposite trend (Melaet et al., 2014).

Operating conditions also play a role in catalytic performance results. Over a series of promoted Fe catalysts, the C_5–C_{15} yield increases with temperature and H_2/CO_2 ratio in the feed within the range of 300°C–350°C and 3–6 mol/mol, respectively. However, methane selectivity also increases with temperature (Rodemerck et al., 2013). The increased C_5–C_{15} yield can be explained by the fact that the endothermic equilibrium-limited RWGS reaction is favoured by higher temperatures and higher H_2/CO_2 ratios, thereby promoting CO-FT and increasing chain growth by the increased CO production.

For mixed CO_x feed on Co-based catalysts, CO_2 mostly acts as CO diluent and increases the selectivity of CH_4 (Riedel et al., 1999). This was observed over different types of supports, implying that it is a characteristic of Co irrespective of support or promoter. Over Co/CNT (carbon nanotube), the hydrogenation of CO is controlled by FT synthesis, whereas that of CO_2 is controlled by methanation. Increasing the CO_2 portion in CO_x promotes the catalyst deactivation and makes the product lighter with higher selectivities of methane. CO_2 acts as an oxidant and oxidizes Co surface, resulting in decreasing activity. The low activity of smaller cobalt particles is attributed to their easier oxidation.

2.4 ISSUES AND CHALLENGES

In view of economics, direct CO_2 hydrogenation is not an economic approach. Basically, at first, a technique for CO_2 capture and separation with high purity must be developed. The cost for hydrogen as feedstock is \$10,000/ton, and the costs for LNG (liquefied natural gas) and methanol (CH_3OH) as the main products are \$770/ton and \$340/ton, respectively. Hence, an expensive feedstock is used in CO_2 hydrogenation to produce such cheap products, which makes this process a non-economically viable. If hydrogen is produced by fully fledged solar power technology, the production cost for CO_2 hydrogenation process will be greatly reduced. Another promising route is CO_2 conversion by solar energy (artificial photosynthesis) (Liu et al., 2019).

2.4.1 CO_2 Capture

In future, CO_2 utilization depends on the availability of captured CO_2, among others. As a consequence, developing technological methods for easy and cheap separation of carbon dioxide for its application as a carbon source is essential.

Basically, CO_2 capture can be classified into four main types: (i) pre-combustion, (ii) post-combustion, (iii) oxy-fuel combustion, and (iv) chemical looping combustion (CLC) capture (Cuellar-Franca and Azapagic, 2015). In the pre-combustion method, the fuel is first reformed to syngas ($CO+H_2$), then treated by WGS, followed by the removal of CO_2 (>20%, which facilitates its separation) and combustion of H_2. Post-combustion capture involves the removal of CO_2 from the flue gas (typically 3%–15% CO_2) as in the conventional energy generation systems. In the oxy-fuel combustion process, the fuel (coal) is burnt with nearly pure O_2 (>95%) mixed with steam or recycled flue gas (RFG).

There are several emerging technologies for CO_2 separation and capture. CLC produces CO_2 without the requirement of costly N_2 separation by avoiding direct contact between (gaseous) fuel and air. This also minimizes NO_x formation (Leung et al., 2014). The oxygen is supplied by a metal oxide, which acts as an oxygen carrier circulating oxygen between air and fuel reactors. The metal oxide should have high oxidation–reduction activity, high melting point, environmental compatibility, and high stability in oxidation–reduction cycles. The oxides of transition metals such as Fe, Co, Cu, Mo, Mn, Cr, and In have been tested, but none of them satisfy all the technical requirements (Al-Mamoori et al., 2017). The technology is still under development with no industrial reference.

For a given capture configuration, a specific separation technology (absorption, membrane separation, etc.) can be adopted for separating CO_2 from a targeted gas stream (fuel gas or flue gas). The technologies differ in their capability in handling different levels of impurities, gas volumes, and CO_2 content, as well as in separation efficiency, which have been reviewed by Takht Ravanchi et al. (2011), Kargari and Takht Ravanchi (2012), Mondal et al. (2012), Spigarelli and Kawatra (2013), Kenarsari et al. (2013), Takht Ravanchi and Sahebdelfar (2014), Songolzadeh et al. (2014), and Leung et al. (2014). The existing and emerging technologies are presented and compared in Table 2.2.

Novel methods have been employed to improve the separation efficiency of the conventional processes. Biological agents like carbonic anhydrase (CA) enzyme enhance CO_2-absorbing properties. CA converts CO_2 to bicarbonate ion very rapidly by accelerating CO_2 hydration although their stability is still a major challenge (Sharma et al., in press). The introduction of hierarchical mesopores into molecular sieve adsorbents improves their CO_2 separation efficiency by reducing the diffusion resistance (Liu et al., 2017b).

TABLE 2.2
Comparison of CO_2 Separation Technologies

Technology	Advantages (or Opportunity)	Disadvantage (or Challenge)
Absorption	• High efficiency (>90%) • Mature technology (commercialized for amine- and alkaline-based processes) • Applicable to low CO_2 partial pressure gas streams (3%–20%) • Equipment corrosion • Absorbent degradation	• Efficiency depends on CO_2 concentration • Energy-intensive absorbent regeneration • Low CO_2 loading capacity of the solvent (0.4–1.2 kg CO_2/kg adsorbent)
Adsorption	• High efficiency (>85%) • Favourable adsorption kinetics	• High energy requirement for regeneration • Low selectivity • Sensitive to oxygen-containing impurities (H_2O, NO_x, SO_x) • No commercial reference
Chemical looping combustion	• N_2 free of CO_2 • No need for separation	• High operational temperature • Oxygen carrier degradation • Under development with no commercial reference
Membrane separation	• High separation efficiency (>80%) • Operational simplicity • Commercial references	• Low fluxes and fouling • Energy intensive for post-combustion • Low selectivity
Cryogenic distillation	• Mature technology	• Application limited to high CO_2 concentration (>90 vol. %) • Need for water removal • Energy intensive • Limited to low temperature
Hydrate-based separation	• High storage capacity • Low energy consumption • Efficiency can be improved by additives	• Low rate of hydrate formation • High pressure operation • Further necessary development

2.4.2 Hydrogen Source

In CO_2 utilization processes, carbon dioxide is used as a carbon source. For hydrogen, a cheap, clean, and sustainable source should be available. Conventionally, about 96% of the large-scale hydrogen production is based on fossil fuels by steam reforming (SR), dry reforming, and partial oxidation (POX) or their combinations (i.e. combined reforming). Besides hydrogen, variable amounts of carbon oxides (mostly CO) are produced, which should be removed, e.g., by WGS followed by CO_2 absorption. In the POX method, coal is gasified with steam and oxygen at high pressure and temperature. This process is called IGCC (integrated gasification combined cycle). From the viewpoint of CO_2 emission, as these technologies are coupled with CCS in which CO_2 emission is reduced, they are favourable (Armaroli and Balzani, 2011).

The chemical looping process (CLP) (Huang and Tan, 2014) is a novel process demonstrated at pilot scale by which hydrogen is produced with *in situ* CO_2 capture. CLC and CLG (chemical looping gasification) are two applications of CLPs. In the CLC process, fuel source is used for the reduction of metal oxide oxygen carriers. On the other hand, by oxidation of the reduced oxygen carrier with air, heat is obtained for electricity. In CLG process, steam and/or CO_2 is used for oxidizing the reduced oxygen carrier after which H_2 and/or CO is obtained. These products can be processed by conventional techniques after which liquid fuels, chemicals, and/or electricity will be generated. In these oxidation–reduction cycles, high-quality heat and high-concentrated hydrogen and CO_2 can be produced from carbonaceous fuels (Zeng et al., 2012; Tong et al., 2013). Despite technical improvements, these technologies utilize fossil fuels, which are not non-renewable sources.

Biomass decomposition and water splitting by solar energy, thermolysis, and biological process are other routes for clean hydrogen production. Among these methods, water electrolysis is more favoured, as electricity is supplied from renewable energy. Biomass decomposition is another valuable route by which H_2 is produced under a net CO_2 reduction condition (Abbasi and Abbasi, 2011; Bičáková and Straka, 2012; Ngoh and Njomo, 2012; Pearson et al., 2012; Li et al., 2013b; Sahebdelfar, 2017).

Since the 1970s, the cycles of thermochemical water splitting have been developed for hydrogen production. These systems are coupled with solar or nuclear energy systems, and their total efficiency is around 50%. Their main drawback is high temperature (>2500°C) requirement for which chemical reagents are used to decrease the reaction temperature (Orhan et al., 2012). There are some challenges for thermolysis process, such as low conversion rates, material corrosion, side reactions, and energy loss of multiple steps. Regarding its cost and efficiency, the thermolysis process is not competitive with other technologies (Graves et al., 2011; Bičáková and Straka, 2012).

Hydrogen can be produced by solar energy, a method that is a solution to environmental and energy problems. Direct solar thermal (namely photolysis) and photoconversion systems are two routes for solar hydrogen production.

Photolysis process is similar to the thermolysis process; i.e., operating temperature is 200°C–2000°C with 70% efficiency. In photolysis, solar radiation warms up water to its dissociation temperature and splits it to hydrogen and oxygen. The main advantages of this method are no pollution generation and no need for fossil fuel. It is worth mentioning that its main limitation to commercialization is the nature of solar energy and thermal energy transfer within a narrow temperature range (Armaroli and Balzani, 2011; Ngoh and Njomo, 2012).

Photoconversion is another process for hydrogen production, which is simple with mild operating conditions. In photocatalytic systems, biomass and water are converted to hydrogen by means of solar energy and photocatalyst (Shimura and Yoshida, 2011).

Water electrolysis is an electrochemical reaction by which hydrogen and oxygen are obtained. Equation (2.16) represents a reduction reaction that occurs on the negatively charged cathode of the electrolyser, whereas Eq. (2.17) represents an oxidation reaction that occurs on the positively charged anode (Goetz et al., 2016):

$$H_2O(l) \rightarrow H_2(g) + \frac{1}{2}O_2(g) \qquad \Delta G_r^0 = +285 \text{ kJ mol}^{-1} \tag{2.15}$$

$$H_2O + 2e^- \rightarrow H_2 + O^{2-} \tag{2.16}$$

$$O^{2-} \rightarrow \frac{1}{2}O_2 + 2e^- \tag{2.17}$$

Worldwide, about 4% of hydrogen is produced by water electrolysis. If electricity is generated from a renewable energy source, this process can be an environmentally friendly. It has been commercialized since the 1890s. The main advantages of this process are its ease of operation and product separation. Moreover, during this process, no GHG is produced and the produced O_2 and H_2 are industrially applicable (Armaroli and Balzani, 2011).

Electrolysis is categorized into two types, namely HTE (high-temperature electrolysis) and LTE (low-temperature electrolysis). HTE method is normally operated above 500°C. Since 1975, SOE (solid oxide electrolysis) cells have been considered as the dominant type of HTE cells, and since the 1960s, AWE (alkaline water electrolysis) cells have been considered as the main type of LTE cells. The main advantages of the LTE method are its fast kinetics and high performance. On the other hand, the main advantage of the HTE method is its high H_2 production rate. The disadvantages of the HTE method are the sintering and agglomeration of electrode materials. AWE cells can easily be replaced and have high durability (approximately 10–20 years). The main advantages of SOE cells are high current density and effective heat utilization (Huang and Tan, 2014).

Water electrolysis is a promising route for H_2 production, but besides electrolyser, a large power source using renewable energy is required as well. Wind, biomass, water, solar energy, and geothermal heat are probable renewable energy resources as they can produce reliable, secure, and affordable electrical energy with low cost (Christopher and Dimitrios, 2012). Due to fluctuation and unavailability of all these renewable resources all the year, their development is restricted (Bajpai and Dash, 2012).

2.4.3 Selection Criteria

Catalytic hydrogenation of carbon dioxide provides a route for a number of oxygenates and hydrocarbons. The selection of a chemical transformation depends on the purposes of the target product such as energy carrier/storage, hydrogen carrier, chemical synthesis, and carbon fixation. The product should also be compatible with the available infrastructure for storage and handling, like transportation facilities.

Hydrogen consumption is common to all transformations and products with lower hydrogen demand per CO_2 reacted and products with less water formation like oxygenates are preferred. Hydrogen should be supplied by water splitting using a renewable energy source (e.g. solar or wind) or from biomass via reforming reaction.

The relative hydrogen loss, j, for a CO_2-based, hydrogen-containing hydrogenation product is defined as the ratio of hydrogen lost as water to hydrogen fed:

$$j = \frac{4C - 2O}{4C + H - 2O} \tag{2.18}$$

where C, H, and O, respectively, denote the number of carbon, hydrogen, and oxygen atoms per molecule. Table 2.3 shows the j values for representative CO_2 hydrogenation products. The numerical value of j ranges from 0 (for formic acid) to 1 (for CO), with smaller values being more desirable. For a given carbon number of a molecule, the order for j is aromatic>olefin>paraffin>alcohol. Thus, j increases with the degree of unsaturation. Furthermore, for the homologous series, it increases with carbon number. This implies that HCOOH followed by methanol are best CO_2-based materials and act as hydrogen carriers in terms of hydrogen transfer efficiency. Furthermore, from their Gibbs free energy of formation from CO_2 and H_2, their decomposition to CO_2 and H_2 to recover the hydrogen is much more favourable near ambient conditions.

For energy storage, the efficiency of the conversion process in converting the input energy to chemical energy is important. This is evaluated by exergic analysis

TABLE 2.3
The j Values of Eq. (2.18) for Representative CO_2 Hydrogenation Products

Product(s)	j value	Stoichiometric H_2/CO_2
HCOOH	0	1
CO	1	1
CH_3OH	0.33	3
C_2H_5OH	0.5	3
CH_3OCH_3	0.5	3
CH_4	0.5	4
C_2H_6	0.57	3.5
C_2H_4	0.67	3
C_6H_6	0.8	2.5

of the transformation process. A low exergic loss is associated with a low environmental impact of the process. Methanation, despite higher hydrogen consumption, has higher exergic efficiency compared to the synthesis of methanol and higher hydrocarbons (Sues et al., 2010). Therefore, it is a better candidate for renewable energy storage. Furthermore, the product can be transported by the existing pipelines for usage.

A hydrogen carrier should have a high hydrogen content and should be easily decomposed to CO-free hydrogen, which is necessary for fuel cell-driven vehicle. Formic acid is desirable for this application. Methanol can also produce hydrogen by SR over Cu-based catalyst.

For transformations utilizing CO_2 as a C_1 feedstock, the hydrogenation product should act as a building block or can easily be transformed into such compounds. Methane is not suitable as it uses a large amount of hydrogen and is chemically very stable and inert. In contrast, lower olefins and aromatics are the basic building blocks of the petrochemical industry. Methanol and DME are also reactive molecules and known chemical intermediates for the production of olefins and aromatics via methanol-to-hydrocarbon transformation over molecular sieve catalysts (Yaripour et al., 2015; Cogate, 2019).

For fuel application, which is potentially the largest sector, easy handling and high energy density are advantageous. Liquid products with little or no oxygen content, like FT products (and also DME), are best selections.

Overall, methanol and DME are the most versatile and general-purpose products obtained using the well-known available technologies. They are multipurpose and multisource chemicals (Michailos et al., 2019). Their versatility leads to the elaboration of the methanol economy by the Nobel laureate George Olah as a substitute to oil and gas economy for sustainable supplying energy carrier and raw material (Olah et al., 2009). Although the use of hydrogen as energy storage medium has higher exergic efficiency than methanol, the storage and use of H_2 are more complicated and cost-intensive due to fugitive nature of hydrogen (Quadrelli et al., 2011). Thus, the methanol economy is also an alternative to the hydrogen economy.

Methanol economy is based on recycling of anthropogenic CO_2 by combining the capturing, storage, and chemical recycling of CO_2. It relies on renewable and abundant CO_2 and H_2O. The energy required is supplied by alternative energies such as solar, wind, and geothermal energies. Water splitting for supplying renewable hydrogen and catalytic hydrogenation of carbon dioxide to methanol and derived products are the pivotal elements of the methanol economy (Goeppert et al., 2014).

On the other hand, methane can be used for the storage of off-peak electricity in integration with infrastructure for natural gas.

2.5 RECENT ADVANCES

The hydrogenation of carbon dioxide to methanol and methane has already been demonstrated at pilot scale or commercial scale.

Lurgi (now Air Liquide) company demonstrated CO_2 hydrogenation to methanol over Süd Chemie (now Clariant) Cu/Zn/Al catalyst at 260°C and 60 bar at pilot scale in 1994. The catalyst showed slight deactivation (Artz et al., 2018).

The CAMERE (carbon dioxide hydrogenation to methanol via reverse water–gas shift reaction) process developed by Korean Institute of Energy and Research (KIER) and Korea Gas Corporation (KOGAS) is a two-step process for methanol synthesis. In the first step at >600°C, CO_2 is partly converted to CO (>60%) by RWGS over a $ZnAl_2O_4$ catalyst. After water removal, the $CO/CO_2/H_2$ mixed gas is converted to methanol over $Cu/ZnO/ZrO_2/Al_2O_3/SiO_2$ catalyst jointly developed with RITE (Research Institute of Innovative Technologies for the Earth) at 250°C–300°C and 5–8 MPa. The two-step system reduces methanol synthesis reactor volume by a factor of 25% (Centi and Perathoner, 2014). The main role of the RWGS reactor is adjusting CO/H_2 ratio that can minimize water production in the methanol reactor, increase methanol yield, and reduce gas reflux ratio. The bench-scale plant has 50–100 kg day^{-1} methanol capacity. CAMERE process is superior to direct CO_2 hydrogenation process but its catalysts and high hydrogen cost need further research (Quadrelli et al., 2011).

A pilot with a capacity of 100 ton/year was constructed by Mitsui Chemicals in 2009. In this plant, effluent CO_2 of a petrochemical plant is used as feedstock and the waste by-product of coke furnaces is used as hydrogen. Besides, photocatalysts were developed by Mitsui Chemicals for H_2O splitting to obtain hydrogen from renewable sources. The first commercial plant by Mitsui is intended to be 600,000 ton $year^{-1}$ to provide feed for downstream olefin and aromatic plants (Yi et al., 2015).

In 2010, the first commercial CO_2 recycling plant was built by CRI (Carbon Recycling International) located in Iceland, and it uses gamogenesis CO_2 and electrolytic H_2 by employing geothermal energy, which renders its operation economical. The wind is also used as an energy source. This plant has been in operation since 2011 with 4000 ton $year^{-1}$ capacity. It was announced in 2014 that with industrial, research, and academic partners, a plant will be instructed for recycling CO_2 from the coal-fired power plant. This plant is named as 'The George Olah Renewable Methanol Plant' (Hertrich and Beller, 2019).

NITE (National Institute of Technology and Evaluation) and RITE, Japan, operated 50 kg day^{-1} pilot with novel $Cu/ZnO/ZrO_2/Al_2O_3/SiO_2$ catalyst (Ushikoshi et al., 1998). Operating at 200–270°C and 30–50 bar, high methanol selectivity (99.7%) can be achieved.

Another process is Carnol that has three sections (Steinberg, 1997):

- CO_2 capture from flue gas of a coal-fired power plant,
- CO_2 conversion (in the presence of H_2) to methanol,
- Methanol use as a fuel in cars.

In the Carnol process, carbon dioxide exited from stack gas of coal-fired power plants is extracted in order to have carbon sources, and methane is thermally decomposed at >800°C to have hydrogen source. Generally, Carnol is a carbon-neutral process, as all the carbon present in methane is turned to solid carbon and no CO_2 emission is observed. It is worth mentioning that the only disadvantage of the Carnol process is its high-temperature requirement.

In Osaka, Mitsui Chemical Inc. built a 100 ton $year^{-1}$ pilot plant by which CO_2 is directly converted to methanol. In this pilot plant, their factory exhaust is used as a carbon source. Renewable energy like sunlight is used to split water for hydrogen production. Of course, this technology is not commercialized yet, due to huge steam and electricity consumption and the lack of long-term stability (Huang and Tan, 2014).

Hydrogenation of carbon dioxide to methane (SNG) has been shown great growth in Europe and especially in Germany since 2009. It has been conducted in pilot scale by several companies because of its potential in excess electricity storage and has been reviewed by Bailera et al. (2017). In Germany, with the highest focus, the attention roots from the fact that the country intends to rely 100% on renewable energies. As wind energy and solar energy are fluctuating and intermittent, long time and high capacity are required. The power to gas (PTG) tackles this problem.

Tohoku University of Technology (Japan) has been a pioneer in PTG with Sabatier reaction (prototype 1996) and installed possibly the first pilot with relatively small scale (<10 kW) started in 2003. Hydrogen (4 N m^3h^{-1}) is obtained by hydrolysing seawater to produce 1 N m^3h^{-1} CH_4 (Bailera et al., 2017).

The Desert Research Institute (DRI, USA) has a small pilot with a maximum 60% CO_2 conversion over PK-7R (20–25 wt.% Ni) Haldor-Topsoe methanation catalyst when operating under $H_2/CO_2 = 4$ at 300°C–350°C (Quadrelli et al., 2011).

In 2012, ZSW (the Centre for Solar Energy and Hydrogen Research) demonstrated a PTG pilot plant that produced 300 $m^3 day^{-1}$ methane. The ZSW 250-kW_{el} plant (Stuttgart) uses a 250 kW alkaline electrolyser. Methanol section comprises two reactor types (tube bundle and plate reactors) operating in combination or separately (Ghaib and Zahrae Ben-Fares, 2018).

In 2013, with the support of ZSW pilot data, in Werlte (Germany), Audi AG demonstrated the first commercial PTG methanation plant (6.3 MW) and worldwide largest SNG plant (325 N m^3h^{-1}) that produced 1000 metric tones of methane per year. In the PTG process, pure H_2 is produced by electrolysis (of alkaline water from wind turbine energy) and a Ni-based catalyst is used in the methanation reactor that operates at 350°C and 30 bar. Carbon dioxide is captured by amine scrubbing from crude biogas. The PTG efficiency is about 54% (without using heat) (Otten, 2014).

The integrated High-Temperature Electrolysis and Methanation for Effective Power to Gas Conversion (HELMETH) project by Sunfire GmbH is aiming to integrate HTE and CO_2 methanation. The methanation module comprises two reactors in series with interstage water removal operating at 300°C and 30 bar using a Ni-based catalyst (Ghaib and Zahrae Ben-Fares, 2018). The heat generated in methanation will be used to vaporize water feeding to the electrolyser that will have a capacity of 15 kW and be operated at 800°C and 15 bar. This will result in the conversion efficiency of more than 85% from renewable electricity to SNG.

The methanation of flue gas (CO_2) is mostly limited to Brandenburg University, Germany. Müller et al. (2017) used an upscaled test station for the investigation of a catalytic CO_2 conversion into methane. Flue gas composition was simulated in a set-up, and real conditions of a power plant were used; 30 N m^3h^{-1} of flue gas and 20 N m^3h^{-1} of hydrogen were used for methane production by the Sabatier process. About 100 kW power is required for H_2 production by electrolysis. In this plant, with synthetic and real flue gas, up to 99% conversion with 100% selectivity is obtained.

The Norwegian RCO_2 AS developed a pilot process for recovering CO_2 from flue gas and converting to methane using renewable hydrogen. Based on the Sabatier reaction, the Tohoku University of Technology and British Petroleum (BP) constructed a pilot plant with very high selectivity in which about 25 vol.% of methane is produced and seawater electrolysis is used for H_2 production (Quadrelli et al., 2011).

2.6 CONCLUSIONS

The use of fossil fuels to supply growing energy demand and their limited resources resulted in attention to carbon dioxide capture and utilization as a C_1 feedstock in the chemical industries to mitigate GHS emissions and to ensure a secure supply of energy in future. The major challenges for exploiting this strategy are low CO_2 concentration in many CO_2-containing streams and large kinetic and thermodynamic barrier in CO_2 conversion. Nevertheless, carbon dioxide can participate in a large number of reactions as a reactant.

In CO_2 hydrogenation, a mixture of CO_2 and H_2 is converted to a variety of chemicals by a thermochemical process. Despite the limited number of products with heterogeneous catalysts, the products comprise important bulk chemicals which can be used as intermediate feedstock, commodity products, and fuels. The reactions are mostly reversible and equilibrium-limited. Therefore, catalysts play a crucial role in the reactions.

The catalysts typically used in CO hydrogenation in syngas chemistry exhibit poor performance in the corresponding reaction for CO_2. However, in cases that the catalyst shows RWGS activity, the distinction is arbitrary. In such a case, CO_2 hydrogenation can be achieved via CO hydrogenation although still modifications in catalyst formulation might be necessary due to excessive water formation, which can damage the catalyst or impact its performance. Tandem catalysts are more frequently used in CO_2 hydrogenation reactions. New classes of catalysts like transition metal carbides have been employed and shown promising results. The selection of a target molecule depends on the goal of reaction (hydrogen transfer, energy transfer, chemical feedstock, etc.), but overall, methanol and DME are more versatile compared to other hydrogenation products. The catalysts for different conversions frequently show similar components (active agent, promoters, and support) and reaction intermediates. Therefore, optimization of catalyst formulation and operating conditions is necessary for acceptable results.

The PTX (power to X, X=methane or methanol) concept for utilizing CO_2 is rapidly developing with several demonstration projects being implemented especially in Europe in the last two decades. Despite this fact, further research and developments are necessary for improving the catalyst formulations, as well as the environmental and economic performance of the processes.

LIST OF ABBREVIATIONS

AWE: Alkaline water electrolysis
ASF: Anderson–Schulz–Flory
BP: British Petroleum

CA:	Carbonic anhydrase
CAMERE:	Carbon dioxide hydrogenation to methanol via reverse water–gas shift reaction
CCS:	Carbon capture and storage
CCU:	Carbon capture and utilization
CCUS:	Carbon capture utilization and storage
CLC:	Chemical looping combustion
CLG:	Chemical looping gasification
CLP:	Chemical looping process
CNT:	Carbon nanotube
CRI:	Carbon Recycling International
DME:	Dimethyl ether
DRI:	Desert Research Institute
ECBM:	Enhanced coal bed methane
EOR:	Enhanced oil recovery
FT:	Fischer–Tropsch
GHG:	Greenhouse gas
GHSV:	Gas Hourly Space Velocity
HELMETH:	High-temperature Electrolysis and Methanation for Effective Power to Gas Conversion
HTE:	High-temperature electrolysis
HTSC:	High-temperature shift conversion
IEA:	International Energy Agency
IGCC:	Integrated gasification combined cycle
IR:	Infra-red
KIER:	Korean Institute of Energy and Research
KOGAS:	Korea Gas Corporation
LNG:	Liquefied natural gas
LTE:	Low-temperature electrolysis
LTSC:	Low temperature shift conversion
MEA:	Monoethanolamine
MOF:	Metal–organic framework
MTH:	Methanol to hydrocarbons
NITE:	National Institute of Technology and Evaluation
POX:	Partial oxidation
PTG:	Power to gas
PTX:	Power to X (X=heat, hydrogen, methane, methanol, or other liquids)
RFG:	Recycled flue gas
RITE:	Research Institute of Innovative Technologies for the Earth
RWGS:	Reverse water–gas shift
SCI-Expanded:	science Citation Index Expanded
SNG:	Synthetic natural gas
SOE:	Solid oxide electrolysis
SR:	Steam reforming
TMC:	Transition metal carbide

TON: Turn-over number
TPD: Temperature-programmed desorption
WGS: Water–gas shift
XPS: X-ray photoelectron spectroscopy

REFERENCES

Abbasi T., S. A. Abbasi, "Renewable hydrogen: Prospects and challenges", *Renew. Sus. Energy Rev.* 15 (2011) 3034–3040.

Ahmad K., S. Upadhyayula, "Greenhouse gas CO_2 hydrogenation to fuels: A thermodynamic analysis", *Env. Prog. Sus. Energy* 38 (2019) 98–111.

Al-Mamoori A., A. Krishnamurthy, A. A. Rownaghi, F. Rezaei, "Carbon capture and utilization update", *Energy Technol.* 5 (2017) 834–849.

Alvarez A., A. Bansode, A. Urakawa, A. V. Bavykina, T. A. Wezendonk, M. Makkee, J. Gascon, F. Kapteijn, "Challenges in the greener production of formates/formic acid, methanol, and DME by heterogeneously catalyzed CO_2 hydrogenation processes", *Chem. Rev.* 117 (2017) 9804–9838.

Araújo O. Q. F., J. L. Medeiros, R. M. B. Alves, *CO_2 Utilization: A Process Systems Engineering Vision*, INTECH, (2014).

Aresta M., A. Dibenedetto, "Carbon dioxide fixation into organic compounds", In Aresta M. (Ed.) "*Carbon Dioxide Recovery and Utilization*" Kluwer Academic Publishers, Dordrecht, (2003).

Aresta M., *"Carbon Dioxide as Chemical Feedstock"*, Wiley-VCH: Weinheim, (2010).

Armaroli N., V. Balzani, "The hydrogen issue", *Chem. Sus. Chem.* 4 (2011) 21–36.

Artz J., T. E. Muller, K. Thenert, "Sustainable conversion of carbon dioxide: An integrated review of catalysis and life cycle assessment", *Chem. Rev.* 118 (2018) 434–504.

Ateka A., P. P. Uriarte, M. Gamero, J. Erena, A. T. Aguayo, J. Bilbao, "A comparative thermodynamic study on the CO_2 conversion in the synthesis of methanol and of DME", *Energy* 120 (2017a) 796–804.

Ateka A., J. Erena, P. P. Uriarte, A. T. Aguayo, J. Bilbao, "Effect of the content of CO_2 and H_2 in the feed on the conversion of CO_2 in the direct synthesis of dimethyl ether over a CuO-ZnO-Al_2O_3/SAPO-18 catalyst", *Int. J. Hydrogen Energy* 111 (2017b) 100–108.

Bailera M., P. Lisbon, L. M. Romeo, S. Espatolero, "Power to gas projects review: Lab, pilot and demo plants for storing renewable energy and CO_2", *Renew. Sus. Energy Rev.* 69 (2017) 292–312.

Bajpai P., V. Dash, "Hybrid renewable energy systems for power generation in stand-alone applications: A review", *Renew. Sus. Energy Rev.* 16 (2012) 2926–2939.

Bičáková O., P. Straka, "Production of hydrogen from renewable resources and its effectiveness", *Int. J. Hydrogen Energy* 37 (2012) 11563–11578.

Catizzone E., G. Bonura, M. Migliori, F. Frusteri, G. Giordano, "CO_2 recycling to dimethyl ether: State-of-the-art and perspectives", *Molecules* 23 (2018) 31–58.

Centi G, S. Perathoner "Perspectives and state of the art in producing solar fuels and chemicals from CO_2", In Centi G, S. Perathoner (Eds.) "*Green CO2: Advances in CO_2 Utilization*", Wiley, New York, pp. 1–24, (2014).

Chakrabarti D., A. de Klerk, V. Prasad, M. K. Gnanamani, W. D. Shafer, G. Jacobs, D. E. Sparks, B. H. Davis, "Conversion of CO_2 over a Co-based Fischer-Tropsch catalyst", *Ind. Eng. Chem. Res.* 54 (2015) 1189–1196.

Christopher K., R. Dimitrios, "A review on exergy comparison of hydrogen production methods from renewable energy sources", *Energy Env. Sci.* 5 (2012) 6640–6651.

Cogate M. R., "Methanol-to-olefins process technology: Current status and future prospects", *Pet. Sci. Technol.* 37 (2019) 559–565.

Cuellar-Franca R. M., A. Azapagic, "Carbon capture, storage and utilization technologies: A critical analysis and comparison of their life cycle environmental impacts", *J. CO_2 Util.* 9 (2015) 82–102.

Dang S., P. Gao, Z. Liu, X. Chen, C. Yang, H. Wang, L. Zhong, S. Li, Y. Sun, "Role of zirconium in direct CO_2 hydrogenation to lower olefins on oxide/zeolite bi-functional catalysts", *J. Catal.* 364 (2018) 382–393.

Dang S., H. Yang, P. Gao, H. Wang, X. Li, W. Wei, Y. Sun, "A review of research progress on heterogeneous catalysts for methanol synthesis from carbon dioxide hydrogenation", *Catal. Today* 330 (2019) 61–75.

Frontera P., A. Macario, M. Ferraro, P. L. Antonucci, "Supported catalysts for CO_2 methanation: A review", *Catal.* 7 (2017) 59–86.

Ganesh I., "Conversion of carbon dioxide into methanol- a potential liquid fuel: Fundamental challenges and opportunities (a review)", *Renew. Sus. Energy Rev.* 31 (2014) 221–257.

Ghaib K., K. Nitz, F. Z. Ben-Fares, "Chemical methanation of CO_2: A review", *Chem. Bio. Chem.* 3 (2016) 266–275.

Ghaib K., F. Zahrae Ben-Fares, "Power-to-Methane: A state-of-the-art review", *Renew. Sus. Energy Rev.* 81 (2018) 433–446.

Gnanamani M. K., G. Jacobs, V. R. R. Pendyala, W. Ma, B. H. Davis, "Hydrogenation of carbon dioxide to liquid fuels". In: G. Centi, S. Perathoner (Eds) "*Green Carbon Dioxide: Advances in CO_2 Utilization*", 1st Ed. Wiley, New York, pp. 99–118 (2014).

Goeppert A., M. Czaun, J. P. Jones, P. G. K. Suriya, G. A. Olah, "Recycling of carbon dioxide to methanol and derived products- closing the loop" *Chem. Soc. Rev.* 43 (2014) 7995–8048.

Goetz M., J. Lefebvre, F. Moers, A. M. D. Koch, F. Graf, S. Bajohr, R. Reimert, T. Kolb, "Renewable Power-to-Gas: A technological and economic review", *Renew. Energy* 85 (2016) 1371–1390.

Graves C., S. D. Ebbesen, M. Mogensen, K. S. Lackner, "Sustainable hydrocarbon fuels by recycling CO_2 and H_2O with renewable or nuclear energy". *Renew. Sus. Energy Rev.* 15 (2011) 1–23.

Gunasekar G. H., K. D. Jung, S. Yoon, "Hydrogenation of CO_2 to formate using a simple, recyclable, and efficient heterogeneous catalyst", *Inorg. Chem.* 58 (2019) 3717–3723.

Gunasekar G. H., K. Park, K. D. J. S. Yoon, "Recent developments in the catalytic hydrogenation of CO_2 to formic acid/formate using heterogeneous catalysts" *Inorg. Chem. Front.* 3 (2016) 882–895.

Guo L., J. Sun, Q. Ge, N. Tsubaki, "Recent advances in direct catalytic hydrogenation of carbon dioxide to valuable C_{2+} hydrocarbons", *J. Mat. Chem. A*, 6 (2018) 23244–23262.

Halmann M. M., *Chemical Fixation of Carbon Dioxide: Methods for Recycling CO_2 into Useful Products*, CRC Press, Boca Raton, FL, (1993).

Hertrich M. F., M. Beller, "Metal-catalyzed hydrogenation of CO_2 into methanol", In Dixneuf, P. H., J-P. Soule (Eds.), "*Organometallics for Green Catalysis*", Springer, Berlin, (2019).

Hao C., S. Wang, M. Li, L. Kang, X. Ma, "Hydrogenation of CO_2 to formic acid on supported ruthenium catalysts", *Catal. Today* 160 (2011) 184–190.

Havran V., M. P. Dudukovic, C. S. Lo, "Conversion of methane and carbon dioxide to higher value products", *Ind. Eng. Chem. Res.* 50 (2011) 7089–7100.

Heyn R.H. "Carbon dioxide conversion", in Harvorth (Ed.), "*Encyclopedia of Catalysis*", Wiley, 2003.

Huang C. H., C. S. Tan, "A review: CO_2 utilization", *Aero. Air Quality Res.* 14 (2014) 480–499.

Inui T., "Highly effective conversion of carbon dioxide to valuable compounds on composite catalysts", *Catal. Today* 29 (1996) 329–337.

Inui T., T. Yamamoto, M. Inoue, H. Hara, T. Takeguchi, J. Kim, "Highly effective synthesis of ethanol by CO_2-hydrogenation on well balanced multi-functional FT type composite catalysts", *Appl. Catal. A. G.* 186 (1999) 395–406.

Jadhav Prakash S. G., D. V. Bhalchandra M. B. Jyeshtharaj B. Joshi, “Catalytic carbon dioxide hydrogenation to methanol: A review of recent studies”, *Chem. Eng. Res. Des.* 92 (2014) 2557–2567.

Jafarbegloo M., A. Tarlani, A. Wahid Mesbah, J. Muzart, S. Sahebdelfar, “NiO-MgO solid solution prepared by Sol–Gel method as precursor for Ni/MgO methane dry reforming catalyst: Effect of calcination temperature on catalytic performance”, *Catal. Lett.* 146 (2016) 238–248.

Jalama K., “Carbon dioxide hydrogenation over nickel-, ruthenium-, and copper-based catalysts: Review of kinetics and mechanism”, *Catal. Rev. Sci. Eng.* 59 (2017) 95–164.

Jia C., J. Gao, Y. Dai, J. Zhang, Y. Yang, “The thermodynamics analysis and experimental validation for complicated systems in CO_2 hydrogenation process”, *J. Energy Chem.* 25 (2016) 1027–1037.

Jia J., C. Qian, Y. Dong, Y. F. Li, H. Wang, M. Ghoussoub, K. T. Butler, A. Walsh, G. A. Ozin, “Heterogeneous catalytic hydrogenation of CO_2 by metal oxides: Defect engineering-perfecting imperfection”, *Chem. Soc. Rev.* 46 (2017) 4631–4644.

Jiang X., X. Nie, X. Wang, H. Wang, N. Koizumi, Y. Chen, X. Guo, C. Song, “Origin of Pd-Cu bimetallic effect for synergetic promotion of methanol formation from CO_2 hydrogenation”, *J. Catal.* 369 (2019) 21–32.

Kargari A., M. Takht Ravanchi, “Carbon dioxide: Capturing and utilization”, in L. Guoxiang (Ed.), *Greenhouse Gases*, InTech, pp. 1–30, Mar. (2012).

Kenarsari S. D., D. Yang, G. Jiang, S. Zhang, J. Wang, A. G. Russell, Q. Wei, M. Fan, “Review of recent advances in carbon dioxide separation and capture”, *RSC Adv.* 3 (2013) 22739–22773.

Kim D. H., Park J. L., Park E. J., Kim Y. D., S. Uhm, “Dopant effect of barium zirconate-based perovskite-type catalysts for the intermediate-temperature reverse water gas shift reaction”, *ACS Catal.* 4 (2014) 3117–3122.

Kusama H., K. Sayama, K. Okada, H. Arawaka, Preprints 74th Annual Meeting Catalysis Soc. Jpn. (1994) 430.

Leung D. Y. C., G. Caramann, M. M. Maroto-Valer, “An overview of current status of carbon dioxide capture and storage technologies”, *Renew. Sus. Energy Rev.* 39 (2014) 426–443.

Li S., H. Guo, C. Luo, H. Zhang, L. Xiong, X. Chen, L. Ma, “Effect of iron promoter on structure and performance of K/Cu–Zn catalyst for higher alcohols synthesis from CO_2 hydrogenation”, *Catal. Lett.* 143 (2013a) 345–355.

Li Y., S. H. Chan, Q. Sun, “Heterogeneous catalytic conversion of CO_2: A comprehensive theoretical review”, *Nanoscale* 7 (2015) 8663–8683.

Li Z., W. Luo, M. Zhang, J. Feng, Z. Zou, “Photoelectrochemical cells for solar hydrogen production: Current state of promising photoelectrodes, methods to improve their properties and outlook”, *Energy Environ. Sci.* 6 (2013b) 347–370.

Li L., D. Mao, J. Xiao, L. Li, X. Guo, J. Yu, “Facile preparation of highly efficient CuO-ZnO-ZrO_2/HZSM-5 bi-functional catalyst for one-step CO_2 hydrogenation to dimethyl ether: Influence of calcinations temperature”, *Chem. Eng. Res. Des.* 111 (2016) 100–108.

Li W., H. Wang, X. Jiang, J. Zhu, Z. Liu, X. Guo, C. Song, “A short review of recent advances in CO_2 hydrogenation to hydrocarbons over heterogeneous catalysts”, *RSC Adv.* 8 (2018) 7651–7669.

Li Z., J. Wang, Y. Qu, H. Liu, C. Tang, S. Miao, Z. Feng, H. An, C. Li, “Highly selective conversion of carbon dioxide to lower olefins”, *ACS Catal.* 7 (2017) 8544–8548.

Liang B., J. Ma, X. Su, C. Yang, H. Duan, H. Zhou, S. Deng, L. Li, Y. Huang, “Investigation on deactivation of Cu/ZnO/Al_2O_3 catalyst for CO_2 hydrogenation to methanol”, *Ind. Eng. Chem. Res.* 58 (2019) 9030–9037.

Liu M., G. Gadikota, “Integrated CO_2 capture, conversion, and storage to produce calcium carbonate using an amine looping strategy”, *Energy Fuels*, 33 (2018) 1–35.

Liu Q., P. He, X. Qian, Z. Fei, Z. Zhang, X. Chen, "Enhanced CO_2 adsorption performance on hierarchical porous ZSM-5 zeolite", *Energy Fuels*, 31 (2017b) 13933–13941.

Liu Y., Y. Yang, Q. Sun, Z. Wang, B. Huang, Y. Dai, X. Qin, X. Zhang, "Chemical adsorption enhanced CO_2 capture and photoreduction over a copper porphyrin based metal organic framework", *ACS Appl. Mater. Interfaces*, 5 (2013) 7654–7658.

Liu M., Y. Yi, L. Wang, H. Guo, A. Bogaerts, "Hydrogenation of carbon dioxide to value-added chemicals by heterogeneous catalysis and plasma catalysis" *Catalysts* 9 (2019) 275–311.

Liu S., H. Zhou, Q. Song, Z. Ma, "Synthesis of higher alcohols from CO_2 hydrogenation over Mo-Co-K sulfide-based catalysts", *J. Taiwan Ins. Chem. Eng.* 76 (2017a) 18–26.

Lu R., K. Kawamoto, "Preparation of mesoporous CeO_2 and monodispersed NiO particles in CeO_2, and enhanced selectivity of NiO/CeO_2 for reverse water gas shift reaction", *Mater. Res. Bull.* 53 (2014) 70–78.

Ma J., N. Sun, X. Zhang, N. Zhao, F. Xiao, W. Wei, Y. Sun, "A short review of catalysis for CO_2 conversion", *Catal. Today* 148 (2009) 221–231.

Martin O., A. J. Martin, C. Mondelli, S. Mitchell, T. F. Segawa, R. Hauert, C. Drouilly, D. Curulla-Ferre, J. Perez-Ramirez, "Indium oxide as a superior catalyst for methanol synthesis by CO_2 hydrogenation", *Angew. Chem. Int. Ed.* 55 (2016) 6261–6265.

Melaet G., A. E. Lindeman, G. A. Somorjai, "Cobalt particle size effects in the Fischer-Tropsch synthesis and in the hydrogenation of CO_2 studied with nanoparticle model catalysts on silica", *Top Catal*, 57 (2014) 500–507.

Michailos S., S. McCord, V. Sick, G. Stokes, P. Styring, "Dimethyl ether synthesis via captured CO_2 hydrogenation within the power to liquids concept: A techno-economic assessment", *Energy Convers. Manag.* 184 (2019) 262–276.

Mikkelsen M., M. Jørgensen, F. C. Krebs, "The teraton challenge: A review of fixation and transformation of carbon dioxide", *Energy Environ. Sci.* 3 (2010) 43–81.

Mondal M. K., H. K. Balsora, P. Varshney, "Progress and trends in CO_2 capture/separation technologies: A review", *Energy* 46 (2012) 431–441.

Müller K., F. Rachow, J. Israel, E. C. C. Schwiertz, D. Scmeisser, "Direct methanation of flue gas at a lignite power plant", *Int. J. Env. Sci.* 2 (2017) 425–437.

Muroyama H., Y. Tsuda, T. Asakoshi, H. Masitah, T. Okanishi, T. Matsui, K. Eguchi, "Carbon dioxide methanation over Ni catalysts supported on various metal oxides", *J. Catal.* 343 (2016) 178–184.

Navarro J. C., M. A. Centeno, O. H. Laguna, J. A. Odriozola, "Policies and motivations for the CO_2 valorization through the Sabatier reaction using structured catalysts. A review of the most recent advances", *Catalysts* 8 (2018) 578–602.

Ngoh S. K., D. Njomo, "An overview of hydrogen gas production from solar energy", *Renew. Sus. Energy Rev.* 16 (2012) 6782–6792.

Nieskens D. L. S., D. Ferrari, Y. Liu, R. Kolonko Jr, "The conversion of carbon dioxide and hydrogen into methanol and higher alcohols", *Catal. Commun.* 14 (2011) 111–113.

Norhasyima R. S., T. M. I. Mahlia, "Advances in CO_2 utilization technology: A patent landscape review", *J. CO_2 Util.* 26 (2018) 323–335.

North M., "What is CO_2? Thermodynamics, basic reactions and physical chemistry", in Styring P., E. A. Quadrelli, K. Armstrong (Eds.) *"Carbon Dioxide Utilization: Closing the Carbon Cycle"*, Elsevier, Amsterdam, (2015).

Olah G. A., A. Goeppert, G. K. Surya Prakash, *Beyond Oil and Gas: The Methanol Economy*, 2nd Ed., Wiley-VCH, Weinheim, (2009).

Olah G. A., T. Mathew, A. Goeppert, G. K. Surya Prakash, "Difference and significance of regenerative versus renewable carbon fuels and products", *Top. Catal.* 61 (2018a) 522–529.

Olah G. A., A. Molnár, G. K. Surya Prakash, *Hydrocarbon Chemistry*, 3rd Ed., Wiley, Hoboken, NJ (2018b), Volume 1, Chapter 3, pp. 125–236.

Omae I., "Recent developments in carbon dioxide utilization for the production of organic chemicals", *Coord. Chem. Rev.* 256 (2012) 1384–1405.

Orhan M. F., I. Dincer, M. A. Rosen, M. Kanoglu, "Integrated hydrogen production options based on renewable and nuclear energy sources", *Renew. Sus. Energy Rev.* 16 (2012) 6059–6082.

Otten R. "The first industrial PTG plant - Audi e-gas as driver for the energy turn around", *CEDEC Gas Day* (2014), Verona, Italy.

Otto A., T. Grube, S. Schiebahn, D. Stolten, "Closing the loop: Captured CO_2 as a feedstock in the chemical industry", *Energy Environ. Sci.* 8 (2015) 3283–3297.

Pearson R. J., M. D. Eisaman, J. W. G. Turner, P. P. Edwards, Z. Jiang, V. L. Kuznetsov, K. A. Littau, L. Di Marco, S. R. G. Taylor, "Energy storage via carbon-neutral fuels made from CO_2, water and renewable energy". *Proc. IEEE* 100 (2012) 440–460.

Porosoff M. D., B. Yan, J. G. Chen, "Catalytic reduction of CO_2 by H_2 for synthesis of CO, methanol and hydrocarbons: Challenges and opportunities", *Energy Environ. Sci.* 9 (2016) 62–73.

Preti, D., Resta, C., Squarcialupi, S., Fachinetti, G. "Carbon dioxide hydrogenation to formic acid by using a heterogeneous gold catalyst", *Angew. Chem. Int. Ed.* 50 (2011) 12551–12554.

Quadrelli E. A., Centi G., Duplan J. L., S. Perathoner, "Carbon dioxide recycling: Emerging large-scale technologies with industrial potential" *Chem. Sus. Chem.* 4 (2011) 1194–1215.

Ren J., X. Qin, J. Z. Yang, Z. F. Qin, H. L. Guo, J. Y. Lin, Z. Li, "Methanation of carbon dioxide over Ni-M/ZrO_2 (M=Fe, Co, Cu) catalysts: Effect of addition of a second metal", *Fuel Proc. Technol.* 137 (2015) 204–211.

Riedel T., M. Claeys, H. Schulz, G. Schaub, S. S. Nam, K. W. Jun, M. J. Choi, G. Kishan, "Comparative study of Fischer-Tropsch synthesis with H_2/CO and H_2/CO_2 syngas using Fe- and Co-based catalysts" *Appl. Catal. A. G.* 186 (1999) 201–213.

Rodemerck U., M. Holen, E. Wagner, Q. Smejkal, A. Barkschat, M. Baerns, "Catalyst development for CO_2 hydrogenation to fuels", *Chem. Cat. Chem.* 5 (2013) 1948–1955.

Roy S., A. Cherevotan, S. C. Peter, "Thermochemical CO_2 hydrogenation to single carbon products: Scientific and technological challenges", *ACS Energy Lett.* 3 (2018) 1938–1966.

Saeidi S., N. A. S. Amin, M. R. Rahimpour, "Hydrogenation of CO_2 to value-added products-A review and potential future developments", *J. CO_2 Util.* 5 (2014) 66–81.

Saeidi S., S. Najari, F. Fazlollahi, M. Khoshtinat Nikoo, F. Sefidkon, J. Jaromir Klemese, L. L. Baxter, "Mechanisms and kinetics of CO_2 hydrogenation to value-added products: A detailed review on current status and future trends", *Renew. Sus. Energy Rev.* 80 (2017) 1292–1311.

Sahebdelfar S., M. Takht Ravanchi, "Carbon dioxide utilization for methane production: A thermodynamic analysis", *J. Petrol. Sci. Eng.* 134 (2015) 14–22.

Sahebdelfar S., "Steam reforming of propionic acid: Thermodynamic analysis of a model compound for hydrogen production from bio-oil", *Int. J. Hydrogen Energy* 42 (2017) 16386–16395.

Sai Prasad P. S., J. W. Bae, K. W. Jun, K. W. Lee, "Fischer–Tropsch synthesis by carbon dioxide hydrogenation on Fe-based catalysts", *Catal. Surv. Asia* 12 (2008) 170–183.

Sharma T., Sharma S., Kamyab H., Kumar A., "Energizing the CO_2 utilization by chemo-enzymatic approaches and potentiality of carbonic anhydrases: A review", *J. Clean. Prod.* 247 (2020) 119138. DOI: 10.1016/j.jclepro.2019.119138.

Shen W. J., K. W. Jun, H. S. Choi, K. W. Lee, "Thermodynamic investigation of methanol and dimethyl ether synthesis from CO_2 hydrogenation", *Korean Chem. Eng.* 1 (2000) 210–216.

Shimura K., H. Yoshida, "Heterogeneous photocatalytic hydrogen production from water and biomass derivatives", *Energy Env. Sci.* 4 (2011) 2467–2481.

Singh A. K., S. Singh, A. Kumar, "Hydrogen energy future with formic acid: A renewable chemical hydrogen storage system", *Catal. Sci. Technol.* 6 (2016) 12–40.

Songolzadeh M., M. Soleimani, M. Takht Ravanchi, R. Songolzadeh, "Carbon dioxide separation from flue gases; A technological review emphasizing on reduction in greenhouse gas emissions", *Sci. World J.* 2014 (2014) 1–34.

Spigarelli B. P., S. K. Kawatra, "Opportunities and challenges in carbon dioxide capture", *J. CO_2 Util.* 1 (2013) 69–87.

Steinberg M, "The Carnol process system for CO_2 mitigation and methanol production" *Energy* 22 (1997) 143–149.

Styring P., E. A. Quadrelli, K. Armstrong, *Carbon Dioxide Utilization: Closing the Carbon Cycle*, Elsevier, Amsterdam, 2015.

Su X., J. Xu, B. Liang, H. Duan, B. Hou, Y. Huang, "Catalytic carbon dioxide hydrogenation to methane: A review of recent studies", *J. Energy Chem.* 25 (2016) 553–565.

Su X., X. Yang, B. Zhao, Y. Huang, "Designing of highly selective and high-temperature endurable RWGS heterogeneous catalysts: Recent advances and the future directions", *J. Energy Chem.* 26 (2017) 854–867.

Sues A., M. Jurascik, K. J. Ptasinski, "Exergetic evaluation of 5 biowastes-to-biofuels routes via gasification", *Energy* 35 (2010) 996–1007.

Sun K., Z. Fan, J. Ye, J. Yan, Q. Ge, Y. Li, W. He, W. Yang, C. Liua, "Hydrogenation of CO_2 to methanol over In_2O_3 catalyst", *J. CO_2 Util.* 12 (2015) 1–6.

Sun Q., J. Ye, C. Liu, Q. Ge, "In_2O_3 as a promising catalyst for CO_2 utilization: A case study with reverse water gas shift over In_2O_3". *Greenhouse Gas Sci Technol.* 4 (2014) 140–144.

Swapnesh A., V. C. Srivastava, I. D. Mall, "Comparative study on thermodynamic analysis of CO_2 utilization reactions", *Chem. Eng. Technol.* 37 (2014) 1–14.

Tada S., T. Shimizu, H. Kameyama, T. Haneda and R. Kikuchi, "Ni/CeO_2 catalysts with high CO_2 methanation activity and high CH_4 selectivity at low temperatures", *Int. J. Hydrogen Energy* 37 (2012) 5527–5531.

Tahriri Zangeneh F., S. Sahebdelfar, M. Takht Ravanchi, "Conversion of carbon dioxide to valuable petrochemicals: An approach to clean development mechanism", *J. Nat. Gas Chem.* 20 (2011) 219–231.

Tahriri Zangeneh F., A. Taeb, K. Gholivand, S. Sahebdelfar, "Thermodynamic analysis of propane dehydrogenation with carbon dioxide and side reactions", *Chem. Eng. Commun.* 203 (2016) 557–565.

Takht Ravanchi M., S. Sahebdelfar, "Carbon dioxide capture and utilization in petrochemical industry: Potentials and challenges", *Appl. Petrochem. Res.* 4 (2014) 63–77.

Takht Ravanchi M., S. Sahebdelfar, F. Tahriri Zangeneh, "Carbon dioxide sequestration in petrochemical industries with the aim of reduction in greenhouse gas emissions", *Fron. Chem. Sci. Eng.* 5 (2011) 173–178.

Tong A., D. Sridhar, Z. Sun, H. R. Kim, L. Zeng, F. Wang, D. Wang, M. V. Kathe, S. Luo, Y. Sun, L. S. Fan, "Continuous high purity hydrogen generation from a syngas chemical looping 25 kW_{th} sub-pilot unit with 100% carbon capture". Fuel 103 (2013) 495–505.

Torrente-Murciano L., D. Mattia, M. D. Jones, P. K. Plucinski, "Formation of hydrocarbons via CO_2 hydrogenation- A thermodynamic study", *J. CO_2 Util.* 6 (2014) 34–39.

Tursunov O., L. Kustov, A. Kustov, "A Brief review of carbon dioxide hydrogenation to methanol over copper and iron based catalysts", *Oil Gas Sci. Technol. Rev.* 72 (2017) 30–38.

Ushikoshi K, K. Mori, T. Watanabe, M. Takeuchi, M. Saito, "A 50 kg/day class test plant for methanol synthesis from CO_2 and H2", In Inui, T., M. Anpo, K. Izui, S. Yanagida, T. Yamaguchi, (Eds) "*Studies in Surface Science and Catalysis*", Elsevier: Amsterdam, pp. 357–362, (1998).

Wang G., D. Mao, X. Guo, J. Yu, "Methanol synthesis from CO_2 hydrogenation over CuO-ZnO-ZrO_2-M_xO_y catalysts (M = Cr, Mo and W)", *Int. J. Hydrogen Energy*, 44 (2019) 4197–4207.

Wang W., S. Wang, X. Ma, J. Gong, "Recent advances in catalytic hydrogenation of carbon dioxide", *Chem. Soc. Rev.* 40 (2011) 3703–3727.

Wang W., Y. Zhang, Z. Wang, J. M. Yana, Q. G. Chang, J. Liu, "Reverse water gas shift over In_2O_3-CeO_2 catalysts", *Catal. Today* 259 (2016) 402–408.

Wu H. C., Y. C. Chang, J. H. Wu, J. H. Lin, I. K. Lin, C. S. Chen, "Methanation of CO_2 and reverse water gas shift reactions on Ni/SiO_2 catalysts: The influence of particle size on selectivity and reaction pathway", *Catal. Sci. Technol.* 5 (2015) 4154–4163.

Yang H., C. Zhang, P. Gao, H. Wang, X. Li, L. Zhong, W. Wei, Y. Sun, "A review of the catalytic hydrogenation of carbon dioxide into value-added hydrocarbons", *Catal. Sci. Technol.* 7 (2017) 4580–4598.

Yaripour F., Shariatinia Z., Sahebdelfar S., Irandoukt A., "Effect of boron incorporation on the structure, products selectivity and lifetime of HZSM-5 nanocatalyst designed for application in methanol-to-olefins (MTO) reaction", *Micro. Meso. Materials* 203 (2015) 41–53.

Ye J., C. Liu, Q. Ge, "DFT study of CO_2 adsorption and hydrogenation on the In_2O_3 Surface", *J. Phys. Chem. C.* 116 (2012) 7817–7825.

Yi Q., W. Li, J. Feng, K. Xie, "Carbon cycle in advanced coal chemical engineering", *Chem. Soc. Rev.* 44 (2015) 5409–5445.

Younas M., L. L. Kong, M. J. K. Bashir, H. Nadeem, A. Shehzad, S. Sethupathi, "Recent advancements, fundamental challenges, and opportunities in catalytic methanation of CO_2", *Energy and Fuels* 30 (2016) 8815–8831.

Zhang J., M. Zhang, S. Chen, X. Wang, Z. Zhou, Y. Wu, T. Zhang, G. Yang, Y. Han, Y. Tan, "Hydrogenation of CO_2 into aromatics over a Zn-CrO_x-zeolite composite catalyst", *Chem. Commun.* 55 (2019b) 973–976.

Zhang X., A. Zhang, X. Jiang, J. Zhu, J. Liu, J. Li, G. Zhang, C. Song, X. Guo, "Utilization of CO_2 for aromatics production over ZnO/ZrO_2-ZSM-5 tandem catalyst", *J. CO_2 Util.* 29 (2019a) 140–145.

Zhao B., Y. X. Pan, C. Liu, "The promotion effect of CeO_2 on CO_2 adsorption and hydrogenation over Ga_2O_3", *Catal. Today* 194 (2012) 60–64.

Zeng L., F. He, F. Li, L. S. Fan, "Coal-direct chemical looping gasification for hydrogen production: Reactor modeling and process simulation", *Energy Fuels* 26 (2012) 3680–3690.

Zhou X., T. Su, Y. Jiang, Z. Qin, H. Ji, Z. Guo, "CuO-Fe_2O_3-CeO_2/HZSM-5 bi-functional catalyst hydrogenated CO_2 for enhanced dimethyl ether synthesis", *Chem. Eng. Sci.* 153 (2016) 10–20.

3 Recent Advances in CO_2 BI-Reforming of Methane for Hydrogen and Syngas Productions

Hamidah Abdullah, Chin Sim Yee, Chi Cheng Chong, Tan Ji Siang, Osarieme Uyi Osazuwa, Herma Dina Setiabudi, Dai-Viet N. Vo, and Sumaiya Zainal Abidin

CONTENTS

3.1 INTRODUCTION

Worldwide, fossils fuels are used for power generation, which is needed in residential, business, industrial, transportation, and the service sectors. Fossil fuels constitute up to 86% of the entire energy demand, which is on the increase due to increased population (Abas, Kalair, and Khan 2015). As we are aware, the burning of fossil fuels such as oil, coal, and gas has directly contributed to increasing greenhouse gases, especially carbon dioxide (CO_2) in the atmosphere, and resulted in the disruption of the carbon cycle, thus creating a global warming impact on the environment. Global CO_2 concentrations have rapidly increased after the industrial revolution started, and in July 2019, it was reported that the CO_2 concentration reached 410 ppm (Earth System Research Laboratory), which surpassed the safety threshold of CO_2 concentration in the atmosphere (350 ppm). On the other hand, nearly 80%–90% of the total composition of natural gas and biogas is methane (CH_4) (Kumar, Shojaee, and Spivey 2015). The U.S. Energy Information Administration published in their Annual Energy Outlook 2013 that the natural gas projection shows continued growth, thus making it an affordable and abundant energy resource. Since CO_2 and CH_4 are the major components of greenhouse gases (Nicoletti Gi 2009; Solomon 2007), expeditious measures to utilize these two greenhouse gas emissions are vital.

Dry reforming of methane (DRM) for hydrogen (H_2) and syn gas (H_2:CO) production, which employs the major greenhouse gases (CO_2 and CH_4) as feedstock, is an upcoming technique to use up the greenhouse gases in the environment. H_2 is considered as the main alternative energy for future sustainable energy systems (Sharma and Ghoshal 2015). It becomes an option to generate energy and is used as fuel due to its renewable nature. Using H_2 in energy systems has gained attraction because it can generate energy when fuel cells are used. Most developed countries have implemented the use of H_2 in their energy systems. Asian countries, particularly Korea and Japan, have deployed programmes for residential fuel cell micro-combined heat and power. Meanwhile, fuel cells have been commercialized in the United States (Dodds et al. 2015). Besides, the application of H_2 in the transportation sector is also considered a major focus on fuel cell research and development. Proton exchange membrane (PEM) type of fuel cell device is expected to be used in place of engines that run on internal combustion for the transportation system (Alaswad et al. 2016). Nowadays, a detailed study has been carried out to obtain cost-effectiveness and improve durability while enhancing the H_2 storage capacity before it can be practically used. Moreover, the use of H_2 is more environmentally friendly compared to that of some fossil fuels. Nicoletti et al (2015) carried out a comparative study between H_2 and the selected fossil fuels and reported that the H_2 fuel cells and the H_2-fuelled engines running on internal combustion resulted in reduced atmospheric pollutions (Nicoletti et al. 2015). Therefore, it is expected that H_2 will play a great role in the future scenario of energy sectors.

Syngas (H_2 and CO), on the other hand, is a major intermediary agent between chemical and synthetic production of fuels via catalytic reaction (Peng et al. 2017). It has various production sources, viz coal; natural gas; biomass; or feedstock from hydrocarbon passing through steam reforming, partial oxidation, and gasification. It can also serve as feedstock for the production of chemicals and fuels in the water–gas shift (WGS) reaction, the synthesis of methanol, and Fischer–Tropsch process (Van de Loosdrecht and Niemantsverdriet 2013). However, syngas obtained from coal or natural gas as feedstock is mostly preferred because it is not cost-competitive for transport fuels (Schulsz 1999). Syngas is typically produced using steam reforming of methane (SRM). However, the reaction needs increased temperatures above 800°C–900°C and a large amount of steam. Therefore, DRM is an alternative way to produce syngas. DRM is the reaction between CO_2 and CH_4. Moreover, DRM is preferred since two greenhouse gases are used as reactants. In order to convert syngas to liquid fuels using the Fischer–Tropsch process (FTP), the syngas ratio (H_2/CO) feed is important. The stoichiometric H_2/CO ratio of 2 is preferred in the FTP, and it has been established that this can be obtained by methane bi-reforming (MBR), which is the combination of SRM and DRM.

In summary, MBR for H_2 and syngas production may play a role in dealing with global CO_2 emissions, abundant CH_4, and renewable energy demand. Recently, research and developments on MBR have resulted in a noticeable improvement in the efficiency of the process and durability of the material, thus ensuring that this technology becomes a reliable option for environmental and future use. Thus, this chapter presents recent advances in this technology in terms of catalyst development and catalyst deactivation.

3.2 METHANE BI-REFORMING REACTION (MBR)

MBR, also called combined steam and dry reforming of methane (CSDRM), is a highly valuable, potential reforming process for yielding H_2/CO ratio of 2, which is applicable for hydrocarbon generation, methanol, and other imperative chemical derivatives (Olah et al. 2013; Vo et al. 2012). In comparison with SRM or DRM as a single process (Fayaz et al. 2016; Bahari et al. 2016; Bukhari et al. 2017; Siang et al. 2018a), the implementation of MBR can evade the auxiliary separation and adjustment steps for tuning the practicable H_2/CO ratios for downstream industrial production, thereby reducing the capital cost and process complexity (Olah et al. 2013; Jabbour et al. 2017; Siang et al. 2018b). In MBR, the generated H_2: CO mixture with a ratio of 2, commonly used for a wide range of industrial applications, was termed as 'metgas', as proposed by Olah et al. (2012). Additionally, MBR offers flexibility in H_2/CO ratios to meet the prerequisite for downstream process production via the simple manipulation of feed composition (Jabbour et al. 2017, Siang et al. 2017). However, the quantity of comprehensive investigations towards MBR is relatively lesser than that of other reforming processes due to its complex reaction network involving multiple reactions. In general, the main reforming reactions in MBR are SRM (Eq. 3.2) and DRM (Eq. 3.3), reactions that form carbon like CH_4 decomposition (Eq. 3.4), Boudouard reaction (Eq. 3.5) as well as non-carbon-forming reactions involving reverse WGS (Eq. 3.6) and carbon gasification by H_2O (Eq. 3.7) or CO_2 (also known as reverse Boudouard reaction) (refer to Eqs. 3.1–3.7).

$$3CH_4 + 2H_2O + CO_2 \rightarrow 8H_2 + 4CO \tag{3.1}$$

$$CH_4 + H_2O \rightarrow 3H_2 + CO \tag{3.2}$$

$$CH_4 + CO_2 \rightarrow 2H_2 + 2CO \tag{3.3}$$

$$CH_4 \rightarrow C + 2H_2 \tag{3.4}$$

$$2CO \rightarrow C + CO_2 \tag{3.5}$$

$$H_2 + CO_2 \rightarrow CO + H_2O \tag{3.6}$$

$$C_2 + H_2O \rightarrow H_2 + CO \tag{3.7}$$

Therefore, understanding the mechanistic steps of MBR can beneficially provide crucial insights into effectively optimizing syngas or H_2 yield under the appropriate operating conditions. Hence, the MBR mechanism and the influence of process variables on the MBR reaction activity as well as the recent advances in MBR process are exclusively reviewed in this section.

3.2.1 Methane Bi-Reforming Reaction Mechanism

From the industrial standpoint, a detailed understanding of the fundamental mechanistic MBR routes is crucial for both catalyst and reactor design in terms of scale-up and commercialization. In mechanistic studies of various reforming routes over MgO-supported catalysts, Qin et al. (1996) suggested that both SRM and DRM processes occurred concomitantly in MBR. In their *in situ* CO_2 transient analysis, the formation rate of carbonaceous species originating from CH_x species as a result of CH_4 decomposition (Eq. 3.4) was superior to the rate of deposited carbon produced from Boudouard reaction (Eq. 3.5). In addition, MBR reportedly possesses a reaction kinetics comparable to that of SRM since both have an analogous reaction mechanism. Overall, the adsorbed oxygen atom (from the oxidizing agents, i.e. H_2O or CO_2) reacts with the adsorbed CH_x ($x = 0$, 1, 2, or 3) species to form CO and H_2. The elementary steps of MBR involving adsorption and activation of reactants to form syngas in association with active site (S) are shown in Eqs. 3.8–3.15, which are as follows:

CH_4 activation:

$$CH_4 + 2S \rightarrow H-S + CH_3-S \tag{3.8}$$

$$CH_3-S + 2S \rightarrow 2H-S + CH-S \tag{3.9}$$

$$CH-S + S \rightarrow H-S + C-S \tag{3.10}$$

H_2O and CO_2 dissociation:

$$H_2O + 3S \rightarrow 2H-S + O-S \tag{3.11}$$

$$CO_2 + 2S \rightarrow CO-S + O-S \tag{3.12}$$

Surface reaction:

$$CH_x-S + O-S + (x-1)S \rightarrow xH-S + CO-S \tag{3.13}$$

H_2 and CO production:

$$2H - S \rightarrow H_2 + 2S \quad (3.14)$$

$$CO - S \rightarrow CO + S \quad (3.15)$$

3.2.2 Effect of Process Variables

In a bid to identify the degree of dependence of catalytic activity on process variables, including feed composition of reactants, temperature, and space velocity, various studies regarding system optimization based on varying process parameters have been extensively scrutinized in recent years. The following subsections provide the thorough discussions about the role of each process variable in MBR performance. The recently reported MBR findings are also summarized in Table 3.1 for comparison among catalysts at different operating conditions.

3.2.2.1 Effects of Feed Composition

As aforementioned, one of the most attractive advantages of MBR is that it provides for feasible H_2/CO ratio adjustment via the direct manipulation of reactant composition, thereby avoiding the requirement of auxiliary separation unit. Hence, numerous studies have broadly assessed the impact of feedstock composition on MBR for altering the H_2/CO ratios. In the study of MBR over Ni/SBA-15 catalyst, Singh et al. (2018) studied the influence of composition of feedstock on the reaction activity at 1073 K. Interestingly, they found that higher $CO_2/(CH_4 + H_2O)$ ratios exhibited higher CH_4 conversion and greater CO_2 conversion. They also assigned this observation to the enhanced rate of CO_2 gasification of the carbonaceous species which originated from CH_4 decomposition. In addition, they observed that the decrease in $H_2O/(CH_4 + CO_2)$ ratio led to a decline in H_2/CO ratio (from 2.14 to 1.83) probably because of improved reverse WGS reaction under H_2O-deficient environment. Elsayed et al. (2018) conducted MBR with two different sets of feed composition, i.e. $CH_4/H_2O/CO_2$ ratios of 1/1/1 and 3/2/1 over co-precipitated Pd- and Pt-promoted Ni-Mg/$Ce_{0.6}Zr_{0.4}O_2$. With a growth in both CH_4 and H_2O feeds, CH_4 conversion was significantly decreased, whereas CO_2 conversion reportedly showed an increasing trend. This behaviour was indicative of the kinetic correlations between WGS, reverse WGS, and coke gasification processes, which were in line with the results presented by Singh et al. (2018).

3.2.2.2 Effects of Reaction Temperature

Since MBR and other additional reactions, namely SRM, DRM, and CH_4 decomposition (widely recognized as the main factors responsible for deactivating catalyst), are highly endothermic reactions, optimizing the reaction temperature for improving the MBR performance and preventing severe coke formation is essential for large-scale MBR production. Itkulova et al. (2018) investigated the generation of syngas from both DRM and bi-reforming of methane (BRM) over Co-Pt/Al_2O_3-ZrO_2. In comparison with DRM, they found that reactant conversions in MBR could reach up to 100% at the temperature ranges of 973–983 K. This was a result of the combined

TABLE 3.1
Summary of Process Variables Influencing the Catalytic MBR Performance Over Recently Reported Catalysts in Literature

Catalyst	Reaction Conditions						
	$CH_4/H_2O/CO_2$ ratio	Temperature (°C)	*GHSV(L $g_{cat}^{-1}h^{-1}$)	CH_4 Conversion (%)	CO_2 Conversion (%)	H_2/CO Ratio	References
$Ni_5\%Al_2O_3$	1/0.8/0.4	800	69–138	82–59	77–54	2.10–2.17	Jabbour et al. (2017)
$Ni_{7.5}\%Al_2O_3$	1/0.8/0.4	800	69–138	87–67	81–61	2.16–2.22	Jabbour et al. (2017)
$Ni_{0.1}Al_2O_3$	1/0.8/0.4	800	69–138	88–82	83–77	2.23–2.33	Jabbour et al. (2017)
Ni/Al_2O_3	1/1.2/0.38	800	36–72	96.4–87.8	38.4–25.6	-	Kim et al. (2017)
$Ni/Mg\text{-}Al_2O_3$	1/1.2/0.38	800	36–72	94.4–81.2	37.4–27.1	-	Kim et al. (2017)
$Ni/Ce\text{-}Al_2O_3$	1/1.2/0.38	800	36–72	96.5–83.3	40.5–27.5	-	Kim et al. (2017)
$Ni/Sm\text{-}Al_2O_3$	1/1.2/0.38	800	36–72	96.7–89.9	42.6–34.1	-	Kim et al. (2017)
Ni/SBA-15	2.5–4.5/3/1.5	800	36	80.9–61.5	47.1–58.9	1.69–2.15	Singh et al. (2018)
	4.5/3/1–2	800	36	52.6–63.4	39.1–60.8	2.65–1.89	
	4.5/1.5–3/1.5	800	36	55.1–61.6	73.6–58.9	1.84–2.15	
Pt/Ni-Mg/	1–3/1–2/1	500	136,000[a]	78–33	32–36	1.2–1.9	Elsayed et al. (2018)
$Ce_{0.6}Zr_{0.4}O_2$	3/2/1	450–600	136,000[a]	18.9–60.2	7.2–94.1	-	
	3/2/1	500	86,700–272,000	39.3–13.7	44.3–11.1	-	
Pd/Ni-Mg/	1–3/1–2/1	500	136,000[a]	42–25	10–13	2.9–3.0	Elsayed et al. (2018)
$Ce_{0.6}Zr_{0.4}O_2$	3/2/1	450–600	136,000[a]	14.5–54.1	0–63.1	-	
	3/2/1	500	86,700–272,000	39.8–7.8	41.0–15.5	-	
B-Ni/SBA-15	3/2/1	700–800	36	44.4–71.8	51.0–60.5	2.13–2.92	Siang et al. (2019)
$La_{0.5}Sr_{0.5}NiO_3$	2/1/2	650–900	70[a]	11.3–77.4	8.3–68.7	-	Yang et al. (2018)
$Co\text{-}Pt/Al_2O_3\text{-}ZrO_2$	1/0.2/1	350–700	1,000[a]	6.3–100	15.3–100	-	(Itkulova et al. 2018)

[a] GHSV unit is $h^{-1} = \frac{\text{Volumetric flow rate of feed}}{\text{Volume of the catalyst}}$.

reactions of CH_4 and H_2O coupled with the reaction between CH_4 and CO_2. However, due to the presence of CH_4 and H_2O side reaction in MBR, the conversion of CH_4 was greater than that of CO_2, and this disparity became significant with increasing temperature in MBR. The effect of temperature on MBR was also assessed by Yang et al. (2018), and they reported that an increased temperature increased the reaction rate of the reforming process due to the endothermicity of MBR and thus showed an improvement in reactant conversions. In addition, they found that MBR was less prone to the formation of carbonaceous species. Hence, they surmised that carbon gasification rate by H_2O and CO_2 was far better than CH_4 decomposition. Siang et al. (2019) conducted an evaluation of longevity for MBR at different temperatures with stoichiometric feed composition on boron-doped Ni/SBA-15. CH_4 and CO_2 conversions increased to about 37.9% and 27.3%, respectively, correspondingly with rising temperature from 973 to 1073 K possibly as a result of the endothermicity of MBR. Similar to the findings of Yang et al. (2018), they also proposed that the carbon gasification reaction by H_2O and CO_2 was faster than CH_4 dehydrogenation since no graphite was detected for the post-reaction analysis of the used-up catalyst, as evident in the Raman and X-ray photoelectron spectroscopic (XPS) analyses.

3.2.2.3 Gas Hourly Space Velocity effects

To eliminate the adverse effects of mass and heat transfer limitations, the intrinsic catalytic MBR activity and the effects of gas hourly space velocity (GHSV) have been extensively investigated on various catalysts in MBR process. Kim et al. (2017) synthesized various Ni/M-Al_2O_3 (M = Sm, Mg or Ce) samples and evaluated their performance for MBR at different GHSV values ranging from 36 to 72 L $g_{cat}^{-1}h^{-1}$. Generally, all specimens exhibited a decline in reactant conversions with increasing GHSV, indicating the hindrance of reactant adsorptions on active metallic sites at extremely high GHSV. Jabbour et al. (2017) conducted longevity tests to examine the relationship between GHSV and coke formation on catalyst surface. They revealed that the stability evaluation for $Ni_x\%Al_2O_3$ at 69 L $g_{cat}^{-1}h^{-1}$ generated less carbon amount in comparison with that at 138 L $g_{cat}^{-1}h^{-1}$. They also suggested that only partial active metallic sites participated in catalytic reaction, which minimized the activation of side reaction like CH_4 decomposition, hence lowering the amount of carbonaceous species formed.

3.3 CATALYST FOR METHANE BI-REFORMING REACTION

Catalysts are crucial in determining the mechanism and distribution of product in the BRM reaction. Numerous catalysts such as monometallic catalysts, bimetallic catalysts, and trimetallic catalysts have been reported for BRM reaction. Catalysts currently used for BRM reaction are summarized in Table 3.2. Generally, catalysts with noble metals are manifestly active and possess better stability than the Ni-based catalysts for reforming reaction. Khani et al. (2016) explored the BRM activity over M/$ZnLaO_4$ (M=3%Pt, 3%Ru, 10%Ni). The 3%Ru/$ZnLaAlO_4$ catalyst achieved higher catalytic performance in terms of inhibiting carbon deposition than the 3%Pt/$ZnLaAlO_4$ and 10%Ni/$ZnLaAlO_4$ catalysts. This behaviour can be linked to better synergistic interactions between the active metals and the $ZnLaAlO_4$ support.

TABLE 3.2
Summary of Currently Employed Catalysts in BRM

Catalysts	Preparation Method	$CH_4/CO_2/H_2O$ Ratio	Reactor	T (°C)	GHSV (L gcat^{-1}h^{-1}) or (h^{-1})*	TOS (h)	Initial Performance X_{CO_2} (%)	Initial Performance X_{CH_4} (%)	Initial Performance H_2/CO Ratio	Final Performance X_{CO_2} (%)	Final Performance X_{CH_4} (%)	Final Performance H_2/CO Ratio	D[a] (%)	References
Monometallic Catalysts														
3%Ru/$ZnLaAlO_4$	Incipient wetness impregnation	1:0.75:0.5	Fixed bed	800	12	30	62.0	94.5	2.10	-	-	-	-	Khani et al. (2016)
3%Pt/$ZnLaAlO_4$							78.2	87.5	1.90	-	-	-	-	
10%Ni/$ZnLaAlO_4$							-	91	-	-	-	-	-	
15%Ni–85%MgO	Ni- PS intermediates	2:1:2	Fixed bed	750	160	24	65.0	80.0	2.0	64	76.3	2.0	4.6	Ashok et al. (2018)
						100	65.0	80.0	2.0	40.0	48.0	2.25	40	
15%Ni–85%SiO_2						24	58.0	68	1.9	32	22	2.36	67.6	
15%NiO-MgO	-	3:1.2:2.4	Tubular flow reactor	830	60	320	74.0	72.0	-	74.0	72.0	-	0	Olah et al. (2012)
10%Ni/ZrO_2	Incipient wetness impregnation	5:2:4	Fixed bed	850	60	20	72	92	2.0	56	82	2.0	10.9	Li et al. (2015)
15%Ni/CeO_2	Co-precipitation impregnation	1:0.8:0.4	Fixed-bed microreactor	800	26.5	12	79	97	1.9	-	97	-	0	Roh et al. (2009)
							-	55.7	-	-	47.4	-	14.9	
12%Ni/30%MgO–Al_2O_3	Incipient wetness impregnation	1:0.4:0.8	Microtubular quartz reactor	650	26.5	-	51.0	68.0	-	-	-	-	-	Roh et al. (2007)
				700	26.5	-	71.0	83.0	-	-	-	-	-	Roh et al. (2007)
15%Ni/MgO-Al_2O_3	Co-precipitation	3:1.2:2.2	Fixed bed	775	86,000*	24	72.0	79.0	2.02	62.0	71.0	2.02	10.1	Li and van Veen (2018)
15%Ni–30%SiO_2-55%MgO	Ni- PS intermediates	2:1:2	Fixed bed	750	160	135	63.0	78.0	2.0	62.0	74.0	2.0	5.1	Ashok et al. (2018)
15%Ni/$Ce_{0.8}Zr_{0.2}O_2$	Co-precipitation impregnation	1:0.8:0.4	Fixed-bed microreactor	800	26.5	20	80	96	1.9	-	96	-	0	Roh et al. (2009)
							-	90.8	-	-	81.7	-	10	

(Continued)

TABLE 3.2 (*Continued*)
Summary of Currently Employed Catalysts in BRM

Catalysts	Preparation Method	$CH_4/CO_2/H_2O$ Ratio	Reactor	T (°C)	GHSV (L gcat^{-1}h^{-1}) or (h^{-1})*	TOS (h)	Initial Performance X_{CO_2} (%)	Initial Performance X_{CH_4} (%)	Initial Performance H_2/CO Ratio	Final Performance X_{CO_2} (%)	Final Performance X_{CH_4} (%)	Final Performance H_2/CO Ratio	D[a] (%)	References
10%Ni/SBA-15	Impregnation	3:1:2	Fixed bed	800	36	-	60.0	62.5	2.05	-	-	-	-	Singh et al. (2017)
10%Ni/SBA-15	Impregnation	3:1:2	Fixed bed	800	3610*	-	54	35	1.6	-	-	-	-	Siang et al. (2018b)
5%Ni/SBA-15	Two solvent impregnations	1:0.4:0.8	Micro activity reference catalytic reactor	800	62.7	10	69.0	58.0	-	0	0	-	100	Jabbour et al. (2015)
5%Ni/Aerosil silica							73.0	88.0	-	70.0	80.0	-	9.1	
5%Ni/MN_3 diatom							20.0	20.0	-	0	0	-	100	
1%Ni/$La_2Zr_2O_7$	Modified Pechini method	1:0.31:0.65	Fixed bed	750	98.7	170	49.5	54.0	-	47.5	53.0	-	1.9	Kumar et al. (2016)
10%Ni–1%B/SBA-15	Sequential impregnation	3:1:2	Fixed bed	800	3610*	-	62.3	55.7	2.06	-	-	-	-	Siang et al. (2018b)
10%Ni–3%B/SBA-15							66	63	2.2	-	-	-	-	
10%Ni–5%B/SBA-15							28.9	26.3	2.4	-	-	-	-	
Bimetallic Catalysts														
10%Ni-La/SBA-15	Impregnation	3:1:2	Fixed bed	800	36	-	63.0	64.5	2.19	-	-	-	-	Singh et al. (2017)
10%Ni–2.5%La/$MgAl_2O_4$	Co-impregnation	1:0.4:1.2	Fixed bed	900	4680	5	54.8	77.7	3.10	44.7	68.1	3.10	12.4	Park et al. (2015)
10%Ni–2.5%Ce/$MgAl_2O_4$	Co-impregnation	1:0.4:0.8	Fixed bed	700	530	20	65.9	81.3	2.10	-	-	-	-	Koo et al. (2014)
10%Ni-10%Ce/$MgAl_2O_4$							52.3	74.9	2.20	-	-	-	-	

(*Continued*)

TABLE 3.2 (*Continued*)
Summary of Currently Employed Catalysts in BRM

Catalysts	Preparation Method	$CH_4/CO_2/H_2O$ Ratio	Reactor	T (°C)	GHSV (L $gcat^{-1}h^{-1}$) or (h^{-1})*	TOS (h)	Initial Performance X_{CO_2} (%)	Initial Performance X_{CH_4} (%)	Initial Performance H_2/CO Ratio	Final Performance X_{CO_2} (%)	Final Performance X_{CH_4} (%)	Final Performance H_2/CO Ratio	D^a (%)	References
4%Co–1%Pt/Al_2O_3	Impregnation	1:1:0.2	Flow quartz reactor	720	1,000*	-	71.9	~100	1.2	-	-	-	-	Itkulova et al. (2014).
4.5% Co–0.5%Pt/Al_2O_3		1:1:0.2		750	1,000*	-	68.5	~100	1.0	-	-	-	-	
4.75% Co–0.25%Pt/Al_2O_3		3:3:0.2		800	1,000*	-	~100	~100	1.3	-	-	-	-	
15%Ni–0.5%Ru/Al_2O_3	Simultaneous impregnation	1:0.56:0.4	Computerized catalytic reactor	750	15.4	96	50.0	53.0	2.0	-	-	-	-	A΄lvarez et al. (2015)
15%Ni–0.5%Ru/Al_2O_3	Consecutive impregnation						25.0	28.0	2.1	-	-	-	-	
0.5% Mo_2C–10%Ni/ZrO_2	Modified hydrothermal method	5:2:4	Fixed bed	850	60	20	80	98	1.8	74	98	1.8	0	Li et al. (2015)
5%Ni–1%Mg/α-Al_2O_3	Incipient wetness impregnation	1:1:0.16	Fixed bed	750	20	-	71.0	74.0	1.10	-	-	-	-	Pour and Mousavi (2015)
4.2%Ni–1%Mg/α-Al_2O_3							66	70	1.09	-	-	-	-	
3.4%Ni–1%Mg/α-Al_2O_3							61	65	1.1	-	-	-	-	
2.6%Ni–1%Mg/α-Al_2O_3							57	60	1.1	-	-	-	-	

(*Continued*)

TABLE 3.2 (*Continued*)
Summary of Currently Employed Catalysts in BRM

Catalysts	Preparation Method	$CH_4/CO_2/H_2O$ Ratio	Reactor	T (°C)	GHSV (L $gcat^{-1}h^{-1}$) or (h^{-1})*	TOS (h)	Initial Performance X_{CO_2} (%)	Initial Performance X_{CH_4} (%)	Initial Performance H_2/CO Ratio	Final Performance X_{CO_2} (%)	Final Performance X_{CH_4} (%)	Final Performance H_2/CO Ratio	D[a] (%)	References
1.8%Ni–1%Mg/ α-Al_2O_3							55	57	1.09	-	-	-	-	
1%Ni–1%Mg/ α-Al_2O_3							52.0	54.0	1.09	-	-	-	-	
10%Ni-$Sn_{0.02}$/ CeO_2-Al_2O_3	Sequential impregnation	1:1:1	Computerized catalytic reactor	800	60	24	74.0	74.0	1.25	40.5	62.0	1.57	16.2	Stroud et al. (2018)
Trimetallic Catalysts														
0.16%Pt–1.4% Ni–1.0% Mg/ $Ce_{0.6}Zr_{0.4}O_2$	Incipient wetness impregnation	3:1:2	Fixed bed	500	136,000*	3	33.0	36.0	2.6	31.0	36.0	2.85	0	Elsayed et al. (2018)
0.13% Pd–1.4% Ni–1.0% Mg/ $Ce_{0.6}Zr_{0.4}O_2$						-	13.0	25.0	3.0	-	-	-	-	

[a] Degree of catalyst deactivation, D (%) = [1 − (Final CH_4 conversion/Initial CH_4 conversion)] × 100.

* GHSV unit is $h^{-1} = \frac{\text{Volumetric flow rate of feed}}{\text{Volume of the catalyst}}$.

However, noble metals are scarce and expensive, hence restricting their wide applications in BRM process. Thus, Ni-based catalysts are now frequently studied for BRM reaction due to their availability, effective performance, and reduced cost, which are obvious reasons for this choice.

As given in Table 3.2, Ni-based catalysts with broad support materials such as single metallic oxides, mixed metal oxides, and silica have been extensively used for BRM reaction. The selection of suitable support materials is a crucial step in developing Ni-based catalyst because the support materials play an important role in the physicochemical properties of catalysts. Appropriate selection of supports can boost the dispersion of active metal particles, facilitate the interaction between the metal and the support, and inhibit the formation of coke. Metal oxides such as MgO, SiO_2, ZrO_2, and CeO_2 have attracted considerable attention in BRM due to their interesting physicochemical properties. Ashok et al. (2018) reported the activity of Ni-SiO_2 and Ni-MgO catalysts prepared by Ni-phyllosilicate (PS) intermediates. PS-derived system is a promising process to induce the metal–support interaction using its unique structural properties. Ni/MgO (T = 750°C, X_{CO_2} = 65%, X_{CH_4} = 80%, H_2/CO = 2, D = 4.6%) exhibited high performance when compared with Ni/SiO_2 (T = 750°C, X_{CO_2} = 58%, X_{CH_4} = 68%, H_2/CO = 1.9, D = 67.6%) as a result of the increased metal dispersion, higher basicity, and stronger interaction between the metal and the support. The good performance of NiO/MgO was also outlined by Olah et al. (2012). In their study, the performance of NiO/MgO was evaluated in BRM reaction from pure CH_4, natural gas, and other sources (CH_4 coal bed and hydrates). Stable reactant conversion (X_{CO_2} = 74%, X_{CH_4} = 74%) was observed at the reaction temperature of 830°C and TOS (time on stream) of 320 h. Li et al. (2015) reported the activity of Ni/ZrO_2 in BRM reaction with good catalytic performance (T= 850°C, X_{CO_2} = 72%, X_{CH_4} =92%, H_2: CO = 2) and longevity (T=850°C, TOS = 20 h, D = 10.9%). Moreover, Roh et al. (2009) reported the BRM reaction over 15%Ni/CeO_2 synthesized via co-precipitation and impregnation methods. 15%Ni/CeO_2 synthesized via co-precipitation method showed better performance than that synthesized via impregnation method because of the strong interaction between the metal and the support, which led to a homogenous distribution of the metal.

Apart from single oxide, mixed metal oxides have been studied in BRM reaction. Roh et al. (2007) reported the activity of 12%Ni/30%MgO-Al_2O_3 and 30%Ni/Al_2O_3 catalysts in BRM reaction. In comparison with Ni/Al_2O_3, adding MgO onto Ni/Al_2O_3 enhanced the catalyst activity and stability as a result of the beneficial effects of MgO as a promoter, thereby enhancing steam adsorption in the reforming process. Furthermore, the addition of MgO prevented the coke formation because of the basic properties of MgO. The application of MgO-AlO_3 in BRM reaction was also reported by Li and Veen (2018) for 15%Ni/MgO-Al_2O_3. At 775°C, the conversion of CO_2 and CH_4 was 72% and 79%, respectively, with H_2/CO ratio of 2.02. However, a sequential decline in activity was observed because of the formation of whiskers of carbon encapsulated in Ni particles. Ashok et al. (2018) explored the application of PS-derived 15%Ni–30%SiO_2–55%MgO catalyst for BRM reaction. The catalyst showed stable catalytic performance (T=750°C, X_{CO_2} =63%, X_{CH_4} =78%, H_2/CO=2.0, D=5.1%) due to its fascinating properties, including moderate acidity sites, high basicity strength, and high structural stability. Roh et al. (2009) studied the enhanced performance of mixed metal Ni oxide ($Ni/Ce_{0.8}Zr_{0.2}O_2$) as compared to Ni/single metal oxide (Ni/Al_2O_3 and Ni/CeO_2). The favourable performance of $Ni/Ce_{0.8}Zr_{0.2}O_2$ was related to the homogeneous metal dispersion and nanocrystalline property of cubic $Ce_{0.8}Zr_{0.2}O_2$ support. In addition to single and mixed metal oxides, silica materials have drawn much attention as support material in BRM reaction. Siang et al. (2017) explored the application of mesoporous silica, i.e. SBA-15, in BRM reaction. SBA-15 was selected as the support material due to its high surface area, mesoporous structure, and increased thermal resistance. Ni/SBA-15 showed a moderate activity in BRM reaction with the conversion of CO_2 and CH_4 having the values of 54% and 35%, respectively, at 800°C. In addition, Jabbour et al. (2015) explored the performance of Ni-silica for BRM reaction. Three silica types, namely SBA-15, MN3 diatom, and Aerosil 300, were used as support materials. The catalysts' performance followed the order of Ni/Aerosil>Ni/SBA15>Ni/NM3 diatom. High stability was observed by Ni/Aerosil, whereas Ni/SBA15 and Ni/NM3 catalysts deactivated rapidly. Ni/SBA-15 and Ni/NM3 deactivations were ascribed to the active metallic Ni species oxidation and mobility. Apart from metal oxides and silica materials, Kumar et al. (2016) reported the potential of $La_2Zr_2O_7$ pyrochlore as support material for BRM reaction. The selection of $La_2Zr_2O_7$ pyrochlore due to its interesting structure made it possible to incorporate active metals in the structure, thus providing strong metal–support interaction and improving the catalyst resistivity towards coking. 1%$Ni/La_2Zr_2O_7$ showed promising results in BRM reaction ($T = 750°C$, $X_{CO_2} = 49.5\%$, $X_{CH_4} = 54\%$) with high stability and virtually no carbon deposition during 20h TOS. To overcome the drawbacks of Ni catalysts, promoters are included as an effort to improve the metal–support interface by enhancing a synergistic effect between the metal and the support. Siang et al. (2018b) explored the effect of boron as a promoter for Ni/SBA-15. The influence of various boron loadings (1–5wt%) on properties and catalytic activity was evaluated. The optimal boron loading was obtained at 3wt% with higher catalytic activity ($T = 800°C$, $X_{CO_2} = 66\%$, $X_{CH_4} = 63\%$, $H_2/CO = 2.2$) compared with unpromoted catalyst ($T = 800°C$, $X_{CO_2} = 54\%$, $X_{CH_4} = 35\%$, $H_2/CO = 1.6$). The superior catalytic behaviour of 10%Ni–3%B/SBA-15 was a result of the positive role that boron plays as an enhancer. Additionally, analysing the spent catalyst revealed that boron inhibited graphite carbon being formed and the deposition of carbon reduced up to four times when compared with unpromoted catalyst.

Apart from boron, the addition of secondary metal as an effort to enhance coke resistance and to improve performance has also been studied. Singh et al. (2017) explored the modification of Ni/SBA-15 with La as a promoter. Adding 10wt% La significantly improved the reaction in the forward direction due to the O_2 storage capacity of La_2O_3 phase and its basicity. In addition, Park et al (2015) found that 2.5wt% La-promoted $MgAl_2O_4$ ($T = 900°C$, $X_{CO_2} = 46.6\%$, $X_{CH_4} = 72.7\%$, $H_2/CO = 3.3$) showed better activity and stability than that of unpromoted $MgAl_2O_4$ ($T = 900°C$, $X_{CO_2} = 39.2\%$, $X_{CH_4} = 55.3\%$, $H_2/CO = 3.5$) because La played a positive role in improving Ni dispersion and strengthening the interaction between the metal and the support, thus in turn improving catalytic activity and stability. Apart from La, Koo et al. (2014) studied the effect of Ce addition on $Ni/MgAl_2O_4$ for BRM reaction. An excellent performance was observed by Ni/SBA-15 promoted with Ce, which was ascribed to the interaction between the metal and support, homogeneous metal dispersion with small NiO crystallite size, high reducibility, and effective transfer of surface O_2.

Additionally, Itkulova et al. (2014) reported how Pt (0.25–1 wt%) influenced the behaviour of Co/Al_2O_3. Complete CH_4 conversion was observed for all catalysts because Pt plays a role in the formation of the dispersed bimetallic nanoparticles. In another study, Álvarez et al. (2015) prepared Ru- and Ni-based catalysts supported on Al_2O_3 for BRM activity by two impregnation techniques, namely simultaneous impregnation and consecutive impregnation. The former method impregnated both the Ru and Ni active metals at the same time, whereas the latter method impregnated initially Ni active metals followed by Ru active metals. The samples of Ni-Ru/Al_2O_3 prepared by simultaneous technique achieved the highest catalytic activity and the lowest coke deposition at 750°C and the ratio of reactant at the inlet of CH_4: H_2O: CO_2 was 1:0.56:0.4. The results obtained can be explained by the stronger Ru–Ni interaction observed using simultaneous technique, which enhanced the carbon deposit gasification, as evidenced by the Raman spectroscopic and H_2-temperature programmed reduction analyses. Li et al. (2015) highlighted the behaviour of bimetallic Mo_2C-Ni/ZrO_2 catalyst in the BRM reaction. In their study, the effects of the Mo_2C loadings (0.2–3.0 wt%) were evaluated. The results showed that an appropriate amount of Mo_2C (0.5%) increased the activity of Ni/Zr (8% and 6% improvement in X_{CO} and X_{CH_4}) due to the presence of new active site originated from Mo_2C. Additionally, Pourand Mousavi. (2015) highlighted the influence of adding alkaline earth element (Mg) on the activity of BRM reaction over Ni-Mg/Al_2O_3, with a fixed amount of Mg (1 wt%) and various amounts of Ni (1–5 wt%). The characterization analysis revealed that increasing Ni loading increased the Ni/Mg size to 29.7 nm from its original size of 13.7 nm. The turnover frequency revealed that the catalyst's activity depended on the size of Ni/Mg particles. In contrast, the selectivity of the products and H_2: CO was not dependent on the size of Ni/Mg particles. Stroud et al. (2018) highlighted the performance of bimetallic 10%Ni-$Sn_{0.02}$/CeO_2-Al_2O_3 in BRM reaction. The catalyst showed a stable conversion towards BRM (T = 800°C, X_{CO_2} = 74%, X_{CH_4} = 74%, H_2/CO = 1.25) and high coke resistance. The improved catalyst performance was credited to the presence of Sn and Ce atoms. The presence of Sn atoms occupied the nucleation sites of C in the presence of Ni atoms, thus inhibiting the formation of coke. Meanwhile, adding Ce modified the support's acidic/basic properties, thus improving the catalyst activity.

In certain cases, the addition of tertiary metals is required to enhance the efficiency of the catalyst towards BRM reaction. Elsayed et al. (2018) outlined the activity of 0.16%Pt–1.4%Ni–1.0%Mg/$Ce_{0.6}Zr_{0.4}O_2$ and 0.13%Pd–1.4%Ni–1.0% Mg/$Ce_{0.6}Zr_{0.4}O_2$ in BRM reaction. Pt-Ni-Mg-based catalyst (T=500°C, X_{CO_2}=33%, X_{CH_4}=36%, H_2/CO = 2.6) had higher activity than Pd-Ni-Mg-based catalyst (T=500°C, X_{CO_2} = 13%, X_{CH_4} = 25%, H_2/CO = 3.0). The superior catalytic activity of Pt-Ni-Mg-based catalyst was credited to its high basicity, thus shifting the equilibrium concentrations as a result of disproportionation of CO, thereby reducing coke deposits.

According to the above discussion, it is reasonable to conclude that the BRM reaction significantly hinges on the type of catalyst that consists of the active metal and the support. Support material is selected by evaluating the physicochemical properties of material, which can have an adverse effect on the metal. In addition, to further boost the surface interaction between the metal and the support, promoters and other (secondary/tertiary) metals can also be introduced.

3.4 DEACTIVATION OF CATALYST DURING METHANE BI-REFORMING REACTION

The highly endothermic nature of BRM reaction is ascribed to the stability of CH_4 and CO_2 molecules. Increased reaction temperature required for BRM reaction can deactivate the catalysts through sintering and coking. These carbons can be produced by CH_4 decomposition that occurs at higher temperature (≥ 750°C) and disproportionation/reduction of CO. Figure 3.1 shows that the deactivation occurred *in situ* during the BRM reaction. Since the amount of carbon formed is highly reliant on the reaction operating parameters and the catalyst employed, the catalysts with criteria of resilient to carbonization, oxidative environment, and increased temperature are highly required to commercialize an economically feasible BRM process (Ashok et al. 2018; Appari et al. 2014; Palma et al. 2014; Zhao et al. 2017). Many approaches such as changing or modifying the support, introducing promoters, and employing diverse preparation methods have been carried out to increase the stability of the BRM catalysts (Kumar et al. 2016; Pakhare and Spivey 2014). Table 3.3 summarizes the stability of these modified catalysts for BRM reaction and its corresponding deactivation phenomena. The pertinent detailed review and comparison are delineated in the subsequent subsections.

3.4.1 Changing or Modifying the Support Used in Methane BI-Reforming

Considering the basic property that is required to enhance resistance to carbon formation, Ni-supported $MgAl_2O_4$ and Ni-supported MgO were prepared through

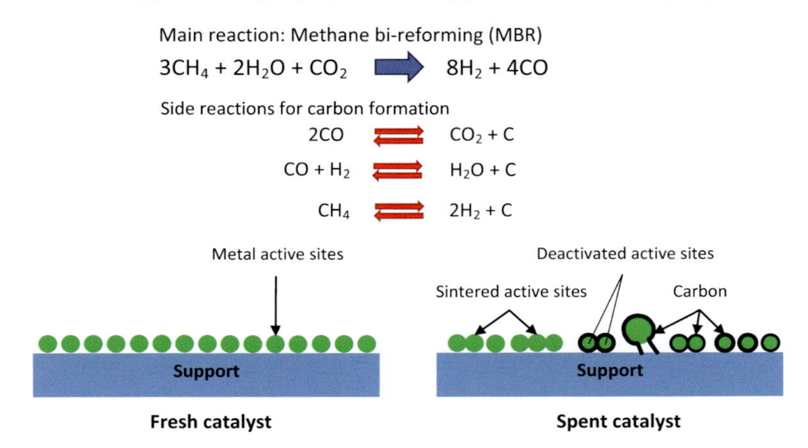

FIGURE 3.1 Graphical representation to illustrate catalyst deactivation occurred during BRM reaction.

TABLE 3.3
Stability of the Modified Catalysts for BRM Reaction and the Type of Deactivations Occurred

Strategies to Prevent Catalyst Deactivation	Type of Catalyst	Preparation Method	Deactivation Phenomena				Degree of Deactivation, %	References
			Coke Formation					
			Amorphous Carbon	Graphitic Carbon	Whisker Carbon	Sintering		
Changing or modifying support	10 wt% $Ni/MgAl_2O_4$	Impregnation	[1]•	[2]••	[3]ND	[4]NA	NA	Koo et al. (2014)
	10 wt% $Ni/MgAl_2O_4$	Impregnation	NA	NA	ND	[5]♣	33.5	Park et al. (2015)
	15 wt% Ni/MgO	Precipitation	ND	ND	ND	♣	40	Ashok et al. (2018)
	15 wt% NiO/MgO	Impregnation	ND	•	ND	NA	0	Olah et al. (2012)
	15 wt% $Ni/MgO\text{-}Al_2O_3$	Precipitation + deposition	•	•	•	♣	10	Li, and van Veen (2018)
	12 wt% $Ni/MgO\text{-}Al_2O_3$	Impregnation	ND	•	ND	ND	6.5	Roh et al. (2007)
	4 wt% Ni/ (pNirb+10.4wt% MgO)	Impregnation	ND	ND	ND	ND	0	Danilova et al. (2015)
	10 wt% Ni/ZrO_2	Modified hydrothermal + impregnation	•	•	ND	♣	11	Li et al. (2015)
	1 wt% $Ni/La_2Zr_2O_7$	Modified Pechini method	•	ND	ND	ND	2	Kumar et al. (2016)
	10 wt% Ni/SBA-15	Impregnation	•	•	ND	ND	NA	Siang et al. (2018b)
	10 wt% Ni/Al_2O_3	Impregnation	ND	••	ND	ND	NA	Stroud et al. (2018)

(*Continued*)

TABLE 3.3 (*Continued*)
Stability of the Modified Catalysts for BRM Reaction and the Type of Deactivations Occurred

Strategies to Prevent Catalyst Deactivation	Type of Catalyst	Preparation Method	Deactivation Phenomena				Degree of Deactivation, %	References
			Coke Formation					
			Amorphous Carbon	Graphitic Carbon	Whisker Carbon	Sintering		
Introducing additives/ promoters	10 wt% Ni–2.5 wt% Ce/$MgAl_2O_4$	Impregnation	•	•	ND	NA	ND	Koo et al. (2014)
	10 wt% Ni–2.5 wt% La/$MgAl_2O_4$	Impregnation	NA	NA	ND	♣	10	Park et al. (2015)
	0.5 wt% Mo_2C–10 wt% Ni/ZrO_2	Impregnation and subsequent reduction carburization	•	•	ND	♣	0	Li et al. (2015)
	0.02 wt% Sn–10 wt% Ni/Ce-Al_2O_3	Impregnation	•	•	ND	ND	9	Stroud et al. (2018)
	0.16%Pt-$Ce_{0.6}Zr_{0.4}O_2$-1.3Ni1.0Mg	Precipitation	ND	ND	ND	ND	0	Elsayed et al. (2018)
	3 wt% B–10 wt% Ni/SBA-15	Impregnation	•	ND	ND	ND	NA	Siang et al. (2018b)

(*Continued*)

TABLE 3.3 (*Continued*)
Stability of the Modified Catalysts for BRM Reaction and the Type of Deactivations Occurred

Strategies to Prevent Catalyst Deactivation	Type of Catalyst	Preparation Method	Deactivation Phenomena				Degree of Deactivation, %	References
			Coke Formation					
			Amorphous Carbon	Graphitic Carbon	Whisker Carbon	Sintering		
Employing diverse catalyst preparation methods	5 wt% Ni/Aerosil silica	Two-solvent deposition	ND	ND	ND	ND	9	Jabbour et al. (2015)
	5 wt% Ni/SBA	Two-solvent deposition	ND	ND	ND	♣	100	Jabbour et al. (2015)
	5 wt% Ni/diatoms MN3	Two-solvent deposition	ND	ND	ND	♣	100	Jabbour et al. (2015)
	15 wt% Ni–35 wt% SiO_2–55 wt% MgO	Ammonia evaporation + hydrothermal	ND	ND	ND	ND	5.1	Ashok et al. (2018)
	15 wt% Ni–60 wt% SiO_2–25 wt% MgO	Ammonia evaporation + hydrothermal	ND	ND	ND	♣	NA	Ashok et al. (2018)

[1] Carbon with amount <10 wt% (wt of carbon/wt of catalyst).
[2] Carbon with amount >10 wt% (wt of carbon/wt of catalyst).
[3] Not detected.
[4] Not available.
[5] Sintering.

impregnation and precipitation methods, respectively (Ashok et al. 2018). Both catalysts, 10 wt% Ni/$MgAl_2O_4$ and 15 wt% Ni/MgO, were sintered and deactivated after TOS for stability evaluation in BRM reaction. In contrast to the 15 wt% Ni/MgO catalyst, the 15 wt% NiO/MgO catalyst synthesized through the impregnation method by Olah et al. (2012) has shown excellent stability, attaining steady CH_4 and CO_2 conversions of 71% and 62%, respectively, throughout the BRM reaction at 830°C, 7 bars, and a GHSV of 600 L $gcat^{-1}h^{-1}$ for the duration of 320 h. Although the formation of carbon did not cause a significant deterioration in performance at lower CH_4-to-H_2O ratio, it is worth mentioning that doubling the amount of steam minimized the carbon formation from 2.35 to 0.47 wt%. Other than being used as the sole support for Ni, MgO has also been used to modify the Al_2O_3 for supporting Ni catalyst for BRM reaction. Li et al. (2018) investigated the BRM reaction using 15 wt% Ni supported on Mg-Al mixed oxide catalyst (15 wt% Ni/MgO-Al_2O_3) prepared using co-precipitation followed by the deposition method. This catalyst exhibited deactivation when it was tested for 24 h TOS at 775°C, while adopting the $CH_4/H_2O/CO_2$ molar ratio of 3:2.2:1.2 and the GHSV of 86 L $gcat^{-1}h^{-1}$. The X_{CH_4} dropped from 79% to 71%, whereas the X_{CO_2} declined from 72% to 62% after 24 h of BRM reaction. The Ni catalyst decreased in activity because of the formation of whiskers of carbon embedded in Ni, leading to the blockage of active sites. Moreover, the increase in Ni particle size from 20.7 to 33.9 nm in the spent catalyst also implied high Ni clogging and sintering *in situ* reaction. On the other hand, there was no sign of sintering for the 12 wt% Ni/MgO-Al_2O_3 catalyst synthesized by Roh et al. (2007) after it was used in the BRM reaction in an 18 h TOS at T = 800°C, $CH_4/H_2O/CO_2$ ratio = 3.0:2.4:1.2, and GHSV = 265 L $gcat^{-1}h^{-1}$. The presence of filamentous coke on the 12 wt% Ni/MgO-Al_2O_3 slightly decreased X_{CH_4} by 6.5%–88% at the end of the activity test. It was proven that Ni/MgO-Al_2O_3 catalyst outperformed the Ni/MgO and Ni/Al_2O_3 catalysts. The basic property of MgO contributed to the enhanced properties of Ni/MgO-Al_2O_3 catalyst such as higher coke resistance, finely dispersed NiO, and strong interaction between Ni and the support.

Danilova et al. (2015) used MgO to modify the support for nickel catalysts, i.e. porous Ni ribbon (pNirb). The catalyst could only be stabilized in BRM reaction if at least 8.6 wt% of MgO underlayer was incorporated to cover the porous Ni ribbon. In the case where insufficient MgO was used, the uncovered porous nickel ribbon was inferred to undergo sintering through surface diffusion mechanism and carbonization, leading to a significant drop in CH_4 conversion during catalyst screening in a flow reactor for 18 h at the operating conditions of T=900°C, $CH_4/H_2O/CO_2$ molar ratio = 3/3.34/1.97, and GHSV of 62.5 L $gact^{-1}h^{-1}$. The best catalyst, i.e. 4 wt% Ni/(pNirb + 10.4 wt% MgO), showed stable and highest activity (X_{CH_4} = 57%) throughout the 18 h with no existing trace of deposited carbon on the spent catalyst. The coke resistance of these catalysts was significantly improved by covering the nickel with MgO underlayer and binding the nickel crystallites epitaxially on the MgO. Besides the basic support, amphoteric metal oxide like ZrO_2 was also used as the support for Ni catalyst employed in BRM reaction by Li et al. (2015). The 10 wt% Ni/ZrO_2 catalyst was subjected to the activity test for 20 h at 850°C with the reactants, $CH_4/H_2O/CO_2$ with a molar ratio of 3:2.4:1.2, and a GHSV of 60 L $gcat^{-1}h^{-1}$. The shell-like coke with higher degree of graphitization on the catalyst encapsulated the Ni particles, preventing the reactants from accessing the active sites and causing a decline of 11% in X_{CH_4} after 20h of the activity test (Li et al. 2015). Contrariwise, the doping of 1wt% Ni onto the $La_2Zr_2O_7$, a highly crystalline mixed basic and amphoteric metal oxide support, has strengthened the interaction of active metal with the support, and this 1wt% $Ni/La_2Zr_2O_7$ catalyst was stable during BRM reaction (Kumar et al. 2016). It showed a consistent activity for BRM reaction temperatures ranging from 700°C to 950°C for 170h. TOS tests were carried out under atmospheric pressure with $CH_4/H_2O/CO_2$ ratio of 3:1.94:0.94 and a GHSV of 104.34L $gcat^{-1}h^{-1}$. In the spent catalyst, Ni sintering was not noticeable and the reduction and oxidation of Ni were still fully reversible. Moreover, the negligible amount of atomic carbon detected on the spent catalysts did not cause catalyst deactivation over the test duration; for instance, X_{CO_2} of 49.5% and X_{CH_4} of 54% were achieved for the BRM at 750°C and remained constant for 24h (Kumar et al. 2016). When compared with other carbons, atomic carbon can be easily oxidized to regenerate the catalyst (Kumar et al. 2015).

3.4.2 Introducing Promoters to Catalyst for Methane Bi-Reforming Reaction

Lanthanum, a rare earth element, was employed as an enhancer of the thermal stability of $MgAl_2O_4$ doped with Ni catalysts (Park et al. 2015). The 10wt% Ni–2.5wt% La/$MgAl_2O_4$ catalyst was synthesized by co-impregnation for bi-reforming of coke oven gas (COG). Operating parameters for the catalytic test were as follows: GHSV of 4680L $gcat^{-1}h^{-1}$ under CH_4:H_2O:CO_2:H_2:CO:N_2 molar ratio of 3:3.6:1.2:6:0.9:0.9, and 40h TOS at 900°C. The conversions of CH_4 and CO_2 catalysed by pristine Ni/$MgAl_2O_4$ catalyst were 83.1% and 57.4%, respectively. These conversions dropped to 55.3% and 39.2%, respectively, upon catalysing the reaction by the aged Ni/$MgAl_2O_4$ catalyst. Also, the conversion of CH_4 and CO_2 catalysed by the aged 10wt% Ni–2.5wt% La/$MgAl_2O_4$ catalyst decreased with lesser extent in comparison with the X_{CH_4} (80.6%) and X_{CO_2} (58%) attained in the reaction catalysed by fresh catalyst, circa 10% and 20%, respectively. Among all the aged catalysts, the catalyst promoted by La showed better dispersion of Ni and lower sizes of crystallite Ni when compared to the unpromoted Ni/$MgAl_2O_4$, indicating the enhanced strong metal–support interactions of La-promoted catalysts that suppress the Ni particles from sintering (Park et al. 2015).

Koo et al. (2014) improved the characteristics of Ni supported on $MgAl_2O_4$ catalysts by adding rare earth oxide and CeO_2 as a promoter. The redox property (Ce^{4+}/Ce^{3+}) of CeO_2 not only increased the Ni dispersion but also provided the active oxygen to gasify the coke on the surface of catalysts. $MgAl_2O_4$ doped using Ni-Ce with Ce/Ni ratio of 0.25 (10wt% Ni–2.5wt% Ce/$MgAl_2O_4$) was proven as the catalyst with the strongest metal–support interaction when compared with the unpromoted Ni/$MgAl_2O_4$ in the BRM reaction.

The catalyst performance was evaluated at various temperatures by a stepwise reduction of 50°C, descending from 700°C to 550°C for 20h at atmospheric pressure with $CH_4/H_2O/CO_2$ ratio of 3:2.4:1.2 and a GHSV of 530L $gcat^{-1}h^{-1}$. At 600°C, a low temperature that favours the whisker-shaped coke formation through Boudouard reaction (CO disproportionation), X_{CH_4} and X_{CO_2} attained in the BRM reaction catalysed by unpromoted 10 wt% $Ni/MgAl_2O_4$ catalyst were 28% and 8%, respectively. The addition of Ce into the promoted catalyst significantly increased the X_{CH_4} by 70%. The 10 wt% Ni–2.5 wt% $Ce/MgAl_2O_4$ catalyst gave the highest efficiency towards eliminating the graphitic carbon in BRM reaction as a result of the superior dispersion of metal, reducibility, and transfer of surface O_2.

CeO_2 was used together with Sn by Stroud et al. (2018) to promote the 10 wt% Ni/Al_2O_3 catalyst. The Sn added would occupy the carbon nucleation sites of Ni atoms, thus decelerating the coke formation process. Considering the detrimental effect on the catalyst activity due to a large amount of Sn that covered the Ni active sites, the optimum molar ratio of Sn/Ni was 0.02. The presence of CeO_2 resulted in the high Ni dispersion, improved storing capacity of O_2, and enhanced acid–base properties of the support, thus leading to better resistance towards carbon deposition. The 0.02 wt% Sn–10 wt% Ni/Ce-Al_2O_3 catalyst was tested in BRM reaction at 700°C and a GHSV of 60 L $gcat^{-1}h^{-1}$ with $CH_4/CO_2/H_2O$ ratio of 3:3:3. After 25 h, the X_{CH_4} declined from approximately 68% to 9%, whereas the X_{CO_2} remained stable and was low at 40%, which might be related to the WGS reaction that was favoured with water as reactant and the reaction catalysed by CeO_2. The used 0.02 wt% Sn–10 wt% Ni/Ce-Al_2O_3 catalyst was confirmed to be not sinter-free. In contrast to the unpromoted catalyst, the carbon deposits on the 0.02 wt% Sn–10 wt% Ni/Ce-Al_2O_3 catalyst were with lower concentration and reduced crystallinity. The better nature of the carbon formed affirmed that the carbon tolerance property of the catalyst improved (Stroud et al. 2018).

Apart from rare earth metal, the addition of noble metal (Pt) could significantly improve its catalytic activity and stability. Noble metal promoters could improve dispersion and reducibility of NiO and subsequently impede the sintering and coking processes (Elsayed et al. 2018; Fu et al. 2018). Elsayed et al. (2018) doped Pt on Ni–Mg/ceria–zirconia catalysts (0.16 wt% Pt-$Ce_{0.6}Zr_{0.4}O_2$–1.3 wt% Ni–1.0 wt% Mg) with the motivation to improve the H_2/CO ratios at lower BRM reaction temperature. A 3 h TOS stability with a GHSV of 136 L $gcat^{-1}h^{-1}$ at unchanged operating conditions was investigated. The feed ratio of 3:2:1 for $CH_4/H_2O/CO_2$ was employed, leading to CH_4 conversion of 37% and CO_2 conversion of 33%. The values remained unchanged with reduced TOS, indicating that the catalyst retained its activity. The spent catalyst was not sintered, and it contained a negligible amount of carbon deposits. Pt can adsorb H_2, hence providing H_2 excess that can enhance the reducibility of CeO_2 to Ce^{3+} by creating O_2 vacancies. Compared to the unpromoted catalyst, the 0.16 wt% Pt-$Ce_{0.6}Zr_{0.4}O_2$–1.3 wt% Ni–1.0 wt% Mg catalyst consisted of additional 25% of basic sites to shift equilibrium of CO disproportionation, thus reducing carbon deposition (Elsayed et al. 2018). The use of scarce and costly rare earth and noble metals as promoter for Ni-based BRM catalyst has restricted its larger-scale application. As an alternative to Pt-like metals, molybdenum carbide, a non-precious transition metal carbide, with its exceptional physicochemical properties was used to promote the 10 wt% Ni/ZrO_2 catalyst for BRM reaction (Li et al. 2015). The modified catalyst with optimized Mo_2C loading, i.e. 0.5 wt% Mo_2C–10 wt% Ni/ZrO_2, was subjected to the activity test for 20 h at 850°C, using atmospheric pressure, $CH_4/H_2O/CO_2$ molar ratio of 3:2.4:1.2, and a GHSV of 60 L $gcat^{-1}h^{-1}$. The 0.5 wt% Mo_2C–10 wt% Ni/ZrO_2 outperformed the unmodified 10 wt% Ni/ZrO_2. The resultant X_{CO_2} and X_{CH_4}

from the BRM reaction catalysed by 0.5 wt% Mo_2C–10 wt% Ni/ZrO_2 were 80% and 98%, respectively, approximately 20%–30% higher than the one obtained using the unmodified catalyst. The attainable conversion also remained constant for the entire test duration. Despite the sintering phenomenon, the better activity of the modified catalyst was ascribed to the enhanced Ni–ZrO_2 interactions that further improved the Ni dispersion and altered the coke morphologies. Furthermore, the quantity and structural build-up of the carbon on the modified catalyst did not differ much from the unmodified one. The two coke structures are amorphous and graphitic carbons. The outstanding stability is resulted from the minimized quantity of graphite carbon on the modified catalyst (Li et al. 2015). Boron promoter, another alternative to rare earth and noble metal oxides, was reported by Siang et al. (2018b) to suppress surface carbon diffusion that formed graphene on the 10 wt% Ni/SBA15. The performance of the best promoted catalyst, 3 wt% B–10 wt% Ni/SBA15, was compared with the performance of the unpromoted catalyst, 10 wt% Ni/SBA15, for the BRM reaction at 700°C–800°C for 10 h, stoichiometric ratio of $CH_4/H_2O/CO_2$ as 3:2:1, and a GHSV of 36 L $gcat^{-1}h^{-1}$. In relation to the unpromoted catalyst, conversions of reactants were improved by 10%–50% depending on the temperature in the BRM reaction catalysed by 3 wt% B–10 wt% Ni/SBA15. Dual carbon types (graphitic and amorphous) were identified in the unpromoted catalyst, whereas only amorphous carbon type was detected in the promoted catalyst. Boron promoter not only suppressed the graphitic carbon formation on the 3 wt% B–10 wt% Ni/SBA15 catalyst but also increased the reactiveness of CNF (carbon nanofilament, a type of amorphous carbon) species and reduced its amount by about four times in comparison with 10 wt% Ni/SBA15 catalyst. Nevertheless, both unpromoted and promoted catalysts were resilient to metallic Ni reoxidation throughout BRM reaction (Siang et al. 2018b).

3.4.3 Employing Diverse Catalyst Preparation Methods for Methane BI-Reforming Reaction

Another approach to obtain a thermally stable and carbon-resistant catalyst for BRM reaction is by inducing strong metal–support interactions during catalyst preparation using diverse methods. The silica-supported Ni (only 5 wt%) catalyst was prepared by Jabbour et al. (2015) using dual solvent technique of deposition with the aim to uniformly disperse the Ni on three silica types: SBA-15 (microporous–mesoporous), Aerosil 300 (mesoporous), and diatom MN3 (macroporous). The longevity of catalysts was tested in the BRM reaction for 10 h at 800°C with the operating parameters such as $CH_4/CO_2/H_2O$ molar ratio of 1:0.4:0.8 and a GHSV of 62 L $gcat^{-1}h^{-1}$. About 5 wt% Ni/Aerosil silica gave an exceptionally improved performance when compared to the 5 wt% Ni/SBA and 5 wt% Ni/diatoms MN3. X_{CH_4} of 88% and X_{CO_2} of 73% were attained in the BRM reaction catalysed by 5 wt% Ni/Aerosil silica, and the performance only dropped 5%–10% after 10 h. On the other hand, X_{CH_4} and X_{CO_2} in the BRM reactions catalysed by 5 wt% Ni/SBA and 5 wt% Ni/diatoms MN3 dropped to zero after 10 h. It was found that the lowest density of silanol functional in Aerosil silica created a platform for intense grafting of Ni on the surface, hence strengthening the interaction of Ni with the support to avoid sintering and oxidation of Ni species during the BRM reaction (Jabbour et al. 2015).

The synthesis of catalyst using PS-derived systems is one of the best solutions due to its unique structural behaviour that can induce an intense interaction of the metal with the support. In order to increase the robustness of PS structures in the steam environment of BRM, Ashok et al. (2018) have inserted Mg-based metal into the framework of Ni-PS structures to suppress their acidic nature. The results of the activity clearly illustrated that 15 wt% Ni–35 wt% SiO_2–55 wt% MgO derived from Ni-PS structure maintained extraordinarily good activity (conversion of CH_4 was 80% and that of CO_2 was 60%) for 140 h in the BRM reaction at 750°C with a GHSV of 200 L $gcat^{-1} h^{-1}$ and a reactant ratio $CH_4/H_2O/CO_2$ of 3:3:1.5. Negligible coke was detected on the spent catalyst at this operating condition. The strong basic strength, improved structural integrity, and moderate acidity during the BRM reaction were responsible for the catalyst performance. Nevertheless, at temperature ≤ 650°C and lower H_2O/CH_4 ratio conditions, carbon deposition of approximately 20 wt% (mostly amorphous in nature) was observed on the 15 wt% Ni–35 wt% SiO_2–55 wt% MgO catalyst. At lower temperature of reaction, the WGS reaction is thermodynamically preferred. The increased amount of CO_2 at lower temperature and insufficient amount of steam at low H_2O/CH_4 condition have resulted in a predominant DRM over SRM and hence increased the coke formation. Despite the significant deactivation of catalysts with high Si/Mg (15 wt% Ni/SiO_2 and 15 wt% Ni/60 wt% SiO_2–25 wt% MgO) during the 24 h reaction and 100 h reaction, respectively, the surface of the catalyst was carbon-free. The deactivation of the 15 wt% Ni/SiO_2 catalyst was ascribed to the decline in the reduced Ni species, whereas the deteriorating activity of the 15 wt% Ni/60 wt% SiO_2–25 wt% MgO was attributable to the remarkable drop in catalyst surface area and pore volume (Ashok et al. 2018).

The active metal supported on mixed oxide catalysts was found to be more stable than the single oxide-supported catalyst for BRM reaction. The complementary effects of basic sites and acidic sites on the supports are of essence to mitigate carbon deposition and sintering while maintaining the catalyst activity. The mixed oxide-supported catalyst showed higher coke resistance and better interaction of the metal with the support, hence increasing the dispersion of the active phase to suppress sintering. Apart from the catalyst support with balanced basic and acidic characteristics, the noble metal and rare earth metal-based promoters could also significantly increase the stability of the catalyst for BRM reaction. This unique property could be linked to their basic properties that hinder coke formation and improve oxygen capacity, which facilitates the coke gasification process. It is noteworthy that a more cost-effective transition metallic carbide like Mo_2C was proven to behave like Pt in promoting the catalyst for BRM reaction. In addition to the catalyst supports and promoters, the method with which the catalyst is synthesized is also a vital factor in ensuring strong links between metal and support of the catalysts adopted in BRM reaction.

In summary, it is obvious that the technology of BRM with CO_2 for generating H_2 and syngas is a major step towards the reduction of globally emitted CO_2 as well as minimizing the excess available CH_4, which is a fallout from high global industrialization. In addition, very recently, the high demand for renewable energy has also led to the need for various techniques and technology geared towards meeting the demand. In the course of these developments, researchers have looked towards improving and enhancing the various techniques of reforming abundantly available greenhouse gases into more useful feedstock to meet the ever-increasing renewable energy demand. The rapid breakthrough in utilizing the bi-reforming technique has resulted in significant strides in the areas of efficiency in the process, improved durability, and usability of materials, thereby making the BRM a very attractive technology in the current time and also for future purposes. This chapter has incorporated ideas of various researchers in the area of recent advances in the bi-reforming technology. The key area that has been tackled in this chapter is in relation to the development of various catalysts used for the bi-reforming process. The major setback today in the world of catalysis particularly when reforming organic compounds like CH_4 is catalyst deactivation. From all the literature reviewed in this work, it is safe to resolve that the efficiency of the BRM process strongly varies with the catalyst efficiency, which is also linked with the catalyst type. This catalyst types are of different forms, which include promoted catalyst, catalyst on support, and perovskite as catalyst. However, in this work, we have narrowed down to studies carried out on catalysts made up of active metal and support. It is worthy to note that apart from the active metal choice, identifying the appropriate support is always key as this goes a long way in distorting the physicochemical properties of the active metal in particular and the entire catalyst structure in general. Various researches have shown that with a good synergy existing between the catalyst materials, the catalyst performance is greatly improved. In addition, the physicochemical properties can also be improved as seen in some work review in this chapter by introducing additional metals called promoters. These promoters improve the metal–support interface, thereby aiding the bi-reforming process. Furthermore, in the bi-reforming process, metals supported on mixed oxides showed more stability when compared to the active metal of single oxide catalyst. The interlinkages between the acidic site and the basic site are vital in minimizing carbon formation, thereby ensuring that the efficiency of the catalyst remains stable throughout the reaction. The mixed oxide-supported catalyst supports high active metal-phase dispersion, which reduces sintering and enhances the catalyst resistance to coking. Other types of catalyst that can improve stability are the rare earth metal-based promoters and noble metals. Several researches have shown that these metals have basic properties required to minimize the formation of carbon during the bi-reforming reaction due to their O_2 capacity, which enhances gasification.

ACKNOWLEDGEMENT

The authors would like to acknowledge the Ministry of Education, Malaysia, for awarding the FRGS research grant vote FRGS/1/2018/TK02/UMP/02/12 (RDU190197) and Universiti Malaysia Pahang (RDU1803118) for financial support.

REFERENCES

Abas, N., Kalair, A., and Khan, N. 2015. Review of fossil fuels and future energy technologies. *Futures* 69 (May): 31–49. doi: 10.1016/j.futures.2015.03.003.

Alaswad, A., Baroutaji, A., Achour, H., Carton, J., Makky, A. A., and Olabi, A. G. 2016. Developments in fuel cell technologies in the transport sector. *International Journal of Hydrogen Energy* 41 (37): 16499–508. doi: 10.1016/j.ijhydene.2016.03.164.

A′lvarez M., A., Centeno, M.A., and Odriozol, J.A. 2015. Ru–Ni catalyst in the combined dry-steam reforming of methane: The importance in the metal order addition. *Topics in Catalysis* 59 (2–4): 303–313.

Appari, S., Janardhanan, V.M., Bauri, R., and S. Jayanti. 2014. Deactivation and regeneration of Ni catalyst during steam reforming of model biogas: An experimental investigation. *International Journal of Hydrogen Energy* 39: 297–304.

Ashok, J., Bian, Z., Wang, Z., and Kawi, S. 2018. Ni-phyllosilicate structure derived Ni-SiO_2-MgO catalysts for bi-reforming applications: acidity, basicity and thermal stability. *Catalysis Science & Technology* 8(6): 1730–1742.

Bahari, M. B., Goo, B. C., and Pham, T. L., et al. 2016. Hydrogen-rich syngas production from ethanol dry reforming on La-doped Ni/Al_2O_3 catalysts: Effect of promoter loading. *Procedia Engineering* 148: 654–661.

Bukhari, S. N., Chin, C. Y., and Setiabudi, H. D., et al. 2017. Tailoring the properties and catalytic activities of Ni/SBA-15 via different TEOS/P123 mass ratios for CO_2 reforming of CH_4. *Journal of Environmental Chemical Engineering* 5: 3122–3128.

Danilova, M. M., Fedorova, Z. A., Kuzmin; V. A., Zaikovskii, V. I., Porsin, A. V. and Krieger T. A. 2015. Combined steam and carbon dioxide reforming of methane over porous nickel based catalysts. *Catalysis Science & Technology* 5: 2761–2768.

Dodds, P. E, Staffell, I., Hawkes, A. D., Li, F., Grünewald, P., McDowall, W., and Ekins, P. 2015. Hydrogen and Fuel Cell Technologies for Heating: A Review. *International Journal of Hydrogen Energy* 40 (5): 2065–2083. doi: 10.1016/j.ijhydene.2014.11.059.

Earth System Research Laboratory, Global Monitoring Division, U.S Department of Commerce, National Oceanic & Atmospheric Administration. 2019. Trends in Atmospheric Carbon dioxide. https://www.esrl.noaa.gov/gmd/ccgg/trends/.

Elsayed, N. H., Maiti, D., and Joseph, B., et al. 2018. Precious metal doped Ni–Mg/Ceria–Zirconia catalysts for methane conversion to syngas by low temperature bi-reforming. *Catalysis Letters* 148(3): 1003–1013.

Fayaz, F., Danh, H. T., and Nguyen-Huy, C., et al. 2016. Promotional effect of Ce-dopant on Al_2O_3-supported Co catalysts for syngas production via CO_2 reforming of ethanol. *Procedia Engineering* 148: 646–653.

Fu, Q., Saltsburg, H., and M. Flytzani-Stephanopoulos. 2003. Active nonmetallic Au and Pt species on ceria-based water-gas shift catalysts. *Science* 301: 935–938.

Itkulova, S. S., Zakumbaeva, G. D., Nurmakanov, Y. Y., Mukazhanova, A. A., and Yermaganbetova, A. K. (2014). Syngas production by bireforming of methane over Co-based alumina-supported catalysts. *Catalysis Today*, 228, 194–198.

Jabbour, K., Kaydouh, M. N., El Hassan, N., El Zakhem, H., Casale, S., Massiani, P., and Davidson, A. 2015. Compared activity and stability of three Ni-silica catalysts for methane bi-and dry reforming. In *2015 International Mediterranean Gas and Oil Conference (MedGO)* (pp. 1–4). IEEE.

Jabbour, K., Massiani, P., and Davidson, A., et al. 2017. Ordered mesoporous "one-pot" synthesized Ni-Mg (Ca)-Al_2O_3 as effective and remarkably stable catalysts for combined steam and dry reforming of methane (CSDRM). *Applied Catalysis B: Environmental* 201: 527–542.

Khani, Y., Shariatinia, Z., and Bahadoran, F. 2016. High catalytic activity and stability of $ZnLaAlO_4$ supported Ni, Pt and Ru nanocatalysts applied in the dry, steam and combined dry-steam reforming of methane. *Chemical Engineering Journal* 299(1): 353–366.

Kim, A. R., Lee, H. Y., and Cho, J. M., et al. 2017. Ni/M-Al_2O_3 (M = Sm, Ce or Mg) for combined steam and CO_2 reforming of CH_4 from coke oven gas. *Journal of CO_2 Utilization* 21:211–218.

Koo, K. Y., Lee, S. H., Jung, U. H., Roh, H. S., and Yoon, W. L. 2014. Syngas production via combined steam and carbon dioxide reforming of methane over Ni–Ce/$MgAl_2O_4$ catalysts with enhanced coke resistance. *Fuel Processing Technology* 119, 151–157.

Kumar, N., Shojaee, M., and Spivey, J. J. 2015. Catalytic bi-reforming of methane: From greenhouse gases to syngas. *Current Opinion in Chemical Engineering* 9 (August): 8–15. doi: 10.1016/j.coche.2015.07.003.

Kumar, N., Roy, A., Wang, Z., L'Abbate, E. M., Haynes, D., Shekhawat, D., and Spivey, J. J. 2016. Bi-reforming of methane on Ni-based pyrochlore catalyst. *Applied Catalysis A: General* 517, 211–216.

Li, M., and van Veen, A. C. 2018. Coupled reforming of methane to syngas ($2H_2$-CO) over Mg-Al oxide supported Ni catalyst. *Applied Catalysis A: General* 550, 176–183.

Li, W., Zhao, Z., Ren, P., and Wang, G. 2015. Effect of molybdenum carbide concentration on the Ni/ZrO_2 catalysts for steam-CO_2 bi-reforming of methane. *RSC Advances* 5(122), 100865–100872.

Nicoletti, G., Arcuri, N., Nicoletti, G., and Bruno, R. 2015. A technical and environmental comparison between hydrogen and some fossil fuels. *Energy Conversion and Management* 89 (January): 205–213. doi: 10.1016/j.enconman.2014.09.057.

Nicoletti Gi, A. G., Il Controllo della, A. F. 2009. CO_2: Misure E Strategie. *Proceedings of Italian Conference ATI, L'Aquila (ITALY).*

van de Loosdrecht, J. NiemantsverdrietJ.W. 2013. Synthesis gas to hydrogen, methanol and synthetic fuels. In *Chemical Energy Storage.*

Olah, G. A., Goeppert, A., and Czaun, M., et al. 2012. Bi-reforming of methane from any source with steam and carbon dioxide exclusively to metgas (CO–$2H_2$) for methanol and hydrocarbon synthesis. *Journal of the American Chemical Society* 135(2): 648–650.

Pakhare, D., and J. Spivey. 2014. A review of dry (CO_2) reforming of methane over noble metal catalysts. *Chemical Society Reviews* 43: 7813–7837.

Palma, S., Bobadilla, L.F., Corrales, A., et al. 2014. Effect of gold on a $NiLaO_3$ perovskite catalyst for methane steam reforming. *Applied Catalysis B: Environmental* 144: 846–854.

Park, J. E., Koo, K. Y., Jung, U. H., Lee, J. H., Roh, H. S., and Yoon, W. L. 2015. Syngas production by combined steam and CO_2 reforming of coke oven gas over highly sinter-stable La-promoted Ni/$MgAl_2O_4$ catalyst. *International Journal of Hydrogen Energy* 40(40): 13909–13917.

Peng, W.X., L.S. Wang, M. Mirzaee, H. Ahmadi, M.J. Esfahani, and S. Fremaux. 2017. Hydrogen and syngas production by catalytic biomass gasification. *Energy Conversion and Management* 135 (March): 270–73. doi: 10.1016/j.enconman.2016.12.056.

Pour, A. N., and Mousavi, M. 2015. Combined reforming of methane by carbon dioxide and water: Particle size effect of Ni–Mg nanoparticles. *International Journal of Hydrogen Energy* 40(38), 12985–12992.

Qin, D., Lapszewicz, J., and Jiang, X. 1996. Comparison of partial oxidation and steam-CO_2 mixed reforming of CH_4 to syngas on MgO-supported metals. *Journal of Catalysis* 159: 140–149.

Roh, H. S., Koo, K. Y., Jeong, J. H., Seo, Y. T., Seo, D. J., Seo, Y. S., and Park, S. B. 2007. Combined reforming of methane over supported Ni catalysts. *Catalysis Letters* 117 (1–2): 85–90.

Roh, H. S., Koo, K. Y., and Yoon, W. L. 2009. Combined reforming of methane over co-precipitated Ni–CeO_2, Ni-ZrO_2 and Ni-$Ce_{0.8}Zr_{0.2}O_2$ catalysts to produce synthesis gas for gas to liquid (GTL) process. *Catalysis Today* 146: 71–75.

Schulz, J. 1999. Short history and present trends of Fischer-Tropsch synthesis. *Applied Catalysis A: General* 3: 186.

Sharma, S., and Ghoshal, S. K. 2015. Hydrogen the future transportation fuel: From production to applications. *Renewable and Sustainable Energy Reviews* 43 (March): 1151–58. doi: 10.1016/j.rser.2014.11.093.

Siang, T. J., Bach, L. G., and Singh, S., et al. 2019. Methane bi-reforming over boron-doped Ni/SBA-15 catalyst: Longevity evaluation. *International Journal of Hydrogen Energy* 44: 20839–20850.

Siang, T. J., Danh, H. T., and Singh, S., et al. 2017. Syngas production from combined steam and carbon dioxide reforming of methane over Ce-modified silica-supported nickel catalysts. *Chemical Engineering Transactions* 56: 1129–1134.

Siang, T. J., Singh, S., and Omoregbe, O., et al. 2018a. Hydrogen production from CH_4 dry reforming over bimetallic Ni–Co/Al_2O_3 catalyst. *Journal of the Energy Institute* 91: 683–694.

Siang, T. J., Pham, T. L., and Van Cuong, N., et al. 2018b. Combined steam and CO_2 reforming of methane for syngas production over carbon-resistant boron-promoted Ni/SBA-15 catalysts. *Microporous and Mesoporous Materials* 262: 122–132.

Singh, S., Bahari, M. B., and Abdullah, B., et al. 2018. Bi-reforming of methane on Ni/SBA-15 catalyst for syngas production: Influence of feed composition. *International Journal of Hydrogen Energy* 43: 17230–17243.

Singh, S., Nga, N. T. A., Pham, T. L. M., Tan, J. S., Phuong, P. T. T., Khan, M. R., and Vo, D. V. N. 2017. Metgas Production from Bi-reforming of methane over La-modified Santa Barbara Amorphous-15 supported Nickel catalyst. *Chemical Engineering Transactions* 56, 1573–1578.

Solomon, S. 2007. The physical basis for climate change. *Contribution of Working Group I to the Fourth Assessment Report of the Intergovernmental Panel on Climate Change, Working Group I.*

Stroud, T., Smith, T. J., Le Saché, E., Santos, J. L., Centeno, M. A., Arellano-Garcia, H., and Reina, T. R. (2018). Chemical CO2 recycling via dry and bi reforming of methane using Ni-Sn/Al_2O_3 and Ni-Sn/CeO2-Al_2O_3 catalysts. *Applied Catalysis B: Environmental* 224, 125–135.

Vo, D. -V. N., Cooper, C. G., and Nguyen, T.-H., et al. 2012. Evaluation of alumina-supported Mo carbide produced via propane carburization for the Fischer–Tropsch synthesis. *Fuel* 93: 105–116.

Yang, E. H., Noh, Y. S., and Hong, G. H., et al. 2018. Combined steam and CO_2 reforming of methane over $La_{1-x}Sr_xNiO_3$ perovskite oxides. *Catalysis Today* 299: 242–250.

Zhao, K., Li, L., Zheng, A., et al. 2017. Synergistic improvements in stability and performance of the double perovskite-type oxides La2-xSrxFeCoO6 for chemical looping steam methane reforming. *Applied Energy* 197: 393–404.

4 Carbonic Anhydrase

An Ancient Metalloenzyme for Solving the Modern Increase in the Atmospheric CO_2 Caused by the Anthropogenic Activities

Claudiu Supuran and Clemente Capasso

CONTENTS

4.1 BACTERIAL CARBONIC ANHYDRASES

The CO_2 hydration/dehydration conversion ($CO_2 + H_2O \rightleftharpoons HCO_3^- + H^+$) is a vital reaction of the bacterial metabolism [1–8]. This facile but physiologically relevant interconversion of carbon dioxide and water into bicarbonate and protons is linked to various metabolic pathways, including photosynthesis and carboxylation reactions, and biochemical processes such as pH homeostasis, electrolyte secretion, and CO_2 and bicarbonate transport [9,10]. Also, CO_2 and HCO_3^- interconversion occurs naturally at a low rate and is precisely tuned in all living organisms to preserve the balance between dissolved inorganic carbon dioxide (CO_2), carbonic acid (H_2CO_3), bicarbonate (HCO_3^-), and carbonate (CO_3^{2-}) [11–14]. The existence of microorganisms depends on the accessibility of the aforementioned metabolites, as they are crucial for the biosynthesis and energy metabolism of microbes [15]. The hydration/dehydration reaction of CO_2 is catalysed by a category of metalloenzymes, recognized as carbonic anhydrases (CAs, EC 4.2.1.1) [4,7,16–18], which are classified into eight genetically separate families (or classes), namely α-, β-, γ-, δ-, ζ-, η-, ϴ, and ι-CAs [4,7,16–19] (see Figure 4.1). The last three classes have been found lately. In 2014, our groups found the η-class in protozoa [20]; however, ϴ- and ι-CAs were discovered in marine diatoms, respectively, in 2016 and 2019 [19,21].

It has been shown to date that only three of eight classes are encoded by the bacterial genome: α-, β-, and γ-CA [7,22–27]. The movement of carbon dioxide and bicarbonate supported by bacterial CAs is shown in Figure 4.2.

The CO_2 and HCO_3^- that are indispensable for bacterial metabolic processes cannot be supplied by the obviously happening non-catalysed CO_2 hydration/dehydration reaction because the speed is too small at physiological pH ($k_{cat\ hydration} = 0.15\ s^{-1}$ and $k_{cat\ dehydration} = 50.0\ s^{-1}$), whereas the catalysed reaction has a rate of 10^4–$10^6\ s^{-1}$ [25,28]. It was speculated that in Gram-negative bacteria, the α-CA, characterized by an N-terminal signal peptide, is capable of converting the atmospheric and/or metabolic CO_2 diffused in the periplasmic space, whereas β- and γ-classes have a cytoplasmic localization and are responsible for CO_2 supply for carboxylase enzyme, pH homeostasis, and other intracellular functions that ensure survival and/or bacterial metabolism [29,30].

Figure 4.3 shows that bacteria demonstrate an intricate distribution pattern of the CA classes. The bacterial genome encodes for the three CA classes (α, β, and γ), but it is also prevalent to distinguish bacteria whose genome encodes only for one or two classes of CA and occasionally for any CA [25,31]. Recently, the amino acid sequence of the β-CAs recognized in the genome of the pathogenic Gram-negative organisms, such as *Helicobacter pylori*, *Vibrio cholerae*, *Neisseria gonorrhoeae*, and *Streptococcus salivarius*, has been shown to have a secretory signal of 18 or more amino acid residues at the N-terminal portion [25,28]. Intriguingly, there is also a short signal peptide at the N-terminus of the carbonic anhydrase (CA) of

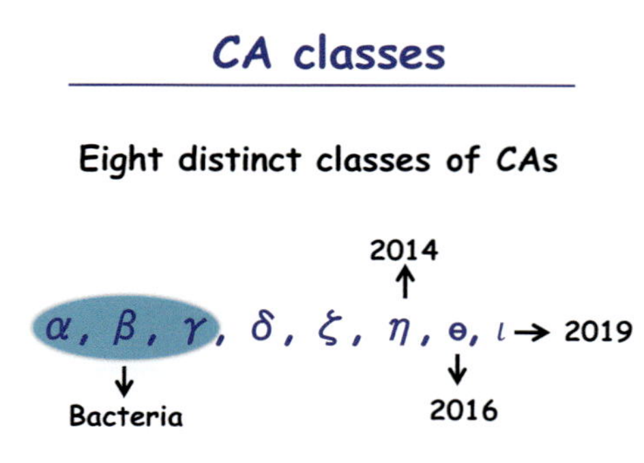

FIGURE 4.1 The α-CAs were found in vertebrates, protozoa, algae, and cytoplasm of green plants and some *Bacteria*; the β-CAs were primarily discovered in *Bacteria*, algae, and chloroplasts of both mono- and dicotyledons, as well as in many fungi and some *Archaea*. The γ-CAs have been discovered in Archaea and some *Bacteria*, whereas δ-, ζ-, ϴ-, and ι-CAs seem to be present only in marine diatoms. The η-class was found in protozoa.

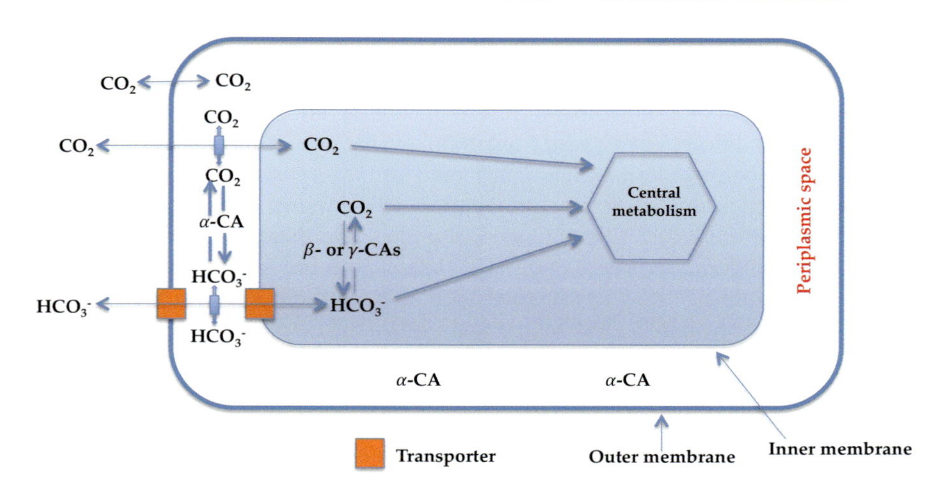

FIGURE 4.2 Transport of carbon dioxide and bicarbonate with the help of bacterial CAs. The α-CA has a periplasmic location and converts the diffused CO_2 into bicarbonate within the periplasmic space, whereas the cytosolic β- and γ-CAs are capable of feeding with CO_2 and HCO_3^-, the central bacterial metabolism.

Distribution of the CA classes

Gram-negative			
Neisseria gonorrhoeae	α	–	–
Helicobacter pylori	α	β	γ
Escherichia coli	–	β	γ
Haemophilus influenzae	–	β	–
Brucella suis	–	β	γ
Salmonella enterica	–	β	–
Vibrio cholerae	α	β	γ
Sulfurihydrogenibium yellowstonense	α	–	γ
Sulfurihydrogenibium azorense	α	–	γ
Porphyromonas gingivalis	–	β	γ
Ralstonia eutropha	α	β	γ
Burkholderia pseudomallei	–	β	γ

Gram-positive			
Mycobacterium tuberculosis	–	β	γ
Clostridium perfringens	–	β	γ
Streptococcus pneumoniae	–	β	γ
Bacillus subtilis	–	β	γ
Leifsonia xyli	–	β	γ
Staphylococcus aureus	–	–	γ
Enterococcus faecalis	–	–	γ

FIGURE 4.3 CA-class distributions in Gram-negative and Gram-positive bacteria.

Methanosarcina thermophile (CAM) protein (γ-CA). Since the signal peptide is essential for translocation in the periplasmic area, the β- or γ-CAs characterized by the presence of a signal peptide may coexist with the α-CAs in the periplasmic area of the Gram-negative bacteria [25,28].

4.2 EXTREMOPHILES AS SOURCE OF STABLE BIOCATALYSTS

Since many bacteria are abundant in environments that are harmful to all other types of living organisms, our groups have concentrated their scientific studies on CAs in thermophilic microorganisms. Extremophiles from Bacteria and Archaea are entirely depicted by microorganisms, whereas in the Eukarya domain, they are primarily algae-forming lichens, fungi, protozoa, including species designed to survive at shallow temperatures (e.g. Antarctic fish) or creatures tolerant to low, high temperatures, and high doses of radiation, like the microscopic invertebrates, known as Tardigrades [32]. The extremophiles populate life-challenging niches and thus are exposed to hot or cold temperatures, high concentrations of salt, as well as acid or alkaline conditions, toxic waste, organic solvents, heavy metals, high pressure, or other environmental situations considered unfavourable for life [33]. In general, five branches characterize extremophiles as follows: (i) thermophiles, bacteria living at temperature ≥45°C, subdivided into moderate thermophiles (45°C–70°C), extreme thermophiles growing optimally at temperatures ≥ 70°C, and hyperthermophiles, organisms growing at very high temperatures (optimal temperatures ≥ 80°C); (ii) psychrophiles, low temperature-adapted bacteria; (iii) acidophiles and alkaliphiles, bacteria living at high acidic or basic pH levels, respectively; (iv) barophiles, bacteria that develop under high pressure; (v) halophiles, bacteria that tolerate high salt concentrations [33,34]. The extremophiles evolved biomolecules adapted to operate under such harsh environmental conditions (temperatures, salinity, pH, pressure, and solvent) to manage these severe habitats. It is important to stress that the mesophilic counterparts cannot survive at all the conditions mentioned above [35,36]. Extremophilic macromolecules at harsh conditions are mainly defined by high stability and optimum enzymatic function with distinctive biotechnological benefits over mesophilic enzymes (optimally active at 25°C–50°C), which are often not well adapted to the severe circumstances of reactions needed in manufacturing procedures. As a consequence, extremophiles are a source of biocatalysts, which can be extensively used in biotechnological purposes, such as esterases/lipases, glycosidases, aldolases, nitrilases/amidases, phosphatases, racemases, thermostable DNA polymerases, and enzymes used in the manufacturing of biofuels or mining processes [36]. The study of the biochemical and physical properties of the ‘extreme’ bacterial CAs has resulted in the finding of molecular characteristics that distinguish them from those of the mesophilic equivalent enzymes, enabling their use in biotechnological areas usually characterized by harmful enzyme activity circumstances [35,37]. As a result, the extreme CAs emerged to be attractive candidates for manufacturing and medical applications such as post-combustion carbon capture, artificial lung development, and biosensors [27,38–49]. CA enzymes can be regarded as a superfamily of biotechnological multitasking molecules because the extreme CAs are capable of combating both the rise of CO_2 in the atmosphere generated by anthropogenic activity and the improvement of human health due to their biomedical concerns.

4.3 EXTREME SspCA AND SazCA

In this regard, our groups explored two CAs recognized in the *Sulfurihydrogenibium yellowstonense* and *Sulfurihydrogenibium azorense* genomes, which are related to the α-CA category [35,50–57]. These CAs were indicated with the acronyms SspCA (from *S. yellowstonense*) and SazCA (from *S. azorense*). The SazCACO_2 hydratase activity ($k_{cat} = 4.40 \times 10^6 s^{-1}$) was the highest between the extreme and other recognized CAs. The human isoform hCA II with a $k_{cat} = 1.40 \times 10^6 s^{-1}$ was the most effective CA up to that moment. Besides, SazCA with a k_{cat}/K_M value of 3.5×10^8 $M^{-1}s^{-1}$ resulted in the second most effective enzyme after superoxide dismutase. SspCA also had an excellent catalytic activity with a k_{cat} of $9.35 \times 10^5 s^{-1}$ and k_{cat}/K_M of 1.1×10^8 $M^{-1}s^{-1}$ for the same reaction (see Table 4.1). SspCA behaviour at elevated temperatures was fascinating, too. SspCA retained an excellent catalytic activity when heated for an extended period (more than 180 min) to 100°C [35,50,52,54–58].

The SspCA's and SazCA's X-ray three-dimensional structures solved by our groups provided the rationale for their thermostability at the molecular level: they have a dimeric structure characterized by elevated compactness, a higher quantity

TABLE 4.1
Comparison of the Biochemical Properties of the Thermostable Bacterial CAs

Organism	Type of Bacteria	Acronym	Family	k_{cat} (s^{-1})	K_{cat}/k_M ($M^{-1}s^{-1}$)
[a]*Sulfurihydrogenibium yellowstonense*	Extreme thermophile	SspCA	α	9.35×10^5	1.1×10^8
[b]*Sulphurihydrogenibium azorense*	Extreme thermophile	SazCA	α	4.40×10^6	3.5×10^8
[c]*Thermovibrio ammonificans*	Extreme thermophile	TaCA	α	1.60×10^6	1.6×10^8
[d]*Persephonellamarina EX-H1T*	Extreme thermophile	PmCA	α	3.2×10^5	3.0×10^7
[e]*Methanobacterium thermoautotrophicum*	Moderate thermophile	Cab	β	1.7×10^4	5.9×10^6
[f]*Methanosarcina thermophila*	Moderate thermophile	Cam	γ	6.8×10^4	3.1×10^6
[g]*Colwellia psychrerythraea*	Psychrophile	CpsCA	γ	6.0×10^5	4.7×10^6
[h]*Pseudoalteromonas haloplanktis*	Psychrophile	PhaCA	γ	1.4×10^5	1.9×10^6
[i]*Nostoc commune*	Psychrophile	NcoCA	γ	9.5×10^5	8.3×10^7
[j]*Pyrococcus horikoshii*	Hyperthermophile	PhoCA	N.D.	N.D.	N.D.

N.D.: not detected.
[a] From Ref. [52,57,58];
[b] From Ref. [50,53];
[c] From Ref. [59];
[d] From Ref. [60];
[e] From Ref. [61];
[f] From Ref. [62];
[g] From Ref. [63,64];
[h] From Ref. [65,66];
[i] From Ref. [67,68];
[j] From Ref. [69].

of secondary structural elements, an enhanced amount of protein surface-charged residues, and a substantial amount of ionic networks with regard to the mesophilic counterpart [35,37].

4.4 SspCA AS A GOOD CANDIDATE IN THE POST-COMBUSTION CARBON DIOXIDE CAPTURE

The rise in the atmospheric CO_2 level caused by the anthropogenic activities and the impacts of this greenhouse gas have encouraged the scientific community to concentrate their studies on different alternatives to reduce CO_2 emissions [70]. An elegant eco-compatible strategy is the post-combustion carbon dioxide capture through the use of 'extreme' CAs (biomimetic strategy), which are preferred in environments characterized by extreme conditions (high temperature, high salinity, extreme pH) [71]. The biomimetic approach is an interesting CO_2 trapping strategy. It enables CO_2 to be converted to water-soluble ions in an eco-compatibility manner with the possibility to use the enzyme for multiple cycles [71]. The recovery of the catalyst from the reaction mixture can be achieved by immobilizing the enzyme on specific supports [72–74]. The word 'immobilized enzymes' refers to proteins that are physically restricted or located within a particular space with preservation of their catalytic activity, and that can be continuously used [75]. Although there is sometimes a reduction in reaction rates, as the enzyme cannot blend loosely with the substrate or a specific conformational change is necessary for the effectiveness of the biocatalyst, there are many examples of enhanced activity and stability due to the immobilization [76]. The need to continually use the biocatalyst leads to the immobilization of the recombinant *Ssp*CA on polyurethane (PU) foam, a pre-polymer of polyethylene glycol [77], onto supported ionic liquid membranes (SMLs), to create a system capable of separating and transforming CO_2 selectively [78]. Besides, the immobilization of the protein on the magnetic particles (MPs) to recover the biocatalyst from the bioreactor effortlessly and practically, e.g. by using a magnet, has also been proposed for SspCA [79].

4.4.1 Immobilization on Polyurethane Foam

The robust SspCA was chosen to produce a three-phase bioreactor (gas, liquid, and solid) packed with the recombinant SspCA immobilized on PU foam, a polyethylene glycol pre-polymer [77]. The calculated specific activity of the PU-SspCA was comparable to that of the free enzyme. The activity of PU-SspCA remained constant for up to a month, whereas that of the free catalyst mildly declined. Once immobilized, the capacity to adsorb CO_2 by PU-SspCA was verified in the laboratory using a three-phase bioreactor [77]. The gas phase was a mixture of N_2 and CO_2 (20% by volume) injected from the bottom of the bioreactor, the aqueous phase (distilled water) was pumped from the top, and the solid phase was the PU-SspCA (enzyme immobilized on PU foam). The CO_2 consumption was tracked using a CO_2 analyser [77]. As a result, the aforementioned laboratory-scale bioreactor showed that the immobilized PU-SspCA converted about 5% or 10% (based on the quantity of immobilized enzyme) of CO_2 from a starting gas mixture of 20%. The CO_2 transformation was about 2% without the biocatalyst. These findings validate the extremophile CA as an outstanding choice for use in post-combustion carbon capture. Russo and colleagues determined the SspCA kinetic constants for the CO_2 hydration reaction and the long-term catalyst strength to high temperatures using the protocol based on the CO_2 absorption experiment in a stirred cell apparatus. In these conditions, SspCA showed a first-order kinetic constant at 25°C of 9.16×10^6 L/(mols) and a half-life of 53 and 8 days at 40°C and 70°C, respectively [80]. The studies were conducted using an alkaline solution of 0.5 M Na_2CO_3/0.5 M $NaHCO_3$ buffer, pH 9.6, as a fluid phase. The SspCA long-term stability at high temperatures is consistent with the working circumstances of a typical absorption device used to capture CO_2 from flue gas and combined with a solvent regeneration and CO_2 retrieval vacuum stripping system.

4.4.2 Immobilization on Membranes

The immobilization of CA onto the membranes makes the selective absorption and separation of CO_2 easier. The immobilized membranes have been widely used to sequester CO_2 from gaseous flows. Generally, the membrane adsorbs CO_2 in alkaline water, and when the CO_2 interacts with the membrane-immobilized CA, bicarbonate is produced in aqueous solution. The use of water affects the liquid membranes employed in immobilization because of its evaporation at relatively low temperatures [81]. To solve this problem, gas separation applications with SLMs are employed, which are characterized by the solvent immobilized inside the porous of the SMLs by capillary forces. Neves et al. provided an excellent idea for the CO_2 extraction from the flue gas flow using the SLMs with the immobilized bovine CA for enhancing the selective transport of CO_2 [82]. The operating temperature of combustion gases can readily exceed the optimum temperature for an enzyme used in the method of CO_2 extraction. That's why our groups impregnated the SLMs with the thermostable SspCA to create a selective system capable to capture and transform the CO_2 [78]. The results showed that the SspCA-SMLs had high CO_2 permeability at high temperatures (up to 100°C) and a reasonable CO_2 conversion [78]. This approach provides novel perspectives into the development of an effective and sustainable strategy to be used in the method of the post-combustion carbon capture.

4.4.3 Binding on Magnetic Particles

For an easy and practical recovery of the catalysts from the bioreactor, e.g. by using a magnet, it has been decided to immobilize SspCA onto the magnetic support. For this purpose, through the carbodiimide activation, the enzyme was covalently immobilized onto the surface of magnetic particles (MPs, Fe_3O_4) formed by co-precipitating ferric sulphate and ferrous chloride with an aqueous ammonia solution [79]. Because of the magnetite's powerful ferromagnetic properties, the biocatalyst can be retrieved from the reaction using a magnet or an electromagnet. The behaviour of the thermostable CA immobilized on the MPs was striking. The activity of the bound SspCA remained constant (100%) at 50°C and 70°C for the entire incubation period (80 h). The long-term stability of the free and bound SspCA or bCA (the mammalian protein) at 25°C was determined, and after an incubation period of 30 days, the residual activity of the bound SspCA was of 100% when compared with that of the free catalyst [74]. SspCA covalently immobilized on the MP surface increased its stability and long-term storage. Furthermore, the biocatalyst can be rapidly recovered from the reaction mixture by an electromagnet [74].

4.4.4 Immobilization on the Bacterial Cell Surface

The strategies described in the aforementioned paragraphs may not be convenient because of the increased costs associated with the purification of biocatalysts and the preparation of the immobilization support. SspCA *in vivo* immobilization can readily solve this restriction [83]. This system, known as the one-step immobilization procedure, overexpresses the *Ssp*CA directly onto the surface of bacterial hosts. It was realized by using the ice nucleation protein (INP) from the Gram-negative bacterium *Pseudomonas syringae* [83]. *Escherichia coli* cells were transformed with a chimeric gene that resulted from the fusion of a signal peptide, the *P. syringae* INP domain (INPN), and the gene encoding for the thermostable α-CA (SspCA) [83]. SspCA was catalytically active and efficiently overexpressed on the external surface of the *E. coli* cells. Moreover, it was stable and active for 15 h at 70°C and 10 days at 25°C [83]. This system drastically reduces the costs needed for the enzyme purification, immobilization, and support preparation when compared with the magnetic Fe_3O_4 particles. Besides, a simple centrifugation step can retrieve the biocatalyst from the reaction mixture. The *in vivo* immobilization can be considered an elegant approach to accomplish the biomimetic capture of CO_2 and other biotechnological systems requiring extremely efficient, thermostable biocatalysts.

4.5 CONCLUSION

SspCA, recognized in the genome of the bacterium *S. yellowstonense*, resulted in a highly active catalyst for CO_2 hydration reaction. It was extremely thermostable, maintaining high catalytic activity even if hot extended (up to 180 min.). Thermostable CAs are proteins that can withstand elevated temperatures and, opposed to mesophilic competitors, can usually withstand high salinity and severe pH. The post-combustion carbon capture can be carried out using the extreme SspCA immobilized onto the PU foam, MPs, SMLs, and bacterial membranes. The *in vivo* immobilization is considered a suitable method for increasing the thermostability of proteins, and the process requires extremely efficient and thermostable catalysts. In summary, the robust CAs play a crucial part in combating the rise of CO_2 in the atmosphere induced by anthropogenic activity or in improving human health as they may be used in certain medical applications (e.g. artificial lungs). For example, to speed up the removal of carbon dioxide from the blood, a mesophilic CA has recently been immobilized on the surface of HFMs (Hollow Fibre Membranes), which are used to transfer CO_2 from artificial lungs [84]. The transfer of CO_2 across the HFMs is improved by increasing the rate of the blood mixing, but the shearing force denatures the mesophilic CA immobilized on the membrane. To overcome this limitation, the robust CAs identified in the genome of thermophilic bacteria can be used to provide efficient artificial lungs.

REFERENCES

1. Del Prete S, De Luca V, De Simone G, Supuran CT, Capasso C. Cloning, expression and purification of the complete domain of the eta-carbonic anhydrase from *Plasmodium falciparum*. *J Enzyme Inhib Med Chem*. 2016;31:54–59.
2. Del Prete S, Vullo D, De Luca V, Carginale V, Osman SM, AlOthman Z, Supuran CT, Capasso C. Cloning, expression, purification and sulfonamide inhibition profile of the complete domain of the eta-carbonic anhydrase from *Plasmodium falciparum*. *Bioorg Med Chem Lett*. 2016;26:4184–4190.
3. Del Prete S, Vullo D, De Luca V, Carginale V, di Fonzo P, Osman SM, AlOthman Z, Supuran CT, Capasso C. Anion inhibition profiles of the complete domain of the eta-carbonic anhydrase from *Plasmodium falciparum*. *Bioorgan Med Chem*. 2016;24:4410–4414.
4. Annunziato G, Angeli A, D'Alba F, Bruno A, Pieroni M, Vullo D, De Luca V, Capasso C, Supuran CT, Costantino G. Discovery of new potential anti-infective compounds based on carbonic anhydrase inhibitors by rational target-focused repurposing approaches. *Chem Med Chem*. 2016;11:1904–1914.
5. Del Prete S, Vullo D, De Luca V, Carginale V, di Fonzo P, Osman SM, AlOthman Z, Supuran CT, Capasso C. Anion inhibition profiles of alpha-, beta- and gamma-carbonic anhydrases from the pathogenic bacterium *Vibrio cholerae*. *Bioorgan. Med Chem*. 2016;24:3413–3417.
6. Abdel Gawad NM, Amin NH, Elsaadi MT, Mohamed FM, Angeli A, De Luca V, Capasso C, Supuran CT. Synthesis of 4-(thiazol-2-ylamino)-benzenesulfonamides with carbonic anhydrase I, II and IX inhibitory activity and cytotoxic effects against breast cancer cell lines. *Bioorgan Med Chem*. 2016;24:3043–3051.
7. Capasso C, Supuran CT. An overview of the carbonic anhydrases from two pathogens of the oral cavity: *Streptococcus mutans* and *Porphyromonas gingivalis*. *Curr Top Med Chem*. 2016;16:2359–2368.
8. Del Prete S, Vullo D, De Luca V, Carginale V, Osman SM, AlOthman Z, Supuran CT, Capasso C. Comparison of the sulfonamide inhibition profiles of the alpha-, beta- and gamma-carbonic anhydrases from the pathogenic bacterium *Vibrio cholerae*. *Bioorgan Med Chem Lett*. 2016;26:1941–1946.
9. Johnson X, Alric J. Interaction between starch breakdown, acetate assimilation, and photosynthetic cyclic electron flow in *Chlamydomonas reinhardtii*. *J Bio Chem*. 2012;287:26445–26452.
10. Tcherkez G, Boex-Fontvieille E, Mahe A, Hodges M. Respiratory carbon fluxes in leaves. *Curr Opinion Plant Bio*. 2012;15:308–314.
11. Smith KS, Ferry JG. Prokaryotic carbonic anhydrases. *FEMS Microbio Rev*. 2000;24:335–366.
12. Maeda S, Price GD, Badger MR, Enomoto C, Omata T. Bicarbonate binding activity of the CmpA protein of the *cyanobacterium Synechococcus* sp. strain PCC 7942 involved in active transport of bicarbonate. *J Bio Chem*. 2000;275:20551–20555.
13. Joseph P, Ouahrani-Bettache S, Montero JL, Nishimori I, Minakuchi T, Vullo D, Scozzafava A, Winum JY, Kohler S, Supuran CT. A new beta-carbonic anhydrase from Brucella suis, its cloning, characterization, and inhibition with sulfonamides and sulfamates, leading to impaired pathogen growth. *Bioorg Med Chem*. 2011;19:1172–1178.
14. Joseph P, Turtaut F, Ouahrani-Bettache S, Montero JL, Nishimori I, Minakuchi T, Vullo D, Scozzafava A, Kohler S, Winum JY, Supuran CT. Cloning, characterization, and inhibition studies of a beta-carbonic anhydrase from *Brucella suis*. *J Med Chem*. 2010;53:2277–2285.

15. Murima P, McKinney JD, Pethe K. Targeting bacterial central metabolism for drug development. *Chem Biol.* 2014;21:1423–1432.
16. Ozensoy Guler O, Capasso C, Supuran CT. A magnificent enzyme superfamily: Carbonic anhydrases, their purification and characterization. *J Enzyme Inhib Med Chem.* 2016;31:689–694.
17. Del Prete S, Vullo D, De Luca V, Carginale V, Ferraroni M, Osman SM, AlOthman Z, Supuran CT, Capasso C. Sulfonamide inhibition studies of the beta-carbonic anhydrase from the pathogenic bacterium *Vibrio cholerae. Bioorg Med Chem.* 2016;24:1115–1120.
18. Del Prete S, De Luca V, De Simone G, Supuran CT, Capasso C. Cloning, expression and purification of the complete domain of the eta-carbonic anhydrase from *Plasmodium falciparum. J Enzyme Inhib Med Chem.* 2016:1–6.
19. Jensen EL, Clement R, Kosta A, Maberly SC, Gontero B. A new widespread subclass of carbonic anhydrase in marine phytoplankton. *ISME J.* 2019;13:2094–2106.
20. Del Prete S, Vullo D, Fisher GM, Andrews KT, Poulsen SA, Capasso C, Supuran CT. Discovery of a new family of carbonic anhydrases in the malaria pathogen *Plasmodium falciparum*: The eta-carbonic anhydrases. *Bioorg Med Chem Lett.* 2014;24:4389–4396.
21. Kikutani S, Nakajima K, Nagasato C, Tsuji Y, Miyatake A, Matsuda Y. Thylakoid luminal theta-carbonic anhydrase critical for growth and photosynthesis in the marine diatom *Phaeodactylum tricornutum. Proc Natl Acad Sci U S A.* 2016;113:9828–9833.
22. Capasso C, Supuran CT. Bacterial, fungal and protozoan carbonic anhydrases as drug targets. *Expert Opin Ther Targets.* 2015;19:1689–1704.
23. Supuran CT, Capasso C. The eta-class carbonic anhydrases as drug targets for antimalarial agents. *Expert Opin Ther Targets.* 2015;19:551–563.
24. Capasso C, Supuran CT. An overview of the selectivity and efficiency of the bacterial carbonic anhydrase inhibitors. *Curr Med Chem.* 2015;22:2130–2139.
25. Capasso C, Supuran CT. An overview of the alpha-, beta- and gamma-carbonic anhydrases from Bacteria: Can bacterial carbonic anhydrases shed new light on evolution of bacteria? *J Enzyme Inhib Med Chem.* 2015;30:325–332.
26. Capasso C, Supuran CT. Sulfa and trimethoprim-like drugs - antimetabolites acting as carbonic anhydrase, dihydropteroate synthase and dihydrofolate reductase inhibitors. *J Enzyme Inhib Med Chem.* 2014;29:379–387.
27. Capasso C, Supuran CT. Anti-infective carbonic anhydrase inhibitors: A patent and literature review. *Expert Opin Ther Pat.* 2013;23:693–704.
28. Supuran CT, Capasso C. New light on bacterial carbonic anhydrases phylogeny based on the analysis of signal peptide sequences. *J enzyme Inhib Med Chem.* 2016;31:1254–1260.
29. Supuran CT. Advances in structure-based drug discovery of carbonic anhydrase inhibitors. *Expert Opin Drug Discov.* 2017;12:61–88.
30. Supuran CT. Structure and function of carbonic anhydrases. *Biochem J.* 2016;473:2023–2032.
31. Supuran CT, Capasso C. An overview of the bacterial carbonic anhydrases. *Metabolites.* 2017;7:56–74.
32. Hygum TL, Fobian D, Kamilari M, Jorgensen A, Schiott M, Grosell M, Mobjerg N. Comparative investigation of copper tolerance and identification of putative tolerance related genes in *Tardigrades. Front Physiol.* 2017;8:95.
33. Taylor MP, van Zyl L, Tuffin IM, Leak DJ, Cowan DA. Genetic tool development underpins recent advances in thermophilic whole-cell biocatalysts. *Microb Biotechnol.* 2011;4:438–448.

34. Zeldes BM, Keller MW, Loder AJ, Straub CT, Adams MWW, Kelly RM. Extremely thermophilic microorganisms as metabolic engineering platforms for production of fuels and industrial chemicals. *Front Microbiol.* 2015;6:1209.
35. Di Fiore A, Capasso C, De Luca V, Monti SM, Carginale V, Supuran CT, Scozzafava A, Pedone C, Rossi M, De Simone G. X-ray structure of the first 'extremo-alpha-carbonic anhydrase', a dimeric enzyme from the thermophilic bacterium *Sulfurihydrogenibium yellowstonense* YO3AOP1. *Acta Crystallogr D Biol Crystallogr.* 2013;69:1150–1159.
36. Coker JA. Extremophiles and biotechnology: Current uses and prospects. *F1000Res.* 2016;5.
37. De Simone G, Monti SM, Alterio V, Buonanno M, De Luca V, Rossi M, Carginale V, Supuran CT, Capasso C, Di Fiore A. Crystal structure of the most catalytically effective carbonic anhydrase enzyme known, SazCA from the thermophilic bacterium *Sulfurihydrogenibium azorense. Bioorg Med Chem Lett.* 2015;25:2002–2006.
38. Supuran CT. CA IX stratification based on cancer treatment: A patent evaluation of US2016/0002350. *Expert Opin Ther Pat.* 2016:1–5.
39. Lomelino C, McKenna R. Carbonic anhydrase inhibitors: A review on the progress of patent literature (2011–2016). *Expert Opin Ther Pat.* 2016;26:947–956.
40. Monti SM, Supuran CT, De Simone G. Anticancer carbonic anhydrase inhibitors: A patent review (2008–2013). *Expert Opin Ther Pat.* 2013;23:737–749.
41. Masini E, Carta F, Scozzafava A, Supuran CT. Antiglaucoma carbonic anhydrase inhibitors: A patent review. *Expert Opin Ther Pat.* 2013;23:705–716.
42. Guzel-Akdemir O, Akdemir A, Pan P, Vermelho AB, Parkkila S, Scozzafava A, Capasso C, Supuran CT. A class of sulfonamides with strong inhibitory action against the alpha-carbonic anhydrase from *Trypanosoma cruzi. J Med Chem.* 2013;56:5773–5781.
43. Aggarwal M, Kondeti B, McKenna R. Anticonvulsant/antiepileptic carbonic anhydrase inhibitors: A patent review. *Expert Opin Ther Pat.* 2013;23:717–724.
44. Carta F, Supuran CT. Diuretics with carbonic anhydrase inhibitory action: A patent and literature review (2005–2013). *Expert Opin Ther Pat.* 2013;23:681–691.
45. Winum JY, Capasso C. Novel antibody to a carbonic anhydrase: Patent evaluation of WO2011138279A1. *Expert Opin Ther Pat.* 2013;23:757–760.
46. Aggarwal M, McKenna R. Update on carbonic anhydrase inhibitors: A patent review (2008–2011). *Expert Opin Ther Pat.* 2012;22:903–915.
47. Carta F, Scozzafava A, Supuran CT. Sulfonamides: A patent review (2008–2012). *Expert Opin Ther Pat.* 2012;22:747–758.
48. Carta F, Supuran CT, Scozzafava A. Novel therapies for glaucoma: A patent review 2007–2011. *Expert Opin Ther Pat.* 2012;22:79–88.
49. Poulsen SA. Carbonic anhydrase inhibition as a cancer therapy: A review of patent literature, 2007–2009. *Expert Opin Ther Pat.* 2010;20:795–806.
50. Akdemir A, Vullo D, De Luca V, Scozzafava A, Carginale V, Rossi M, Supuran CT, Capasso C. The extremo-alpha-carbonic anhydrase (CA) from *Sulfurihydrogenibium azorense*, the fastest CA known, is highly activated by amino acids and amines. *Bioorg Med Chem Lett.* 2013;23:1087–1090.
51. Capasso C, De Luca V, Carginale V, Cannio R, Rossi M. Biochemical properties of a novel and highly thermostable bacterial alpha-carbonic anhydrase from *Sulfurihydrogenibium yellowstonense* YO3AOP1. *J Enzyme Inhib Med Chem.* 2012;27:892–897.
52. De Luca V, Vullo D, Scozzafava A, Carginale V, Rossi M, Supuran CT, Capasso C. Anion inhibition studies of an alpha-carbonic anhydrase from the thermophilic bacterium *Sulfurihydrogenibium yellowstonense* YO3AOP1. *Bioorg Med Chem Lett.* 2012;22:5630–5634.

53. De Luca V, Vullo D, Scozzafava A, Carginale V, Rossi M, Supuran CT, Capasso C. An alpha-carbonic anhydrase from the thermophilic bacterium *Sulphurihydrogenibium azorense* is the fastest enzyme known for the CO_2 hydration reaction. *Bioorg Med Chem*. 2013;21:1465–1469.
54. Vullo D, De Luca V, Scozzafava A, Carginale V, Rossi M, Supuran CT, Capasso C. Anion inhibition studies of the fastest carbonic anhydrase (CA) known, the extremo-CA from the bacterium *Sulfurihydrogenibium azorense*. *Bioorg Med Chem Lett*. 2012;22:7142–7145.
55. Vullo D, De Luca V, Scozzafava A, Carginale V, Rossi M, Supuran CT, Capasso C. The first activation study of a bacterial carbonic anhydrase (CA). The thermostable alpha-CA from *Sulfurihydrogenibium yellowstonense* YO3AOP1 is highly activated by amino acids and amines. *Bioorg Med Chem Lett*. 2012;22:6324–6327.
56. Vullo D, De Luca V, Scozzafava A, Carginale V, Rossi M, Supuran CT, Capasso C. The extremo-alpha-carbonic anhydrase from the thermophilic bacterium *Sulfurihydrogenibium azorense* is highly inhibited by sulfonamides. *Bioorg Med Chem*. 2013;21:4521–4525.
57. Vullo D, Luca VD, Scozzafava A, Carginale V, Rossi M, Supuran CT, Capasso C. The alpha-carbonic anhydrase from the thermophilic bacterium *Sulfurihydrogenibium yellowstonense* YO3AOP1 is highly susceptible to inhibition by sulfonamides. *Bioorg Med Chem*. 2013;21:1534–1538.
58. Alafeefy AM, Abdel-Aziz HA, Vullo D, Al-Tamimi AM, Al-Jaber NA, Capasso C, Supuran CT. Inhibition of carbonic anhydrases from the extremophilic bacteria *Sulfurihydrogenibium yellostonense* (SspCA) and *S. azorense* (SazCA) with a new series of sulfonamides incorporating aroylhydrazone-, [1,2,4]triazolo[3,4-b][1,3,4]thiadiazinyl- or 2-(cyanophenylmethylene)-1,3,4-thiadiazol-3(2H)-yl moieties. *Bioorg Med Chem*. 2014;22:141–147.
59. James P, Isupov MN, Sayer C, Saneei V, Berg S, Lioliou M, Kotlar HK, Littlechild JA. The structure of a tetrameric alpha-carbonic anhydrase from *Thermovibrio ammonificans* reveals a core formed around intermolecular disulfides that contribute to its thermostability. *Acta crystallogr Section D, Biol Crystal*. 2014;70:2607–2618.
60. Kanth BK, Jun SY, Kumari S, Pack SP. Highly thermostable carbonic anhydrase from *Persephonella marina* EX-H1: Its expression and characterization for CO_2 sequestration applications.. *Process Biochem*. 2014;49.
61. Smith KS, Ferry JG. A plant-type (beta-class) carbonic anhydrase in the thermophilic methanoarchaeon *Methanobacterium thermoautotrophicum*. *J Bacteriol*. 1999;181:6247–6253.
62. Alber BE, Colangelo CM, Dong J, Stalhandske CM, Baird TT, Tu C, Fierke CA, Silverman DN, Scott RA, Ferry JG. Kinetic and spectroscopic characterization of the gamma-carbonic anhydrase from the methanoarchaeon *Methanosarcina thermophila*. *Biochemistry*. 1999;38:13119–13128.
63. De Luca V, Vullo D, Del Prete S, Carginale V, Osman SM, AlOthman Z, Supuran CT, Capasso C. Cloning, characterization and anion inhibition studies of a gamma-carbonic anhydrase from the Antarctic bacterium *Colwellia psychrerythraea*. *Bioorgan Med Chem*. 2016;24:835–840.
64. Vullo D, De Luca V, Del Prete S, Carginale V, Scozzafava A, Osman SM, AlOthman Z, Capasso C, Supuran CT. Sulfonamide inhibition studies of the gamma-carbonic anhydrase from the Antarctic bacterium *Colwellia psychrerythraea*. *Bioorgan Med Chem Lett*. 2016;26:1253–1259.
65. Vullo D, De Luca V, Del Prete S, Carginale V, Scozzafava A, Capasso C, Supuran CT. Sulfonamide inhibition studies of the gamma-carbonic anhydrase from the Antarctic bacterium *Pseudoalteromonas haloplanktis*. *Bioorgan Med Chem Lett*. 2015;25:3550–3555.

66. De Luca V, Vullo D, Del Prete S, Carginale V, Scozzafava A, Osman SM, AlOthman Z, Supuran CT, Capasso C. Cloning, characterization and anion inhibition studies of a new gamma-carbonic anhydrase from the Antarctic bacterium *Pseudoalteromonas haloplanktis*. *Bioorgan Med Chem*. 2015;23:4405–4409.
67. De Luca V, Del Prete S, Carginale V, Vullo D, Supuran CT, Capasso C. Cloning, characterization and anion inhibition studies of a gamma-carbonic anhydrase from the Antarctic cyanobacterium *Nostoc commune*. *Bioorgan Med Chem Lett*. 2015;25:4970–4975.
68. Vullo D, De Luca V, Del Prete S, Carginale V, Scozzafava A, Capasso C, Supuran CT. Sulfonamide inhibition studies of the gamma-carbonic anhydrase from the Antarctic cyanobacterium *Nostoc commune*. *Bioorgan Med Chem*. 2015;23:1728–1734.
69. Jeyakanthan J, Rangarajan S, Mridula P, Kanaujia SP, Shiro Y, Kuramitsu S, Yokoyama S, Sekar K. Observation of a calcium-binding site in the gamma-class carbonic anhydrase from *Pyrococcus horikoshii*. *Acta Crystallogr Section D, Biol Crystal*. 2008;64:1012–1019.
70. He Q, Silliman BR. Climate change, human impacts, and coastal ecosystems in the anthropocene. *Curr Biol*. 2019;29:R1021–R1035.
71. Savile CK, Lalonde JJ. Biotechnology for the acceleration of carbon dioxide capture and sequestration. *Curr Opin Biotechnol*. 2011;22:818–823.
72. Homaei AA, Sariri R, Vianello F, Stevanato R. Enzyme immobilization: An update. *J Chem Biol*. 2013;6:185–205.
73. Noh G, Docherty S, Lam E, Huang X, Mance D, Alfke J, Copéret C. CO_2 Hydrogenation to CH_3OH on supported Cu nanoparticles: Nature and role of Ti in bulk oxides vs isolated surface sites. *J Phys Chem*. 2019;2–17.
74. Tanvi S, Swati S, Hesam K, Ashok K. Energizing the CO_2 utilization by chemo-enzymatic approaches and potentiality of carbonic anhydrases: A review 2020; 247:1–13.
75. Tosa T, Mori T, Fuse N, Chibata I. Studies on continuous enzyme reactions. I. Screening of carriers for preparation of water-insoluble aminoacylase. *Enzymologia*. 1966;31:214–224.
76. Mohamad NR, Marzuki NH, Buang NA, Huyop F, Wahab RA. An overview of technologies for immobilization of enzymes and surface analysis techniques for immobilized enzymes. *Biotechnol Biotechnol Equip*. 2015;29:205–220.
77. Migliardini F, De Luca V, Carginale V, Rossi M, Corbo P, Supuran CT, Capasso C. Biomimetic CO_2 capture using a highly thermostable bacterial alpha-carbonic anhydrase immobilized on a polyurethane foam. *J Enzyme Inhib Med Chem*. 2014;29: 146–150.
78. Abdelrahim MYM, Martins CF, Neves LA, Capasso C, Supuran CT, Coelhoso IM, Crespo JG, Barboiu M. Supported ionic liquid membranes immobilized with carbonic anhydrases for CO_2 transport at high temperatures. *J Mem Sci*. 2017;528:225–230.
79. Perfetto R, Del Prete S, Vullo D, Sansone G, Barone CMA, Rossi M, Supuran CT, Capasso C. Production and covalent immobilisation of the recombinant bacterial carbonic anhydrase (SspCA) onto magnetic nanoparticles. *J Enzyme Inhib Med Chem*. 2017;32:759–766.
80. Russo ME, Olivieri G, Capasso C, De Luca V, Marzocchella A, Salatino P, Rossi M. Kinetic study of a novel thermo-stable alpha-carbonic anhydrase for biomimetic CO_2 capture. *Enzyme Micro Technol*. 2013;53:271–277.
81. Trachtenberg MC, Cowan RM, Smith DA, Horazak DA, Jensen MD, Laumb JD, Vucelic AP, Chen H, Wang L, Wu X. Membrane-based, enzyme-facilitated, efficient carbon dioxide capture. *Energy Procedia*. 2009;1:353–360.
82. Neves LA, Afonso C, Coelhoso IM, Crespo JG. Integrated CO_2 capture and enzymatic bioconversion in supported ionic liquid membranes. *Sep Pur Tech*. 2012;97:34–41.

83. Del Prete S, Perfetto R, Rossi M, Alasmary FAS, Osman SM, AlOthman Z, Supuran CT, Capasso C. A one-step procedure for immobilising the thermostable carbonic anhydrase (SspCA) on the surface membrane of *Escherichia coli*. *J Enzyme Inhib Med Chem*. 2017;32:1120–1128.
84. Arazawa DT, Oh HI, Ye SH, Johnson CA, Jr., Woolley JR, Wagner WR, Federspiel WJ. Immobilized carbonic anhydrase on hollow fiber membranes accelerates CO(2) removal from blood. *J Memb Sci*. 2012;404–404:25–31.

5 Engineering of Microbial Carbonic Anhydrase for Enhanced Carbon Sequestration

Anand Giri, Veerbala Sharma, Shabnam Thakur, Tanvi Sharma, Ashok Kumar, and Deepak Pant

CONTENTS

5.1 INTRODUCTION

Carbonic anhydrase (CA) is a zinc-containing enzyme that presents in both eukaryotes and prokaryotes and rapidly catalyses the hydration reaction of CO_2 to HCO_3^- and proton (H^+), and vice versa (Di Fiore et al., 2015). As per the latest measurement of CO_2 by NOAA in 2017, the concentration of CO_2 in the atmosphere was about 400 ppm, and it is rising by 2 ppm annually (NOAA, 2017) due to rapid industrial development, uncontrolled CO_2 emissions from various industries, burning of fossil fuels, and uncontrolled human growth. CO_2 has a significant impact on climate change and global warming. The increased CO_2 concentration in the atmosphere since the post-industrial era can correlate with increasing global surface temperatures (Giri and Pant, 2018). Over the last century (1906–2005), the average global temperature data revealed that the average temperature increases by 0.7°C ± 0.2°C over that period (Jansen et al., 2007). In the last decade, the sequestration of CO_2 by microbial CA have been attracted much attention as an alternative way to carbon management technology due to high reaction rate of CA (10^4–$10^6 s^{-1}$) and high tolerance rate towards high CO_2 concentration (Bose and Satyanarayana, 2017). However, low thermal stability, poor activity, and reusability of CA have some industrial limitations over its optimum utilization. Protein engineering has been widely applied to alter the existing protein structure or enzyme in order to improve its catalytic properties for various environmental, industrial, and pharmaceutical applications. Exploiting CA activity by CA engineering is an attractive technique to speed up its diverse application in biomimetic route for CO_2 sequestration from the combustion fuel gases with its improved environmental compatibility, potential economic viability, activity, and stability (Jo et al., 2016; Giri and Pant, 2019a). Recent advancement of experimental and computational CA engineering can lead to the developments of the large-scale industrial biocatalyst. Various proteins have been successfully engineered by several methods such as site-directed mutagenesis, directed evolution, protein immobilization, biological and chemical modifications, active site-directed substitutions, ProSAR: amino acid contributions, and rational design of substrate specificity (Kazlauskas and Bornscheuer, 2009; Kumar et al., 2018; Verma et al., 2019). Directed evolution is an effective technique for the major changes in protein active site with efficient biocatalytic performances (Alvizo et al., 2014; Boone et al., 2013; Zhang et al., 2016; Zuo et al., 2018). Other methods like ProSAR: amino acid contributions are fast and provide an easier way to alter enzymes with optimal activity (Lippow and Tidor, 2007; Yao et al., 2015; Chen et al., 2015; Zhang et al., 2017). This chapter focuses on mainly microbial CA engineering techniques and briefly addresses the environmental applications of the immobilized CAs. Methods involving the modification and expression of the CA gene, immobilization, amino acid modification, engineering in *de novo* disulphide bond in free microbial CA, and CA immobilization are significantly promising techniques (Kumar et al., 2015; Zhang et al., 2016; Wu et al., 2016; Chen et al., 2016; Kumar et al., 2018a, 2018b) and are cost-effective processes for the overall CO_2 mitigation (Kambar and Ozdemir, 2010). Several studies of engineered recombinant microbial CA (Del Prete et al., 2012; Joseph et al., 2010), directed evolution (Alvizo et al., 2014; Jo et al., 2016), and immobilization (Sharma et al., 2011; have expressed and modified CA, which is an attempt to overcome its activity, stability, and recovery issue. CA engineering involves the following three steps: (i) engineering strategies or selecting protein changes such as directed evolution, rational design; (ii) making these changes (mutagenesis); and (iii) screening or selection (Kazlauskas and Bornscheuer, 2009). Choosing different strategies can result in various benefits or drawbacks in each of these steps (Figure 5.1).

5.2 STRUCTURE AND FUNCTION OF MICROBIAL CA

Microbial CA belongs to three genetically different classes α, β, and γ that contain a tetrahedrally active catalytic Zn (II) ion in the active site. Three imidazole groups of histidines (His-94, 96, and 119) and a hydroxyl molecule in α- and γ-CA are attached with the metal ion in the active site. In the β-CA, one histidine and two

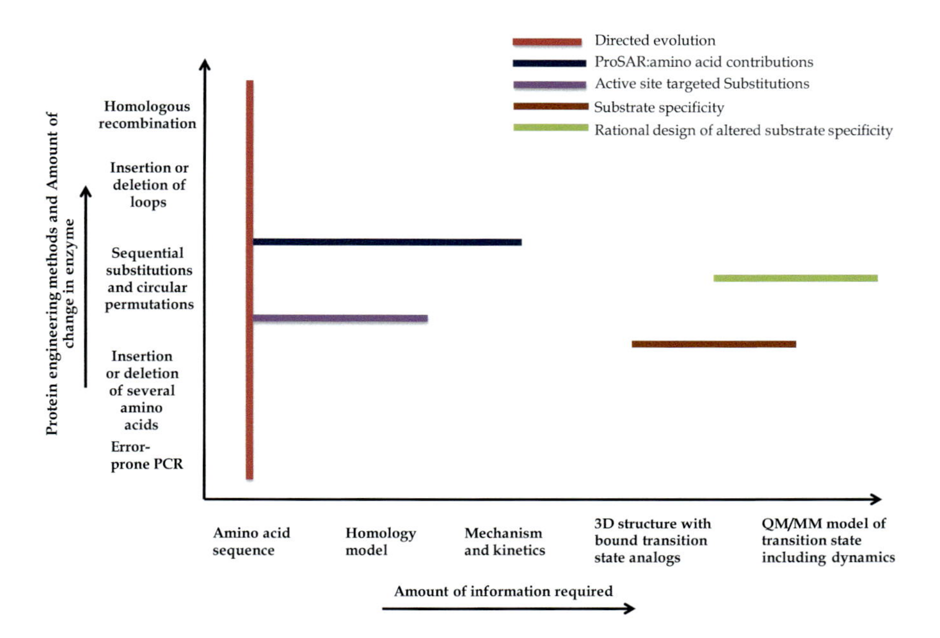

FIGURE 5.1 Proposed CA engineering methods based on the amount of available information of target CA.

cysteine residues are attached with the metal ion present in the active site (Capasso and Supuran, 2015). γ-CA not only contains Fe(II) ion but can also actively bind with Zn(II) or Co(II) ions (Supuran and Capasso, 2017). Leu-198, Val-(121, 143, 207), and Trp-209, a cluster of hydrophobic amino acids, are responsible for carbon dioxide binding and participate in the proton shuttling process (Ferry, 2010). A cluster of hydrophilic amino acids of CA (Asn-62, His-64, Tyr-7, and Asn-67) are responsible for the orientation of CO_2, during CO_2 catalysis (Domsic and McKenna, 2010). CAs from *Sulfurihydrogenibium yellowstonense*, *Thermovibrio ammonificans*, *Neisseria gonorrhoeae*, and *Sulfurihydrogenibium azorense* are the only known three-dimensional structures that belong to α-CA (Di Fiore et al., 2013; Supuran and Capasso, 2017), and the X-ray crystal structure of CAs is determined from many bacterial species such as *Mycobacterium tuberculosis*, *Citrobacter freundii*, *Pseudomonas spp.*, *Haemophilus influenzae*, *Vibrio cholerae*, *Salmonella enterica*, and *Escherichia coli* (Ferraroni et al., 2015; Giri et al., 2018; Giri and Pant, 2019a). Only one CAM (Carbonic Anhydrase Methanosarcina) from methane-producing archaea *Methanosarcina thermophila* has been crystallized until now (Kisker et al., 1996).

5.2.1 Catalytic Activity of CA

Despite their differences in structure, CAs have a common mechanism of action. The catalytic reaction of CA is a two-step ping-pong reaction that catalyses CO_2, which is expressed as follows (Supuran and Capasso, 2017):

$$E-ZnH_2O \leftrightarrow E-ZnOH^- + H^+ \quad (5.1)$$

$$E-ZnOH^- + CO^2 \leftrightarrow E-ZnHCO_3^- \quad (5.2)$$

$$E-ZnHCO_3^- + H_2O \leftrightarrow E-ZnH_2O + HCO_3^- \quad (5.3)$$

In the catalytic mechanism of CA, the histidine residue accepts a proton (H^+) which is formed from the water molecules. Nucleophilic attack by zinc-bounded hydroxyl group on the molecule of carbon dioxide takes place, which results in enzyme–bicarbonate complex formation (Silverman and Lindskog, 1988). In α-CA, His-64 receives a proton from the active site of water molecules by its interference with the molecule of zinc-bounded water molecule (Tripp et al., 2001). Lys-91 and Tyr-131 near the active site cavity of CA isozyme can also influence the k_{cat} and can be targeted to incorporate these imidazole analogues. Kimber et al. (2000) analysed this observation and divided β-class into two subclasses, the 'Cab type' and 'plant type', found from *Methanothermobacter thermoautotrophicum* and *Pisum sativum* CA, respectively.

5.3 MICROBIAL CARBONIC ANHYDRASE ENGINEERING

The development of recombinant technology by the expression of high-yielding CA gene into the targeted microorganism is a promising biotechnological method to enhance carbon sequestration potential of microbial CA. The methods of CA engineering have been effective towards CO_2 sequestration.

5.3.1 Genetic Engineering

Recently, genetic engineering of CA has gained considerable research interest because of climate change and energy crisis. The natural capability of CO_2 fixation by CA can be genetically engineered to achieve a desired result with improved activity and/or stability and generate high-value bioproducts (Ducat et al., 2011; Chen et al., 2012). Jo et al. (2013) expressed CA from *N. gonorrhoeae* (ngCA) in the periplasm of *E. coli* and successfully demonstrated the whole-cell catalyst as an effective catalyst for CO_2 sequestration (Figure 5.2). In another study, engineered yeast (*Saccharomyces cerevisiae*) is also used for the enhancement of CO_2 hydration reaction (Barbero et al., 2013). Recombinant CAs from different microorganisms such as *Hahellachejuensis*, *N. gonorrhoeae*, and *Thalassiosira weissflogii* are expressed in *E. coli*, which after expression results in high pH and temperature stability, doubled esterase activity, and accelerated rate of calcite crystal formation (Ki et al., 2013). Fan et al. (2011) studied ice nucleation protein (INP) from *Pseudomonas syringae* as surface-anchoring support for the expression ofα-CA from *Helicobacter pylori* on the outer membrane of *E. coli*, which served as resourceful biocatalyst for CO_2 sequestration with significantly increased CO_2 removal rate. Del Prete et al. (2012) purified α-CA from *V. cholerae* (VchCA) and cloned, which is four times more active than the other microbial CAs. α-Type CA protein (CAA1) of *Mesorhizobium loti* coded by *msi040* gene is expressed in nitrogen-fixing and free-living bacteria for CO_2 fixation in associated root nodules (Kalloniati et al., 2009).

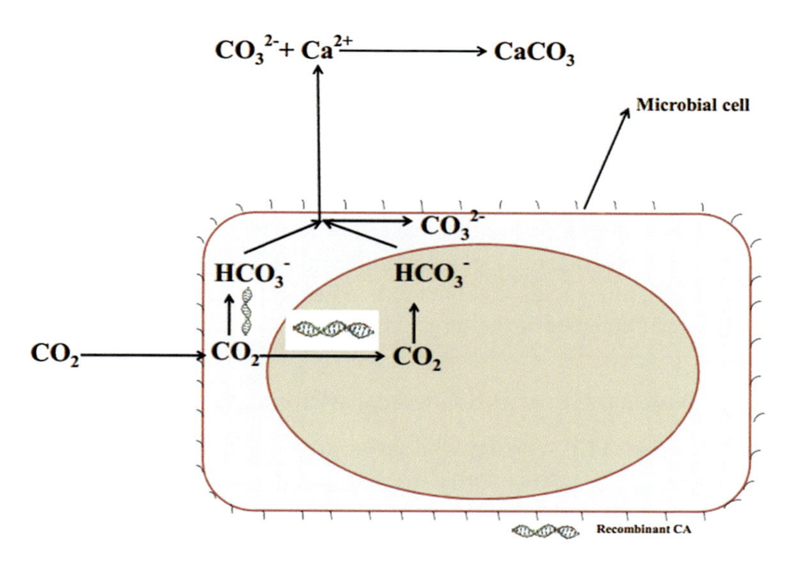

FIGURE 5.2 Design concept for the CO_2 mitigation by genetic engineering of CA.

5.3.2 Directed Evolution

Another strategy is the directed evolution, which causes changes in the catalytic site to give enzyme variants with catalysing capabilities, and the creation of highly stable β-CA originating from *Desulfovibrio vulgaris* is the example, which accelerates the CO_2 absorption above 100°C in alkaline solvents (Alvizo et al., 2014). The structure and active site of many CAs like α-CA from *S. azorense* can alter the proton transfer rate of enzyme. A disulphide bonding is conserved in human carbonic anhydrases (hCAs) (IV, VI, IX, XII, and XIV) along with β-CA from many microbes like *N. gonorrhoeae* and is responsible for the stability of CA (Aggarwal et al., 2013). The engineering of disulphide bond present in the microbial CA shows an enhancement of thermostability and a resistance to denaturation as compared to native CA (Mårtensson et al., 2002). Introduction of the designed, *de novo* disulphide bond into CA originating from *N. gonorrhoeae* (*ng*CA), which later expressed in *E. coli*, shows an improved kinetic and thermodynamic stability (Dombkowski et al., 2013). Gould and Tawfik (2005) suggested that two mutations of human CA (Ala65Val and Thr200Ala) through the directed evolution can increase a promising function of CA in the industrial applications. CA obtained from *D. vulgaris* (DvCA) accelerates CO_2 absorption at 100°C and in the presence of alkaline solvent (Fox et al., 2007; Alvizo et al., 2014; Kumar et al., 2018c; 2019; Sharma et al., 2019). Directed evolution can be carried out in different ways; for example, *in vitro* recombination-based directed evolutions focused on the generation of gene libraries and the process of increasing the enzyme activity should be developed. The directed evolution computational methods are also rapidly used for improving and creating new enzymes, thus enhancing their catalytic function (Hibbert and Dalby, 2005). The substitutions of the hydrophobic residues into hydrophilic substitutes in CA retain the high catalytic efficiency and stabilize the denaturing temperature (Boone et al., 2013). Parra-Cruz et al. (2018) used molecular dynamics simulations to study the association between thermostability and structural flexibility of the bacterial α-CA and designed the disulphide bonds and mutants with increased stability. Random, targeted mutagenesis, and recombination strategies are mainly used for the directed evolution of protein by producing sequence libraries. Random mutagenesis has taken the place of amino acid substitutions to explore beneficial mutations. However, targeted mutagenesis has taken the place of conformational changes in protein for the randomization of amino acids. Recombination introduces a wide number of changes in a simultaneous sequence of the targeted protein (Bloom et al., 2005).

5.3.3 CA Immobilization and Chemical Modification

The immobilization of CA onto the solid surface and the chemical modification of CA are eco-friendly and effective approaches successfully used to increase its catalytic property for various industrial and biotechnological applications. Chemically modified and immobilized CA can be capable of enhancing carbon dioxide absorption from a gas stream and enhancing carbon dioxide sequestration and mineral formation (Cowan and Fernandez-Lafuente, 2011). The incorporation of aldehyde can lead to the best protein modifications; various studies on the incorporation of aldehyde in *in vivo* and in *vitro* conditions were carried out because aldehyde can easily react with lysine residues (Pant et al., 2017; Giri and Pant, 2019b). Bootorabi et al. (2008) studied the reaction of CA with acetaldehyde after monitoring its activity and found that acetaldehyde binding with CA may affect the hydrogen-bonding patterns, may result in conformational changes of CA, and may change the acid–base properties within the active site and the surface of an enzyme, which in turn leads to the decreased CA activity. The various cross-linking agents such as bis(N-hydroxysuccinimide) ester, dialdehyde, bis-imidate, diacid chloride, malondialdehyde, glutaraldehyde, and dimethyl suberimidate are useful for the preparation and modification of α-, β-, and γ-CA polypeptides and exhibit improved catalytic properties under carbon-capture process conditions (Novick et al., 2013). The modification of cysteine residue of CA leads to an enhancement in the proton transfer, whereas unmodified cysteine residue shows threefold higher CO_2 hydration rate (Elder et al., 2004).

The binding of different ligands within the active site of CA depends on (i) coordination sphere of CA active site, including Zn^+ atoms; (ii) hydrophobic sites of targeted CA; (iii) hydrophilic sites of CA; and (iv) nucleophilicity (Giri and Pant, 2019b). Cysteine residues of *β*-CA can be modified by the thiol group as a nucleophilic reagent (Kim et al., 2008). The catalytic rate of CA in CO_2 hydration process can be enhanced by various cross-linking agents (Tu et al., 1990). Stefanucci et al. (2018) studied the activation of *β*- and *γ*-CA from the pathogenic bacteria *M. tuberculosis*, *V. cholera*, and *Burkholderia pseudomallei* by the incorporation of acidic amino acids as tripeptide. CA from *T weissflogii* is also significantly activated by different amines and amino acids (Angeli et al., 2018).

CA immobilization provides an effective way to improve the operational stability of enzymatic technology in carbon capture or sequestration process. Based on a large number of studies, enzyme immobilization can be done by different methods using supports such as solid surface, polymeric materials, and on or within soft poly molecular assemblies (Cao, 2005; Mateo et al., 2007; Wu et al., 2016; Kumar et al., 2019). Currently, CA immobilization is an active research area due to its large number of applications. The simplest CA immobilization method is the physical adsorption on the soft or solid surface like colloidal gold sols used for the adsorption of CA molecule and retention of its enzymatic activity (Jesionowski et al., 2014; Drozdov et al., 2016). The CA was successfully immobilized on spherical SBA-15 silica particles by different approaches, such as adsorption, cross-linking, and covalent binding, and studied for the sequestration of CO_2 (Vinoba et al., 2011). The results demonstrated that the silver nanoparticle-conjugated hCA showed ~25-fold higher CO_2-capture efficiency and the highest operational stability than the free hCA (Vinoba et al., 2012). Chitosan, polyurethane (PU) foam, glass, activated carbon, hydrogels, and silica beads are the supports that have been used in the immobilization of CA (Zhang et al., 2011; Wanjari et al., 2011; Forsyth et al., 2013). Recently, bioinspired silica has been used for enzyme immobilization as a green method due to its mild processing conditions, improved stability, high activity, and excellent immobilization efficiency (Patwardhan, 2011) (Table 5.1).

Overall, the immobilized CA acts as a green catalyst for reducing carbon dioxide emissions compared to other alkali or a soluble biocatalyst process. Other than these methods, the recent development of CA engineering is via chemical modifications involving (i) the modification of natural or unnatural amino acids (i.e. site-selective

TABLE 5.1
Example of Microbial CA Immobilization and Function

Sr. No.	CA Used	Support Material and Methods	Results	References
1	Microbial CA	Silica or SiO_2–ZrO_2 composite nanoparticles	Improved CA stability	Zhang et al. (2013)
2	*Bacillus pumilus* TS1 CA	Surfactant-modified (SDS) silylated chitosan	Longer storage stability, increased K_m and V_{max}	Yadav et al. (2010)
3	CA from *Bacillus pumilus, Pseudomonas fragi*, and *Micrococcus lylae*	Surfactant-modified silylated chitosan	Improved storage stability	Prabhu et al. (2011)
4	CA from *Bacillus pumilus*	Chitosan beads	Increased K_m value	Wanjari et al. (2011)
5	Recombinant(α-CA) from *Sulfurihydrogenibium yellowstonense*	Magnetic Fe_3O_4 nanoparticles (MNP)	Enhanced stability and storage of CA	Perfetto et al. (2017)
6	α-CA from *Sulfurihydrogenibium yellowstonense*	PU foam	Long-term stability	Migliardini et al. (2014)

incorporation) and (ii) the recognition-driven modification, which also improves the catalytic activity of microbial CA (Sakamoto and Hamachi, 2018).

5.4 CARBON SEQUESTRATION PERFORMANCES OF ENGINEERED CA

Microbial CA engineering refers to an induced change in the sequence of proteins to achieve the desired result with improved properties in terms of increased activity and/or stability, ease of handling, and enzyme separation during its industrial applications. The role of engineered microbial CA in carbon sequestration is given in Table 5.2.

5.5 ENGINEERING PROMOTERS

Promoter engineering attempts to lead the vibrant range necessary to promote gene expression with fine-tuning, high-strength over expressions, and synthetic circuit design for metabolic and synthetic biology engineering in any host organism. It attempts to amend transcriptional capacity by mutating, enhancing, or otherwise altering the promoter DNA sequence. Site-selective mutagenesis (SSM) by error-prone PCR (ep-PCR), saturation mutagenesis of nucleotide spacer regions, viral promoters, and hybrid promoter engineering are the main promoter strategies towards enzyme engineering (Blazeck and Alper, 2013). Fan et al. (2011) used two different promoters (T7 and Tac) with two different vectors, pKK223-3 and pET22b(+), to create fusion genes and to express them on the *E. coli*. The smallest INP-N fused protein acts as a surface-anchoring motif and shows the highest level of expression and activity of CA. The T7 promoter gives a high level of protein expression than the Tac promoter. Saturation mutagenesis, a directed method, involves a nucleotide substitution at the specific region and retention of consensus regions of the promoter system. However, only variable regions are mutated during promoter engineering of CA (Patwardhan et al., 2009).

5.6 CONCLUSION

Research and development in the CA engineering for CO_2 sequestration and valuable eco-friendly products has become a great attraction in recent years. The favourable characteristic of microbial CA is considered as an efficient and promising eco-friendly alternative for reducing atmospheric CO_2. For convenient handling of CA, facile separation, efficient reuse and recovery, and enhancement of stability and activity, CA engineering provides a vital role. CA engineering like immobilization of CA may also display much better functional properties with improved specificity and/or selectivity. Physical, ionic, and covalent interactions are found between the support system and CA. Like synthetic resin in organic polymer, biopolymer provides immobilized CA with specific kinetic properties of CA. Recombinant engineering of CA is a good option for increasing the catalytic activity and stability of CA for CO_2 sequestration process. Recombinant CA is also a promising technique

TABLE 5.2
Role of Engineered Microbial CA in Carbon Sequestration

Type and CA Source	Engineering Method	Description	Resulted Role	References
α-CA, *Neisseria gonorrhoeae* (*ng*CA)	Periplasmic expression	*ng*CA expressed in *E. coli*	Increased formation of calcium carbonate and improved thermostability	Jo et al. (2013)
α-CA, *Neisseria Gonorrhoeae* (*ng*CA)	Site-directed mutagenesis	Selected variants expressed in *E. coli*	Enhanced kinetic and thermodynamic stability.	Jo et al. (2016)
α-CA, *Hahellachejuensis* (HC-αCA)	Gene expression	Recombinant CA expressed in *E. coli*	The refolded HC-a CA displayed high pH and temperature stability, and doubled esterase activity	Ki et al. (2013)
α-CA, *Neisseria gonorrhoeae* (*ng*CA)	Gene expression	Recombinant CA expressed in *E. coli*	High CO_2 hydration activity with accelerated rate of calcite crystal formation	Kim et al. (2012)
δ-CA, *Thalassiosiraweissflogii* (TWCA1)	Gene expression	Expressed in *E. coli*	Recombinant protein assay possessing both CO_2 hydration and increased esterase activity	Lee et al. (2013)
β-CA, *Desulfovibrio vulgaris*	Directed evolution	Use of protein sequence activity relationship algorithm to create new variants of CA	Increased thermostability and alkali tolerance and enhanced rate of CO_2 absorption 25-fold compared with the non-evolved CA	Alvizo et al. (2014)
CA from *Serratia sp.*	Gene expression	CA gene from *Serratia sp.* expressed in *E. coli*	Increased stability	Srivastava et al. (2015)
CAs of *Sulfurihydrogenibium yellowstonense* (*SspCA*)	Immobilization	PU foam	Increased thermostability at 70°C	Capasso et al. (2012)

for the conversion of atmospheric CO_2 to eco-friendly products for long-term uses. Further advancements and engineering in microbial CA overexpression systems can produce faster and higher levels of CA for environmental applications.

ACKNOWLEDGEMENT

The author AG gratefully acknowledges CSIR, New Delhi Government of India, for the grant of Senior Research Fellowship (SRF).

REFERENCES

Aggarwal, M., Boone, C. D., Kondeti, B., & McKenna, R. (2013). Structural annotation of human carbonic anhydrases. *Journal of Enzyme Inhibition and Medicinal Chemistry*, *28*(2), 267–277.

Alvizo, O., Nguyen, L.J., Savile, C.K., Bresson, J.A., Lakhapatri, S.L., Solis, E.O.P., Fox, R.J., Broering, J.M., Benoit, M.R., Zimmerman, S.A., &Novick, S. J. (2014). Directed evolution of an ultrastable carbonic anhydrase for highly efficient carbon capture from flue gas. *Proceedings of the National Academy of Sciences*, *111*(46), 16436–16441.

Angeli, A., Alasmary, F.A., Del Prete, S., Osman, S.M., AlOthman, Z., Donald, W.A., Capasso, C., & Supuran, C.T. (2018). The first activation study of a δ-carbonic anhydrase: TweCAδ from the diatom *Thalassiosira weissflogii* is effectively activated by amines and amino acids. *Journal of Enzyme Inhibition and Medicinal Chemistry*, *33*(1), 680–685.

Barbero, R., Carnelli, L., Simon, A., Kao, A., Monforte, A. D. A., Riccò, M., ... Belcher, A. (2013). Engineered yeast for enhanced CO_2 mineralization. *Energy & Environmental Science*, *6*(2), 660–674.

Blazeck, J., & Alper, H. S. (2013). Promoter engineering: Recent advances in controlling transcription at the most fundamental level. *Biotechnology Journal*, *8*(1), 46–58.

Bloom, J. D., Meyer, M. M., Meinhold, P., Otey, C. R., MacMillan, D., & Arnold, F. H. (2005). Evolving strategies for enzyme engineering. *Current Opinion in Structural Biology*, *15*(4), 447–452.

Boone, C. D., Habibzadegan, A., Tu, C., Silverman, D. N., & McKenna, R. (2013). Structural and catalytic characterization of a thermally stable and acid-stable variant of human carbonic anhydrase II containing an engineered disulfide bond. *Acta Crystallographica Section D: Biological Crystallography*, *69*(8), 1414–1422.

Bootorabi, F., Jänis, J., Valjakka, J., Isoniemi, S., Vainiotalo, P., Vullo, D., Supuran, C.T., Waheed, A., Sly, W.S., Niemelä, O., & Parkkila, S., 2008. Modification of carbonic anhydrase II with acetaldehyde, the first metabolite of ethanol, leads to decreased enzyme activity. *BMC Biochemistry*, *9*(1), 32.

Bose, H., & Satyanarayana, T. (2017). Microbial carbonic anhydrases in biomimetic carbon sequestration for mitigating global warming: Prospects and perspectives. *Frontiers in Microbiology*, *8*, 1615.

Cao, L. (2006). *Carrier-Bound Immobilized Enzymes: Principles, Application and Design*. John Wiley & Sons.

Capasso, C., & Supuran, C. T. (2015). An overview of the alpha-, beta-and gamma-carbonic anhydrases from Bacteria: Can bacterial carbonic anhydrases shed new light on evolution of bacteria? *Journal of Enzyme Inhibition and Medicinal Chemistry*, *30*(2), 325–332.

Capasso, C., De Luca, V., Carginalea, V., Caramuscioc, P., Cavalheiroc, C. F., & Canniod, R. (2012). Characterization and properties of a new thermoactive and thermostable carbonic anhydrase. *Chemical Engineering*, *27*, 271–276.

Chen, J., An, Y., Kumar, A., & Liu, Z. (2017). Improvement of chitinase Pachi with nematicidal activities by random mutagenesis. *International Journal of Biological Macromolecules*, *96*, 171–176.

Chen, L., Jiang, H., Cheng, Q., Chen, J., Wu, G., Kumar, A., ... Liu, Z. (2015). Enhanced nematicidal potential of the chitinase pachi from Pseudomonas aeruginosa in association with Cry21Aa. *Scientific Reports*, *5*, 14395.

Chen, P. H., Liu, H. L., Chen, Y. J., Cheng, Y. H., Lin, W. L., Yeh, C. H., & Chang, C. H. (2012). Enhancing CO_2 bio-mitigation by genetic engineering of cyanobacteria. *Energy & Environmental Science*, *5*(8), 8318–8327.

Cowan, D. A., & Fernandez-Lafuente, R. (2011). Enhancing the functional properties of thermophilic enzymes by chemical modification and immobilization. *Enzyme and Microbial Technology*, *49*(4), 326–346.

Del Prete, S., Isik, S., Vullo, D., De Luca, V., Carginale, V., Scozzafava, A., Supuran, C.T. & Capasso, C., 2012. DNA cloning, characterization, and inhibition studies of an α-carbonic anhydrase from the pathogenic bacterium Vibrio cholerae. *Journal of Medicinal Chemistry, 55*(23), 10742–10748.

Di Fiore, A., Alterio, V., Monti, S. M., De Simone, G., & D'Ambrosio, K. (2015). Thermostable carbonic anhydrases in biotechnological applications. *International Journal of Molecular Sciences*, *16*(7), 15456–15480. doi:10.3390/ijms160715456.

Di Fiore, A., Capasso, C., De Luca, V., Monti, S. M., Carginale, V., Supuran, C. T., ... De Simone, G. (2013). X-ray structure of the first extremo-α-carbonic anhydrase', a dimeric enzyme from the thermophilic bacterium *Sulfurihydrogenibium yellowstonense* YO3AOP1. *ActaCrystallographica Section D: Biological Crystallography*, *69*(6), 1150–1159.

Dombkowski, A. A., Sultana, K. Z., & Craig, D. B. (2014). Protein disulfide engineering. *FEBS Letters*, *588*(2), 206–212.

Domsic, J. F., & McKenna, R. (2010). Sequestration of carbon dioxide by the hydrophobic pocket of the carbonic anhydrases. *Biochimica et Biophysica Acta (BBA)-Proteins and Proteomics*, *1804*(2), 326–331.

Drozdov, A. S., Shapovalova, O. E., Ivanovski, V., Avnir, D., & Vinogradov, V. V. (2016). Entrapment of enzymes within sol–gel-derived magnetite. *Chemistry of Materials*, *28*(7), 2248–2253.

Ducat, D. C., Way, J. C., & Silver, P. A. (2011). Engineering cyanobacteria to generate high-value products. *Trends in Biotechnology*, *29*(2), 95–103.

Elder, I., Han, S., Tu, C., Steele, H., Laipis, P. J., Viola, R. E., & Silverman, D. N. (2004). Activation of carbonic anhydrase II by active-site incorporation of histidine analogs. *Archives of Biochemistry and Biophysics*, *421*(2), 283–289.

Fan, L. H., Liu, N., Yu, M. R., Yang, S. T., & Chen, H. L. (2011). Cell surface display of carbonic anhydrase on *Escherichia coli* using ice nucleation protein for CO_2 sequestration. *Biotechnology and Bioengineering*, *108*(12), 2853–2864.

Ferraroni, M., Del Prete, S., Vullo, D., Capasso, C., & Supuran, C. T. (2015). Crystal structure and kinetic studies of a tetrameric type II β-carbonic anhydrase from the pathogenic bacterium Vibrio cholerae. *Acta Crystallographica Section D: Biological Crystallography*, *71*(12), 2449–2456.

Forsyth, C., Yip, T. W., & Patwardhan, S. V. (2013). CO_2 sequestration by enzyme immobilized onto bioinspired silica. *Chemical Communications*, *49*(31), 3191–3193.

Fox, R. J., Davis, S. C., Mundorff, E. C., Newman, L. M., Gavrilovic, V., Ma, S. K., ...Grate, J. (2007). Improving catalytic function by ProSAR-driven enzyme evolution. *Nature Biotechnology*, *25*(3), 338.

Giri, A., Banerjee, U. C., Kumar, M., & Pant, D. (2018). Intracellular carbonic anhydrase from *Citrobacter freundii* and its role in bio-sequestration. *Bioresource Technology*, *267*, 789–792.

Giri, A., & Pant, D. (2018) Carbon management and greenhouse gas mitigation. In: Saleem Hashmi (eds) *Reference Module in Materials Science and Materials Engineering*. Elsevier. https://doi.org/10.1016/b978-0-12-803581-8.11041-0

Giri, A., & Pant, D. (2019a) CO_2 management using carbonic anhydrase producing microbes from western Indian Himalaya. *Bioresource Technology Reports*, *8*, 100320.

Giri, A., & Pant, D. (2019b) Carbonic anhydrase modification for carbon management. *Environmental Science and Pollution Research*, 27, 1294–1318.

Gould, S. M., & Tawfik, D. S. (2005). Directed evolution of the promiscuous esterase activity of carbonic anhydrase II. *Biochemistry*, *44*(14), 5444–5452.

Hibbert, E. G., & Dalby, P. A. (2005). Directed evolution strategies for improved enzymatic performance. *Microbial Cell Factories*, *4*(1), 29.

Jansen, E., Overpeck, J., Briffa, K. R., Duplessy, J.-C., Joos, F., Masson-Delmotte, V., et al. (2007). Palaeo climate. In: Solomon, S., Qin, D., Manning, M., Chen, Z., Marquis, M., Averyt, K. B., et al., editors. *Climate change the physical science basis. Contribution of Working Group I to the Fourth Assessment Report of the Intergovernmental Panel on Climate Change*. Cambridge/New York: Cambridge University Press.

Jesionowski, T., Zdarta, J., & Krajewska, B. (2014). Enzyme immobilization by adsorption: A review. *Adsorption*, *20*(5–6), 801–821.

Jo, B. H., Kim, I. G., Seo, J. H., Kang, D. G., & Cha, H. J. (2013). Engineered *Escherichia coli* with periplasmic carbonic anhydrase as a biocatalyst for CO_2 sequestration. *Applied and Environmental Microbiology*, *79*(21), 6697–6705.

Jo, B. H., Park, T. Y., Park, H. J., Yeon, Y. J., Yoo, Y. J., & Cha, H. J. (2016). Engineering de novo disulfide bond in bacterial α-type carbonic anhydrase for thermostable carbon sequestration. *Scientific Reports*, *6*, 29322.

Joseph, P., Turtaut, F., Ouahrani-Bettache, S., Montero, J. L., Nishimori, I., Minakuchi, T., … Supuran, C. T. (2010). Cloning, characterization, and inhibition studies of a β-carbonic anhydrase from Brucellasuis. *Journal of Medicinal Chemistry*, *53*(5), 2277–2285.

Kalloniati, C., Tsikou, D., Lampiri, V., Fotelli, M.N., Rennenberg, H., Chatzipavlidis, I., Fasseas, C., Katinakis, P., & Flemetakis, E. (2009). Characterization of a Mesorhizobium loti α-type carbonic anhydrase and its role in symbiotic nitrogen fixation. *Journal of bacteriology*, *191*(8), 2593–2600.

Kambar, B., & Ozdemir, E. (2010). Thermal stability of carbonic anhydrase immobilized within polyurethane foam. *Biotechnology Progress*, *26*(5), 1474–1480.

Kazlauskas, R. J., & Bornscheuer, U. T. (2009). Finding better protein engineering strategies. *Nature Chemical Biology*, *5*(8), 526.

Ki, M. R., Min, K., Kanth, B. K., Lee, J., & Pack, S. P. (2013). Expression, reconstruction and characterization of codon-optimized carbonic anhydrase from *Hahellachejuensis* for CO_2 sequestration application. *Bioprocess and Biosystems Engineering*, *36*(3), 375–381.

Kim, I. G., Jo, B. H., Kang, D. G., Kim, C. S., Choi, Y. S., & Cha, H. J. (2012). Biomineralization-based conversion of carbon dioxide to calcium carbonate using recombinant carbonic anhydrase. *Chemosphere*, *87*(10), 1091–1096.

Kim, Y., Ho, S. O., Gassman, N. R., Korlann, Y., Landorf, E. V., Collart, F. R., & Weiss, S. (2008). Efficient site-specific labeling of proteins via cysteines. *Bioconjugate Chemistry*, *19*(3), 786–791.

Kimber, M. S., & Pai, E. F. (2000). The active site architecture of Pisumsativum β-carbonic anhydrase is a mirror image of that of α-carbonic anhydrases. *The EMBO Journal*, *19*(7), 1407–1418.

Kisker, C., Schindelin, H., Alber, B. E., Ferry, J. G., & Rees, D. C. (1996). A left-hand beta-helix revealed by the crystal structure of a carbonic anhydrase from the archaeon *Methanosarcinathermophila*. *The EMBO Journal*, *15*(10), 2323–2330.

Kumar, A., Kim, I. W., Patel, S. K., & Lee, J. K. (2018c).Synthesis of protein-inorganic nanohybrids with improved catalytic properties using Co_3 (PO_4) 2. *Indian Journal of Microbiology*, *58*(1), 100–104.

Kumar, A., Park, G. D., Patel, S. K., Kondaveeti, S., Otari, S., Anwar, M. Z., ... Sohn, J. H. (2019). SiO_2 microparticles with carbon nanotube-derived mesopores as an efficient support for enzyme immobilization. *Chemical Engineering Journal*, *359*, 1252–1264.

Kumar, A., Wu, G., & Liu, Z. (2018a). Synthesis and characterization of cross linked enzyme aggregates of serine hydroxyl methyltransferase from Idiomerinaleihiensis. *International Journal of Biological Macromolecules*, *117*, 683–690.

Kumar, A., Wu, G., Wu, Z., Kumar, N., & Liu, Z. (2018b. Improved catalytic properties of a serine hydroxymethyltransferase from Idiomarinaloihiensis by site directed mutagenesis. *International Journal of Biological Macromolecules*, *117*, 1216–1223.

Kumar, A., Zhang, S., Wu, G., Wu, C. C., Chen, J., Baskaran, R., & Liu, Z. (2015). Cellulose binding domain assisted immobilization of lipase (GSlip–CBD) onto cellulosic nanogel: Characterization and application in organic medium. *Colloids and Surfaces B: Biointerfaces*, *136*, 1042–1050.

Lee, R. B. Y., Smith, J. A. C., & Rickaby, R. E. (2013). Cloning, expression and characterization of the δ-carbonic Anhydrase of *Thalassiosira weissflogii* (Bacillariophyceae). *Journal of Phycology*, *49*(1), 170–177.

Lippow, S. M., & Tidor, B. (2007). Progress in computational protein design. *Current Opinion in Biotechnology*, *18*(4), 305–311.

Mårtensson, L. G., Karlsson, M., & Carlsson, U. (2002). Dramatic stabilization of the native state of human carbonic anhydrase II by an engineered disulfide bond. *Biochemistry*, *41*(52), 15867–15875.

Mateo, C., Palomo, J. M., Fernandez-Lorente, G., Guisan, J. M., & Fernandez-Lafuente, R. (2007). Improvement of enzyme activity, stability and selectivity via immobilization techniques. *Enzyme and Microbial Technology*, *40*(6), 1451–1463.

Migliardini, F., De Luca, V., Carginale, V., Rossi, M., Corbo, P., Supuran, C. T., & Capasso, C. (2014). Biomimetic CO_2 capture using a highly thermostable bacterial α-carbonic anhydrase immobilized on a polyurethane foam. *Journal of Enzyme Inhibition and Medicinal Chemistry*, *29*(1), 146–150.

NOAA (2017). Trends in Atmospheric Carbon Dioxide [online]. Available at: https://www.co2.earth/daily-co2. Earth Systems Research Laboratory.

Novick, S., & Alvizo, O. (2013). U.S. Patent No. 8,354,262. Washington, DC: U.S. Patent and Trademark Office.

Pant, D., Sharma, V., Singh, P., Kumar, M., Giri, A., & Singh, M. P. (2017). Perturbations and 3R in carbon management. *Environmental Science and Pollution Research*, *24*(5), 4413–4432.

Parra-Cruz, R., Jäger, C. M., Lau, P. L., Gomes, R. L., & Pordea, A. (2018). Rational design of thermostable carbonic anhydrase mutants using molecular dynamics simulations. *The Journal of Physical Chemistry B*, *122*(36), 8526–8536.

Patwardhan, R. P., Lee, C., Litvin, O., Young, D. L., Pe'er, D., & Shendure, J. (2009). High-resolution analysis of DNA regulatory elements by synthetic saturation mutagenesis. *Nature Biotechnology*, *27*(12), 1173.

Patwardhan, S. V. (2011). Biomimetic and bioinspired silica: Recent developments and applications. *Chemical Communications*, *47*(27), 7567–7582.

Perfetto, R., Del Prete, S., Vullo, D., Sansone, G., Barone, C. M., Rossi, M., ... Capasso, C. (2017). Production and covalent immobilisation of the recombinant bacterial carbonic anhydrase (SspCA) onto magnetic nanoparticles. *Journal of Enzyme Inhibition and Medicinal Chemistry*, *32*(1), 759–766.

Prabhu, C., Wanjari, S., Puri, A., Bhattacharya, A., Pujari, R., Yadav, R., Das, S., Labhsetwar, N., Sharma, A., Satyanarayanan, T. & Rayalu, S. (2011). Region-specific bacterial carbonic anhydrase for biomimetic sequestration of carbon dioxide. *Energy & Fuels*, *25*(3), 1327–1332.

Sakamoto, S., & Hamachi, I. (2018). Recent progress in chemical modification of proteins. *Analytical Sciences*, *35*. 5–27.

Sharma, A., Bhattacharya, A., & Shrivastava, A. (2011). Biomimetic CO_2 sequestration using purified carbonic anhydrase from indigenous bacterial strains immobilized on biopolymeric materials. *Enzyme and Microbial Technology*, *48*(4–5), 416–426.

Sharma, T., Sharma, S., Kamyab, H., & Kumar, A. (2019). Energizing the CO_2 utilization by Chemo-enzymatic approaches and potentiality of carbonic anhydrases: A review. *Journal of Cleaner Production*, 247, 119138.

Silverman, D. N., & Lindskog, S. (1988). The catalytic mechanism of carbonic anhydrase: Implications of a rate-limiting protolysis of water. *Accounts of Chemical Research*, *21*(1), 30–36.

Srivastava, S., Bharti, R. K., Verma, P. K., & Shekhar Thakur, I. (2015). Cloning and expression of gamma carbonic anhydrase from Serratia sp. ISTD04 for sequestration of carbon dioxide and formation of calcite. *Bioresource Technology*, *188*, 209–213.

Stefanucci, A., Angeli, A., Dimmito, M.P., Luisi, G., Del Prete, S., Capasso, C., Donald, W.A., Mollica, A. & Supuran, C.T. (2018). Activation of β-and γ-carbonic anhydrases from pathogenic bacteria with tripeptides. *Journal of Enzyme Inhibition and Medicinal Chemistry*, *33*(1), 945–950.

Supuran, C. T., & Capasso, C. (2017). An overview of the bacterial carbonic anhydrases. *Metabolites*, *7*(4), 56.

Tripp, B. C., Smith, K., & Ferry, J. G. (2001). Carbonic anhydrase: New insights for an ancient enzyme. *Journal of Biological Chemistry*, *276*(52), 48615–48618.

Tu, C., Paranawithana, S.R., Jewell, D.A., Tanhauser, S.M., LoGrasso, P. V., Wynns, G.C., Laipis, P.J., Silverman, D. N. (1990). Buffer enhancement of proton transfer in catalysis by human carbonic anhydrase III. *Biochemistry*, *29*(27), 6400–6405.

Verma, R., Kumar, A., & Kumar, S. (2019). Synthesis and characterization of cross-linked enzyme aggregates (CLEAs) of thermostable xylanase from Geobacillus thermodenitrificans X1. *Process Biochemistry*, *80*, 72–79.

Vinoba, M., Bhagiyalakshmi, M., Jeong, S. K., Yoon, Y. I., & Nam, S. C. (2011). Capture and sequestration of CO_2 by human carbonic anhydrase covalently immobilized onto amine-functionalized SBA-15. *The Journal of Physical Chemistry C*, *115*(41), 20209–20216.

Vinoba, M., Bhagiyalakshmi, M., Jeong, S. K., Yoon, Y. I., & Nam, S. C. (2012). Carbonic anhydrase conjugated to nanosilver immobilized onto mesoporous SBA-15 for sequestration of CO_2. *Journal of Molecular Catalysis B: Enzymatic*, *75*, 60–67.

Wanjari, S., Prabhu, C., Yadav, R., Satyanarayana, T., Labhsetwar, N., & Rayalu, S. (2011). Immobilization of carbonic anhydrase on chitosan beads for enhanced carbonation reaction. *Process Biochemistry*, *46*(4), 1010–1018.

Wu, G., Zhan, T., Guo, Y., Kumar, A., & Liu, Z. (2016). Asn336 is involved in the substrate affinity of glycine oxidase from Bacillus cereus. *Electronic Journal of Biotechnology*, *19*(4), 26–30.

Yadav, R., Wanjari, S., Prabhu, C., Kumar, V., Labhsetwar, N., Satyanarayanan, T., Kotwal, S. & Rayalu, S. (2010). Immobilized carbonic anhydrase for the biomimetic carbonation reaction. *Energy & Fuels*, *24*(11), 6198–6207.

Zhang, K., Guo, Y., Yao, P., Lin, Y., Kumar, A., Liu, Z., … Zhang, L. (2016). Characterization and directed evolution of BliGO, a novel glycine oxidase from *Bacillus licheniformis*. *Enzyme and Microbial Technology*, *85*, 12–18.

Zhang, S., Han, Y., Kumar, A., Gao, H., Liu, Z., & Hu, N. (2017). Characterization of an L-phosphinothricin resistant glutamine synthetase from Exiguobacterium sp. and its improvement. *Applied Microbiology and Biotechnology*, *101*(9), 3653–3661.

Zhang, S., Lu, H., & Lu, Y. (2013). Enhanced stability and chemical resistance of a new nanoscale biocatalyst for accelerating CO_2 absorption into a carbonate solution. *Environmental Science & Technology*, *47*(23), 13882–13888.

Zhang, S., Zhang, Z., Lu, Y., Rostam-Abadi, M., & Jones, A. (2011). Activity and stability of immobilized carbonic anhydrase for promoting CO_2 absorption into a carbonate solution for post-combustion CO_2 capture. *Bioresource Technology*, *102*(22), 10194–10201.

Zuo, W., Nie, L., Baskaran, R., Kumar, A., & Liu, Z. (2018). Characterization and improved properties of Glutamine synthetase from Providencia vermicola by site-directed mutagenesis. *Scientific Reports*, *8*(1), 15640.

6 Electrochemical CO_2 Reduction Reaction on Nitrogen-Doped Carbon Catalysts

Mahima Khandelwal

CONTENTS

6.1 INTRODUCTION

In the past few decades, the substantial growth in population and technologies led to the drastic increase in energy consumption. As a result, we have seen a massive consumption/combustion of fossil fuel resources (coal, natural gas, and petrol) in order to meet the energy demands. Excessive use of fossil fuels resulted in the production of harmful gases such as CO_2, SO_x, and NO_x. The substantial increase in the level of greenhouse gases like CO_2 in the atmosphere led to the drastic climate changes, which has become a matter of concern for human society and environment (Canadell et al. 2007, Chu and Majumdar 2012, Fernandes et al. 2019). Therefore, it has become an urgent need to recycle the atmospheric CO_2 in order to make a cleaner environment. To solve this issue, various new strategies have been emerged, such as thermo-, bio-, photo-, and electrochemical methods, to mitigate CO_2 and turn into useful low-cost value-added chemical fuels, i.e. carbon monoxide (CO), ethylene (C_2H_4), methane (CH_4), formic acid (HCOOH), methanol (CH_3OH), and ethanol (C_2H_5OH) (Kondratenko et al. 2013, Sharma et al. 2020). Among the various CO_2 reduction strategies, an electrochemical CO_2 reduction is an effective approach, which has recently gained considerable interest from researchers for the conversion of CO_2 into practical low-cost chemical fuels because of its distinctive advantages. Some of the advantages of electrochemical CO_2 reduction are as follows: (i) it uses electrons as clean reductants under mild conditions and makes use of environmentally friendly aqueous electrolytes; (ii) it is simple and effective to couple with carbon-free electricity sources such as solar, wind, and hydroelectric; (iii) it operates with high reaction rates and exhibits high efficiency under ambient conditions; (iv) these systems are modular, compact, and easily scalable (Costentin et al. 2013). In CO_2 reduction reaction (CO_2RR), the cathode materials (electrocatalysts) play a pivotal role in the product selectivity and efficiency, reaction mechanism, and electrokinetics as compared to other energy-related processes such as oxygen reduction reaction (ORR), oxygen evolution reaction (OER), and hydrogen evolution reaction (HER) (Fernandes et al. 2019, Kuang and Zheng 2016, Jiao et al. 2015, Gao, Wang, et al. 2019, Khandelwal, Chandrasekaran, et al. 2018). Hitherto, different kinds of electrocatalysts have been employed for the electrochemical CO_2RR, such as carbonaceous materials, transition metals, metal oxides, and nanostructured 2D materials (Wang et al. 2019, Hou et al. 2019, Feng et al. 2017, Liu, Guo, et al. 2017). Among all, carbon-based materials have gained immense interest for CO_2RR due to their unique physicochemical properties, and several reviews can be found focusing on the performance of carbon-based materials towards CO_2RR (Wang et al. 2019, Hou et al. 2019). However, the effect of heteroatom doping on the carbon materials has not been largely covered for CO_2RR. Therefore, this chapter focuses on heteroatom-doped carbon materials specifically on N-doped carbon (NDC) electrocatalysts for CO_2RR. The details of other heteroatom-doped carbons (S, P, B, and F) for CO_2RR are outside the reach of this chapter.

6.2 ELECTROCHEMICAL CO_2 REDUCTION REACTION (CO_2RR)

CO_2 is a chemically inert, linear, and thermodynamically stable molecule having two C=O bonds with the bond dissociation energy of 750 kJ mol^{-1}, which is much larger than C–C (336 kJ mol^{-1}), C–O (327 kJ mol^{-1}), and C–H (411 kJ mol^{-1}) bonds (Wu et al. 2017, Jia et al. 2019). Consequently, CO_2RR has simple thermodynamics, but slow kinetics. There are few challenges related to the electrochemical reduction of CO_2, like the direct CO_2 reduction, implying one-electron transfer for the formation of intermediate ($CO_2^{\cdot -}$), which is thermodynamically not favourable as it occurs at a very high negative potential of approx. −1.90 V *vs.* standard hydrogen electrode (SHE) due to the bending of the extremely stable and linear CO_2 molecule, suggesting that enormous amount of energy is required for electroreduction (Benson et al. 2009, Vasileff et al. 2017). Moreover, CO_2RR is quite difficult in an aqueous medium due to its competition with HER arising because of its similar thermodynamic potentials (Vasileff et al. 2017). Further, the poor solubility of CO_2 in aqueous medium limits the reaction rate and results in poor turnover frequency (TOF) since the CO_2RR proceeds *via* the adsorption of CO_2 molecule at the electrode–electrolyte interface (Li and Oloman 2005). Therefore, it is of utmost importance to develop the catalysts that can suppress the formation of hydrogen as much as possible and facilitate the CO_2 reduction into value-added products with low overpotential, high selectivity with maximum Faradaic efficiency (FE), and long stability. CO_2RR proceeds *via* different multistep pathways involving 2–18 electrons resulting in various reduction products such as CO, CH_3OH, HCOOH, C_2H_5OH, CH_4, and C_2H_4. The electron pathway and the end reduction product formation depend on the number of factors such as the electrocatalyst properties, electrolyte pH, temperature, pressure, CO_2 concentration, electrolyte, and electrode potential (Mao et al. 2015). The thermodynamically driven CO_2 electrochemical half-reactions along with their standard electrode potential against SHE in aqueous medium at pH 7.0 can be found in recent reviews (Fernandes et al. 2019, Jia et al. 2019, Schneider et al. 2012).

The electrocatalytic performance of a catalyst for CO_2RR has been evaluated using various parameters such as onset potential (E_{onset}), current density (j), FE, and Tafel slope. The details about these parameters can be found in the review by Fernandes et al. (2019). Hitherto, different transition metals were studied as catalysts for the electrochemical CO_2RR, which have been categorized into three main groups according to the product selectivity (Hori et al. 1994). The first group containing Sn, Hg, and Pb electrodes are selective for the $HCOOH/HCOO^-$ production. The second group containing Au, Ag, Pd, and Zn electrodes are selective for the CO production. The last group containing Cu electrode is capable of transforming CO_2 into hydrocarbons and alcohols as major products such as C_2H_4 and C_2H_5OH (Hori et al. 1994, Sun et al. 2017). But these metal catalysts have low selectivity, low stability, and high overpotential (Fernandes et al. 2019, Lim et al. 2014). Therefore, metal-free carbon catalysts have been emerged as a better option in place of metal catalysts due to their relatively larger surface area, high electrical conductivity, remarkable thermal and chemical stability, mechanical strength, and adjustable surface chemistry. Additionally, carbon materials are relatively cheap, environmentally benign, and can be produced in large quantities at a much lower cost. Despite having the remarkable features, pristine carbon material suffers from poor CO_2RR activity due to its inability to activate the CO_2 molecules. Therefore, heteroatom doping in the carbon frameworks has been proved to be an effective approach for enhancing the electrocatalytic activity of carbon catalysts for CO_2RR (Liu, Ali, et al. 2017, Duan et al. 2017). The heteroatom doping (N, S, B, P, and F) in carbon structure alters the charge neutrality and spin density of carbon atoms and induces charge redistribution (Wang et al. 2012). The induced strain and defect sites due to doping enable the adsorption of CO_2 molecules and the activation of reaction intermediates in heteroatom-doped carbon materials, which contributes to the enhancement of electrocatalytic activity towards CO_2 reduction. Apart from this, the doping of heteroatoms in carbon materials also modulates their texture properties, which may increase the surface area and pore size and improves their wettability of the carbon materials. These features help in CO_2 diffusion and ion transfer, which contribute towards electrocatalysis.

Among the various explored heteroatoms such as B, P, N, S, and F (Fernandes et al. 2019, Sreekanth et al. 2015, Liu, Ali, et al. 2018, Xie et al. 2018), nitrogen is so far the most suitable dopant for carbon materials due to its similar size and relatively larger electronegativity (3.04) to that of carbon atom (2.55). The N-doping in carbon frameworks has been identified in four different bonding configurations, namely pyridinic-N (398.1–399.3 eV), pyrrolic-N (399.8–401.2 eV), quaternary-N/graphitic-N (401.1–402.7 eV), and N-oxides (402.0–406.0 eV), by X-ray photoelectron spectroscopy (XPS), as has been shown in Figure 6.1 (Wang et al. 2012, Khandelwal et al. 2017, Khandelwal, Li, et al. 2018, Jurewicz et al. 2003). Pyridinic-N and pyrrolic-N are mainly observed as defect sites, which are present at the edges of the carbon framework contributing to one and two p electrons to π system, respectively.

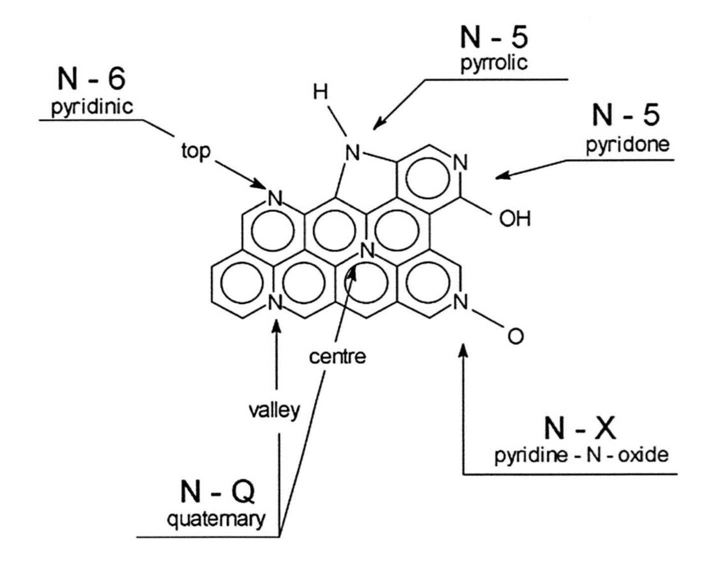

FIGURE 6.1 Different bonding configurations of nitrogen atom in carbon lattice. (Reprinted with permission from Jurewicz et al. (2003) Copyright 2003, Elsevier.)

On the other hand, graphitic-N replaces the carbon atom in the hexagonal rings of carbon framework (Wang et al. 2012, Jurewicz et al. 2003). The electrocatalytic activity of NDC materials towards CO_2RR is mainly associated with the presence of different N-moieties. The presence of these different N-moieties disturbs the charge neutrality of carbon atoms, induces the redistribution of charge, and consequently facilitates the adsorption of CO_2 molecules and activation for CO_2RR. Moreover, the different configurations of nitrogen in carbon matrix introduce the Lewis basicity on the surface of carbon and modulate the electronic conductivity, pore structure, and surface wettability of carbon materials (Wang et al. 2012, Khandelwal et al. 2017, Khandelwal, Li, et al. 2018). As a result, NDC materials exhibit an enhanced electrocatalytic performance for CO_2RR with different activity and product selectivity.

This chapter systematically summarizes the development of NDC catalysts for enhancing the electrochemical performance towards CO_2RR. In the following section, we have focused mainly on the synthesis of NDC electrocatalysts employing different synthetic routes. Specifically, the emphasis has been made on the different experimental conditions, which leads to the tunable amount of total N-contents, different N-bonding configurations, and modulated surface properties such as surface area, pore size, and total pore volume of NDC materials for electrochemical reduction of CO_2. Thereafter, we have made an elaborative discussion about the electrochemical performance of various NDC catalysts for CO_2RR focusing on the primarily N-doped active sites. Furthermore, we have also shed some light on the improved interactions between the different N-bonding configurations present in NDC materials and CO_2 molecule, which facilitates the adsorption capacity of CO_2 molecule and thereby enhances the CO_2 reduction into different value-added products with high selectivity and activity.

Finally, the current challenges and future prospects of NDC materials have been discussed for electrochemical CO_2 reduction. Overall, this chapter aims to provide an overview of the advanced NDC materials for the electrochemical CO_2RR. We believe that this chapter will help to provide a better understanding to the readers in this research area.

6.3 SYNTHESIS OF N-DOPED CARBON CATALYSTS FOR CO_2RR

Generally, the synthesis of heteroatom-doped carbon materials, especially NDCs, has been reported by the following two methods: (i) *in situ* doping (i.e. nitrogen precursor and carbon material were treated together) and (ii) doping after the synthesis of carbon material (i.e. the 'as-synthesized' carbon material was post-treated with the nitrogen precursor). Both protocols lead to the doping of nitrogen into the carbon matrix. For *in situ* doping, chemical vapour deposition (CVD) is a widely employed method because it has several advantages such as controlled morphology and product quality. CVD method involves the vaporization of carbon and nitrogen sources, which leads to their deposition on the substrate. For example, Wu et al. (2015) reported the synthesis of N-doped carbon nanotubes (NCNTs) by liquid CVD approach, having a total N-content of 5 at.% employing a liquid precursor (acetonitrile (ACN)) and dicyandiamide with ferrocene under Ar/H_2 atmosphere at 850°C. Researchers have reported that the experimental conditions such as employed N-doping precursor and growth conditions influence the total N-doping content and graphitic content. In this context, Sharma et al. (2015) reported the synthesis of N-doped CNTs (NCNTs) using three different N-containing precursors, namely ACN, dimethylformamide (DMF), and triethylamine (TEA), at different growth temperatures (750°C, 850°C, and 950°C) for modulating the total nitrogen content. Their study revealed that NCNTs synthesized by ACN as N-doping precursor at 750°C, 850°C, and 950°C exhibited different total N-content (at.%) of 1.9, 4.9, and 3.9, respectively. However, NCNTs synthesized by employing DMF and TEA at the growth temperature of 850°C resulted in almost similar total N-content of 2.8 and 2.4 at.%, respectively. These results significantly demonstrated the role of different growth temperatures and N-doping precursors in the total N-doping content. The growth conditions not only influence the total N-content, but also monitor the ratio of sp^2-carbon phase to sp^3-carbon phase in NDC materials, which was manifested by Wanninayake et al. (2020). They synthesized nitrogen-doped ultra-nanocrystalline thin films of the diamond by a microwave-assisted CVD technique. The carbon host (graphitic (sp^2) and diamond-like (sp^3)) structure was monitored by varying CH_4 and Ar compositions in the source gas mixture during microwave-assisted CVD growth. It was observed that the N-doping in sp^2-carbon (graphitic) shows an enhanced catalytic performance compared to the sp^3-carbon (diamond). Despite all the aforementioned advantages of CVD approach, this approach has also some disadvantages such as harsh experimental conditions, high cost, and contamination with metal impurities (Fernandes et al. 2019, Liu, Ali, et al. 2017). Therefore, other methods such as laser ablation, arc discharge, microwave irradiation, N_2 plasma, and multistep hard-template protocols (Fernandes et al. 2019, Wang et al. 2012, Gao, Xie, et al. 2019) have also been used for the preparation of NDCs as an electrocatalyst. Apart from these methods, direct carbonization/pyrolysis of carbon sources with nitrogen-rich precursors such as melamine, urea, dicyandiamide, pyridine, and quinoline is one of the widely adopted methods for obtaining NDCs (Wang et al. 2012, Gao, Xie, et al. 2019, Chen et al. 2018, Liu, Li, et al. 2015). In a recent study by Gao, Xie, and Han, et al. (2019), NDC nanosheets were obtained by employing the two-step thermal annealing approach using glucose and melamine as carbon and nitrogen sources, respectively (Figure 6.2). They have demonstrated the significant role of annealing temperature in modulating the total surface nitrogen content in a carbon matrix, which decreases significantly from 6.75 to 2.54 at.%. as the temperature increases from 800°C to 1100°C. They have also mentioned that with an increased pyrolysis temperature, the relative percentage of pyridinic-N (28.12%–14.88%) and pyrrolic-N (39.22%–12.32%) decreased, whereas for graphitic-N, it was found to be increased from ~ 21.00% to 57.00%. In contrast to this, Chen et al. (2018) reported a decrease in only pyrrolic-N content at the increased temperature in N-doped fullerene derivative (N-C61) synthesized by pyrolysing commercially available [6,6]-phenyl-C61-butyric acid methyl ester (PC61BM as a carbon source) with urea (N-source). Moreover, they observed that the pyrrolic-N content was diminished completely as the carbonization temperature raised to 1000°C. Contrary to the reports discussed above, researchers have also varied the ratios of N-rich precursor to carbon source for adjusting the N-dopant content and its configurations. For example, Liu, Yang and Huang, et al. (2018) reported 3D N-doped graphene nanoribbon network (N-GRW) by pyrolysis/carbonization of melamine (nitrogen-rich precursor) and L-cysteine (carbon source). In N-GRW catalysts, the ratio of melamine to L-cysteine was changed from 1:1 to 8:1 in order to tune the N-doping total content and at.% of N individual content. From the experimental outcomes and theoretical calculations, they have concluded that among various N-doping configurations, pyridinic-N is serving as an active site for the CO_2RR.

The other most extensively employed approach for the synthesis of NDCs is post-treatment synthesis where the pristine carbon source has been treated with nitrogen-rich precursor at high temperatures. Thus, the as-obtained NDC catalysts having

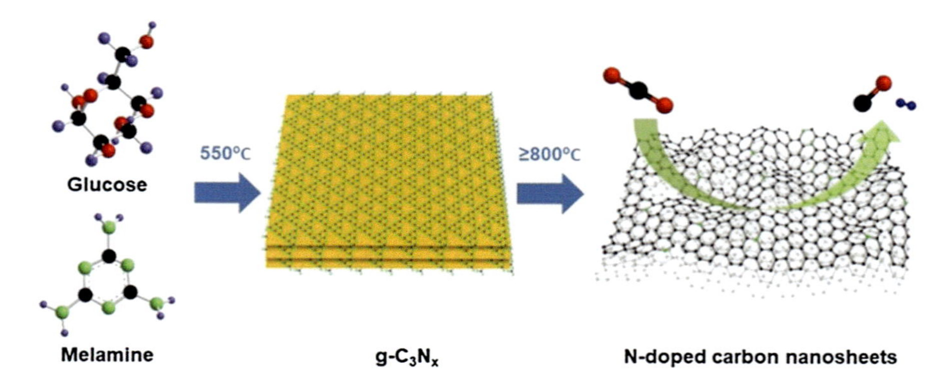

FIGURE 6.2 Schematic illustration for the synthesis of nitrogen-doped CSs and the process of electrochemical CO_2RR. (Reprinted with permission from Gao, Xie, et al. (2019). Copyright 2019, American Chemical Society.)

various N-moieties show different electrocatalytic activity and product selectivity for CO_2RR. Among the different bonding configurations of nitrogen (pyridinic-, pyrrolic-, and graphitic-N), it has been mostly observed that the catalyst having a high density of pyridinic-N sites shows promising electrocatalytic activity for CO_2 reduction. Keeping this in mind, Ma et al. (2019) recently reported NCNTs with a high amount of pyridinic-N by pyrolysis of phenanthroline as a heterocyclic N-rich precursor with multiwalled CNT. Specifically, phenanthroline has been used as a precursor to increase the concentration of pyridinic-N. Therefore, it is evident that the used N-precursors play a significant role in tuning the N-configurations as well as the total N-content. Li, Xiao, et al. (2019) performed the N-doping on the as-synthesized cubic carbon foam with flowing ammonia in the furnace for different times ranging from 30 to 120 min under constant temperature.

Apart from performing the nitrogen treatment on traditional carbon materials for CO_2RR such as graphene, CNTs, fullerene, and nanodiamonds, other carbon materials have also been explored, which are derived from metal–organic framework (MOF), biomass, and fossil fuels (Wang et al. 2018, Li et al. 2017, Yao et al. 2019, Li, Wang, and Xiao, et al. 2019). For instance, Wang et al. (2018) synthesized NDC catalyst by the pyrolysis of MOF, zeolitic imidazolate framework (ZIF-8) with a nitrogen-rich organic linker under an inert atmosphere at different pyrolysis temperatures (700°C–900°C) followed by the acid washing. Different pyrolysis temperatures were used in order to control the at.% of different N-contents (Figure 6.3). Biomass-derived carbons have also been reported from various sources such as wheat flour (Li et al. 2017) and plant (Typha) (Yao et al. 2019), which resulted in N-doped porous carbon structures. Yao et al. (2019) reported the synthesis of N-doped nanoporous carbon from cheap and renewable biomass (Typha) by hydrothermal treatment followed by calcination in NH_3, which demonstrated high surface area, pore volume, and pyridinic-N content. Moreover, the effect of calcination temperature on porosity and N-content was investigated. Similarly, Li, Wang and Xiao, et al. (2019) derived the nitrogen-doped porous carbon from bituminous coal, and subsequently, the ammonia etching was performed for incorporating N-dopants and generating hierarchical porous structure, which is advantageous for enhancing mass transport of gaseous substrates towards the surface of the electrocatalyst.

The above synthetic strategies allow us to modulate the total N-content, at.% of different N-bonding configurations, morphologies, and porosity in NDCs by changing the experimental conditions such as carbonization temperature and reaction time and by employing different N-precursors. The presence of different N-bonding configurations as easily accessible active sites in NDC structure makes easier for CO_2 molecule to adsorb and activate, which consequently results in enhanced electrocatalytic performance for CO_2RR.

6.3.1 Electrochemical Performance of N-Doped Carbon Catalysts for CO_2RR

The electrochemical performance of NDC catalyst towards CO_2 reduction has been measured by various parameters such as onset potential, FE of the products at different potentials, current density, Tafel slope value, and stability. Therefore, in this

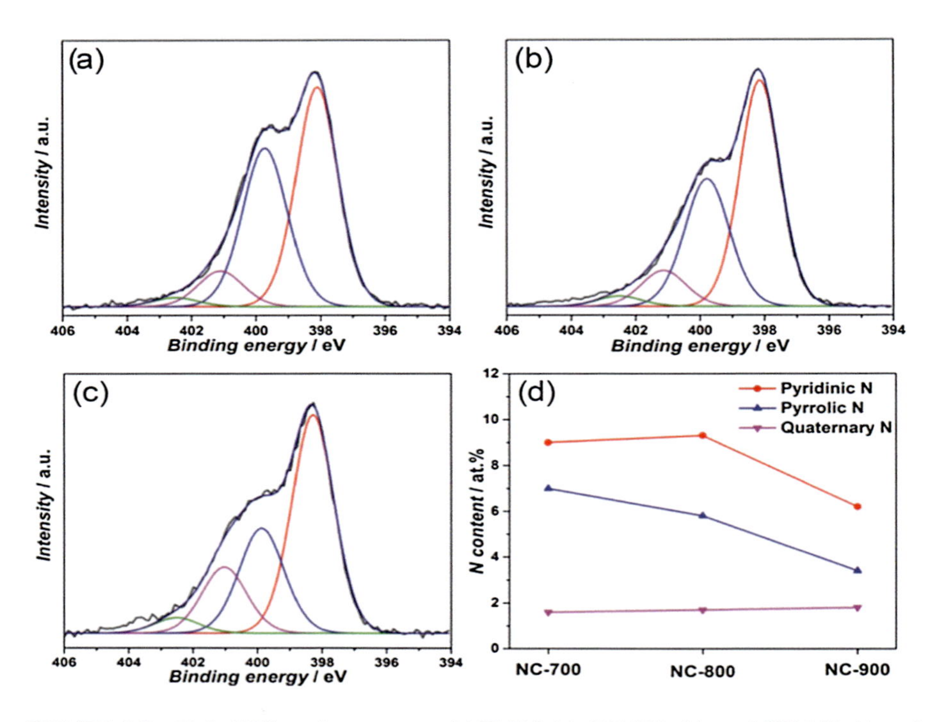

FIGURE 6.3 N 1s XPS region spectra of NC-700 (a), NC-800 (b), and NC-900 (c) and N-species distribution in NC-900 (d). (Reprinted with permission from Wang et al. (2018). Copyright 2018, American Chemical Society.)

section, we have presented a comprehensive and updated literature/example on the electrochemical performance of NDC catalysts for CO_2RR. Table 6.1 summarizes the performance of various NDC materials applied as electrocatalysts for CO_2RR (Wu et al. 2015, Sharma et al. 2015, Chen et al. 2018, Liu, Yang, and Huang, et al. 2018, Ma et al. 2019, Wang et al. 2018, Yao et al. 2019, Kumar et al. 2013, Zhang et al. 2014, Liu, Chen, et al. 2015, Sun et al. 2016, Wu et al. 2016, Cui et al. 2017, Li, Fechler, et al. 2018, Zhu et al. 2019, Hursán et al., 2019, Li, Wang, et al. 2019, Kuang et al. 2019, Li, Xiao, et al. 2018, Wang et al. 2016). Furthermore, the electrochemical performance of some of these NDCs towards CO_2RR has been discussed in detail in chronological order.

In the pioneering work by Kumar et al. (2013), the synthesis of metal-free low-cost N-doped carbon nanofibres (N-CNFs) by pyrolysing heteroatomic polyacrylonitrile (PAN) polymer (N-source) for CO_2RR was reported. At low overpotential (0.17 V *vs.* SHE), N-CNFs reduced CO_2 to CO with a high FE of 98% at −0.573 V *vs.* SHE. Furthermore, N-CNFs exhibited exceptionally high current density compared with bulk Ag (approx. 13 times), 5-nm Ag nanoparticles (4 times), and carbon films (2 times). The high performance of N-CNFs towards CO_2 reduction was attributed to the corrugated surface, enhanced catalytic active sites due to doped-N, and high work function, which leads to the high binding energy of reaction intermediates to the surface of catalyst (N-CNFs). In order to examine the role of the nitrogen atom in

TABLE 6.1
Summary of Electrochemical Performance of Various NDC Catalysts towards CO_2RR

NDC Electrocatalysts	Electrolyte	E_{onset}[a] (V *vs.* RHE)	$J_{major\ product}$[b] (mA cm^{-2})	FE_{max}[c] (Potential (V *vs.* RHE))	Tafel Slope (mV dec^{-1})	Ref(s).
NCNTs	0.1 M $KHCO_3$	−0.30	−1.0 (−0.8 V *vs.* RHE)	CO: 80.0% (−0.78)	203	Wu et al. (2015)
NCNT-ACN-850	0.1 M $KHCO_3$	−0.70	-	CO; 80% (−1.05)	160	Sharma et al. (2015)
N-C61-800	0.5 M $KHCO_3$	−0.42	12.2 (−1.1 V *vs.* RHE)	$HCOO^-$; 91.2% (−0.90)	146	Chen et al. (2018)
N-GRW (GM2)	0.5 M $KHCO_3$	-	-	CO; 87.6 % (−0.40)	129.8	Liu, Yang, et al. (2018)
NCNT-NH_3	0.5 M $NaHCO_3$	−0.40	9.7 (−0.7 V *vs.* RHE)	CO: 96.2% (−0.70)	92	Ma et al. (2019)
NC-900	0.1 M $KHCO_3$	-	−1.1 (−0.93 V *vs.* RHE)	CO; 78 % (−0.93)	-	Wang et al. (2018)
NC-900	0.5 M $NaHCO_3$	−0.30	−0.40 (−0.5 V *vs.* RHE)	CO; 82 % (−0.50)	127	Yao et al. (2019)
N-CNF	[EMIM]BF_4	-	-	CO; 98% (−0.573 V *vs.* SHE)	-	Kumar et al. (2013)
PEI-NCNT	0.1 M $KHCO_3$	-	2.2[d] (−1.8 V *vs.* SCE)	$HCOO^-$; 85% (−1.8 V *vs.* SCE)	134	Zhang et al. 2014
NDD_L/Si RA	0.5 M $NaHCO_3$	−0.36	-	CH_3COO^-; 91.2–91.8 % (−0.8–1.0)	77.1	Liu, Chen, et al. 2015
NGM-1/CP	[Bmim]BF_4, 3 wt% H_2O	-	1.42[e] (−1.4 V *vs.* SHE)	CH_4; 93.5 ± 1.2 % (−1.40 V *vs.* SHE)	-	Sun et al. (2016)
N-GQDs	1 M KOH	−0.26	46 (−0.86 V *vs.* RHE)	C_2H_4; 31 % (−0.75)	198	Wu et al. (2016)
CN-H-CNT	0.1 M $KHCO_3$	−0.20	-	CO; ~88% (−0.50)	124	Cui et al. (2017)
SaU-900	0.1 M $KHCO_3$	−0.37	-	CO; 22% (−0.85)	-	Li, Fechler, et al. (2018)
1D/2D NR/CS-900	0.5 M $KHCO_3$	−0.207	3.78[e] (−0.55 V *vs.* RHE)	CO; 94.2 % (−0.45)	65	Zhu et al. (2019)
NC-27	0.5 M $KHCO_3$	−0.27 ± 0.3	−1.85 (−0.8 V *vs.* RHE)	CO; 76 % (−0.60)	-	Hursán et al. (2019)

(Continued)

TABLE 6.1 (*Continued*)
Summary of Electrochemical Performance of Various NDC Catalysts towards CO_2RR

NDC Electrocatalysts	Electrolyte	E_{onset}[a] (V *vs.* RHE)	$J_{major\ product}$[b] (mA cm^{-2})	FE_{max}[c] (Potential (V *vs.* RHE))	Tafel Slope (mV dec^{-1})	Ref(s).
CNPC-1100	0.1 M $KHCO_3$	-	−1.0 (−0.6 V *vs.* RHE)	CO; 92 % (−0.60)	163	Li, Wang, and Xiao, et al. (2019)
MNC-D	0.1 M $KHCO_3$	−0.28	−6.8 (−0.58 V *vs.* RHE)	CO; ~92% (−0.58)	138	Kuang et al. (2019)
WNCNs-1000	0.1 M $KHCO_3$	−0.40	−1.15 (−0.60 V *vs.* RHE)	CO; ~84% (−0.49)	132	Li, Xiao, and Hao, et al. (2018)
N-graphene	0.5 M $KHCO_3$	−0.30	−7.5 (−0.84 V *vs.* RHE)	$HCOO^-$; 73% (−0.84)	135	Wang et al. (2016)

[a] E_{onset} = onset potential.
[b] $J_{major\ product}$ = partial current density normalized by electrode area at a particular potential.
[c] FE_{max} = maximum Faradaic efficiency of major product.
[d] $J_{major\ product}$ = current density normalized by electroactive surface area.
[e] $J_{major\ product}$ = total current density normalized by electrode area.

the CO_2 reduction mechanism, XPS core-level spectra were recorded before and after 9 h of the electrochemical reaction. N 1s core-level spectrum of N-CNFs revealed the presence of three N-bonding configurations, namely pyridinic-N (398.5 eV), quaternary-N (401.1 eV), and N-oxides (402.2 eV) with the fractions of 25.8%, 36.7%, and 37.5%, respectively. However, N 1s spectrum of N-CNFs after 9 h of electrochemical reaction showed a decrease in the peak area of N-oxides from 37.5% to 10.0% with the appearance of an additional peak corresponding to pyridonic-N (400.1 eV), which clearly indicated the conversion of N-oxides to pyridonic-N. However, the peak areas for pyridinic-N and quaternary-N remained unchanged. These results ruled out the possibility of nitrogen being directly involved in the CO_2 reduction. Therefore, the conclusion was made that the CO_2 reduction was anticipated by the positive charges on the carbon atoms in CNFs due to the presence of electronegative N-atoms.

Later in 2014, Zhang et al. (2014) reported metal-free NCNTs by ammonia plasma treatment, which were further treated with polyethylenimine (PEI) layer (PEI-NCNT). The as-synthesized electrocatalyst was found to be effective for the selective reduction of CO_2 to $HCOO^-$ in aqueous media. The increase in N-content was observed after PEI deposition on NCNT (11.3 at.%) as compared to that of NCNT (7.6 at.%). PEI-NCNT electrode exhibited much higher current density (mA cm^{-2}) and FE for $HCOO^-$ (7.2; 85%) as compared to PEI-CNT (3.8; 8%) and NCNT electrodes (3.0; 59%) at −1.8 V *vs.* saturated calomel electrode (SCE). The PEI film that acted as a co-catalyst was added to NCNT for enhancing the electrocatalytic activity. It is worth noting that the improvement in catalytic activity was only apparent in PEI-NCNT and not for undoped CNT. Based on these experimental results, it was proposed that CO_2 was adsorbed by basic N-binding sites in NCNT and reduced to $CO_2^{\cdot-}$, which was stabilized by PEI layer *via* hydrogen-bonding interactions (NCNT-N-C(O)O$^-$...H-N-PEI), thus resulting in lower E_{onset} values for reducing CO_2 to $CO_2^{\cdot-}$.

Wu et al. (2015) synthesized NCNTs as a facile electrocatalyst for the CO_2 reduction to CO. The as-synthesized NCNTs catalyst exhibited a maximum FE_{CO} of 80% at low overpotential (−0.26 V *vs.* reversible hydrogen electrode (RHE)) (Figure 6.4). The excellent electrocatalytic performance of NCNTs was found due to the enhanced electrical conductivity and the incorporation of pyridinic-N as the most efficient catalytically active site accompanied by pyrrolic and quaternary-N. Moreover, the density functional theory (DFT) study revealed that the formation of the key intermediate (COOH*) lowered the energy barrier for the potential-limiting step. Furthermore, the strong COOH* binding but weak adsorption of CO contributed towards the selectivity of CO.

Subsequently, Sharma et al. (2015) synthesized NCNTs in order to exploit the active centres generated due to defects and to uncover the role of defects towards the selectivity of NCNTs. Different NCNTs were synthesized by ACN, DMF, and TEA at different temperatures (750, 850°C, and 950°C). Among the various synthesized samples, NCNT-ACN-850 exhibited highest N-content (4.9 at.%) and highest electrocatalytic activity with a maximum FE for CO (80% at −1.05 V) and lowest onset potential (−0.70 V) as compared to NCNT-TEA-850 (FE; 25% at −1.05 V; onset potential; −1.0 V) and NCNT-DMF-850 (FE; 40% at −1.05 V; onset potential; −0.8 V). The detailed analyses about N-functionalities demonstrated pyridinic (1.1 at.%) and

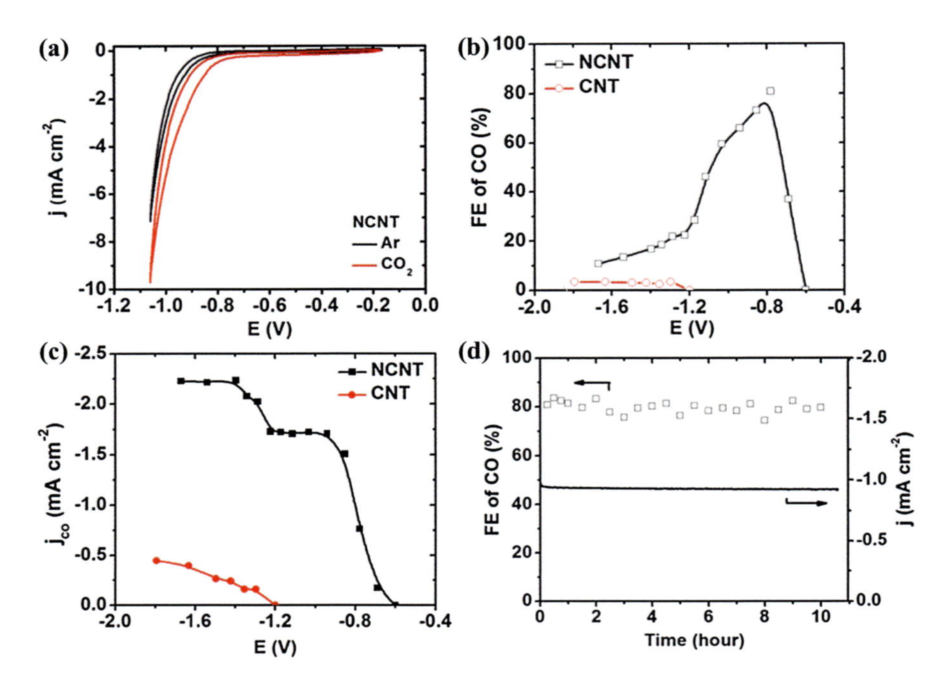

FIGURE 6.4 Cyclic voltammetry curves of NCNTs in Ar- and CO_2-saturated 0.1 M $KHCO_3$ electrolyte at 50 mV s^{-1} (a). Dependence of FE of CO on applied cell potential during electrocatalysis of CO_2 reduction for both NCNT and CNT catalysts (b). Partial current density of CO *vs.* applied cell potential for NCNT and CNT catalysts (c). Stability performance of NCNTs for CO_2 reduction operated at a potentiostatic mode of −0.8 V for 10 h (d). (Reprinted with permission from Wu et al. (2015). Copyright 2015, American Chemical Society.)

graphitic-N (3.5 at.%) to be the highest content as defects, which were accountable for lowest overpotential, and their maximum FE_{CO} was found to be −0.18 V and 80%, respectively, in NCNT-ACN-850 compared to that in pristine CNTs, while the effect of pyrrolic N-moiety on the CO_2 reduction activity was negligible. The experimental observations about the contribution of graphitic-N and pyridinic-N towards the enhancement of catalytic activity were further analysed with the theoretical calculations. DFT calculations exhibited pyridinic-like defects in the graphitic plane having an excessive negative charge and retaining lone pairs that were capable of binding CO_2 and resulted in enhanced catalytic activity. For graphitic-N, the electrons were less available for CO_2 binding as they were situated in the antibonding orbital (π^*). However, for pyrrolic-N due to defects geometry, N-atom moved towards the centre of the tube and made it difficult for the electrons to access for CO_2 binding.

In contrast to the above reports which are selective for the conversion of CO_2 to C1 products, Liu, Chen et al. (2015) reported N-doped nanodiamond/Si rod (NDD_L/Si RA) array as a catalyst for the electrochemical conversion of CO_2 to CH_3COO^- (C2) product using aqueous electrolyte (0.5 M $NaHCO_3$). The as-synthesized electrocatalyst exhibited the E_{onset} of −0.36 V *vs.* RHE having a maximum $FE_{CH_3COO^-}$ of 91.2%–91.8% in −0.8 to −1.0 V (*vs.* RHE) potential range (Figure 6.5). The enhanced

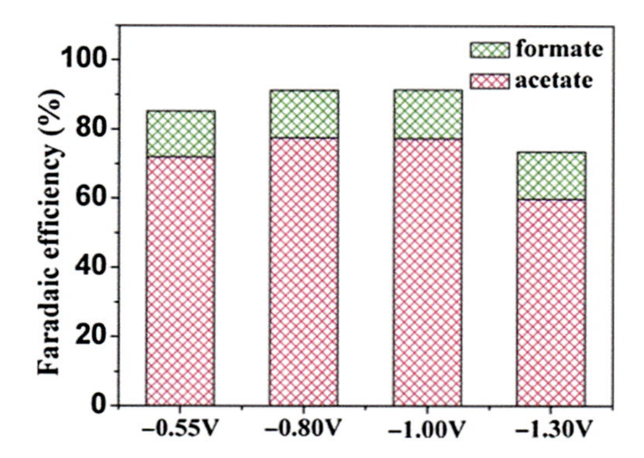

FIGURE 6.5 FE for acetate and formate production for electrochemical reduction of CO_2 on NDD_L/Si RA electrodes at −0.55 to −1.30 V (CO_2-saturated 0.5 M $NaHCO_3$ solution). (Reprinted with permission from Liu, Chen et al. (2015). Copyright 2015, American Chemical Society.)

electrocatalytic activity was attributed to the high enough N-content (N-sp^3C active sites), which induced the defect sites in the carbon structure and polarized the adjacent carbon atoms, thereby accelerating the adsorption of CO_2 molecule.

Sun et al. (2016) reported the electrochemical CO_2 reduction to CH_4 employing N-doped graphene-like carbon materials (NGMs). Different NGMs were prepared by employing different N-precursors, which resulted in varied N-content (3.17–6.52 at.%) measured by XPS. For electrocatalytic measurements, the electrodes were prepared by employing differently synthesized NGMs with carbon papers, and the performance was evaluated in ionic liquids with a trace amount of water content. The catalyst with the highest N-content (NGM-1/CP synthesized using 3-pyridinecarbonitrile) was observed to be highly selective for CH_4 formation with the maximum FE of 93.5%. Moreover, the current density for Cu electrode was found to be lower (six times) under similar experimental conditions. Furthermore, the authors concluded that FE for the formation of CH_4 was highly dependent on the active N-species: as the active N-content increased from 1.8% to 4.8%, FE also exhibited an increase from 20.8% to 93.5%. From the experimental results, the authors inferred that the pyridinic- and pyridonic/pyrrolic-N were the active species that contributed to CO_2 reduction to CH_4. Based on the experimental outcomes, the possible mechanism for the CO_2 electrochemical reduction to C2 product on NGM/CP electrode was proposed by the authors, as has been shown in Figure 6.6. In brief, CO_2 was first adsorbed on to the different N-bonding configurations (pyridinic- and pyridonic/pyrrolic-N), which act as binding sites, and reduced to $CO_2^{\cdot-}$. Subsequently, ionic liquid facilitates the transformation of CO_2 into 2, which upon coupling with CO_2 molecule from solution (Lewis acid) forms 3. Then, the second electron transfer will take place to form CO_{ads} (4), which upon accepting electron and proton gets desorbed/converted into CHO_{ads} (5). In the subsequent final steps (6–8), after accepting additional electrons and protons, CHO_{ads} is converted into CH_4.

Wu et al. (2016) reported N-doped graphene quantum dots (N-GQDs) having thickness and size distribution in the range of 0.7–1.8 nm and 1–3 nm, respectively.

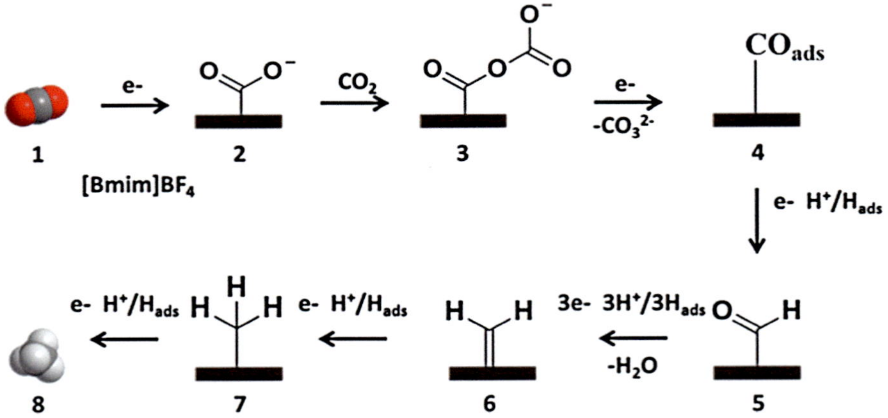

FIGURE 6.6 Schematic diagram for the CO_2 reduction mechanism at NGM/CP electrode. (Reprinted with permission from Sun et al. (2016). Copyright 2016, Royal society of chemistry.)

The as-synthesized N-GQDs exhibited CO and $HCOO^-$ selectivity at a potential of −0.26 V *vs.* RHE. Interestingly, as the more negative potential was applied, C2 and C3 product formations were observed like C_2H_4 and multicarbon oxygenates such as C_2H_5OH, CH_3COO^-, and n-C_3H_7OH along with CH_4, in addition to C1 products (CO and $HCOO^-$) (Figure 6.7). For example, at specific potentials of −0.75 and −0.86 V *vs.* RHE, C_2H_4 and CH_4 were the major products with the highest FE of 31% and 15%, respectively. Moreover, the selectivity of N-GQDs was not only limited to hydrocarbons but also found to be selective for multicarbon oxygenates. For instance, at −0.78 V, the maximum FE was reached to 26% in which 16% was accountable for C_2H_5OH as the major product. It is important to note that the electrocatalytic performance of metal-free N-GQDs was found to be similar to that of metal-based Cu catalyst. In order to understand more about the electrocatalytic activity and product selectivity of the as-synthesized N-GQDs, undoped GQDs and N-doped reduced graphene oxide (N-RGOs) were also synthesized. Undoped GQDs catalysed the reduction of CO_2 at more negative potential having CO and $HCOO^-$ as the major products but with lower FEs than N-GQDs. Moreover, GQDs also exhibited much slower kinetics than N-GQDs, which was indicated by the higher Tafel slope value in case of GQDs (371 mV dec^{-1}) as compared to N-GQDs (198 mV dec^{-1}) (Figure 6.7d). The difference in the catalytic activity was understood due to the presence of N-doping active centres (defects) in N-GQDs. To further understand the detailed role of different N-bonding configurations towards CO_2RR, XPS measurements of N-GQDs were taken before and after the CO_2RR, which revealed a decrease in the relative concentration of pyridinic-N from 65% to 38% and an increase in the relative concentration of pyrrolic/pyridonic-N from 20% to 50%. On the other hand, the relative concentration of quaternary-N remained almost unchanged (15% to 12%). These results suggested the adsorption of CO_2 molecule on pyridinic-N-species because the binding energy value of pyridinic-N was similar to that of pyrrolic-N. However, in case

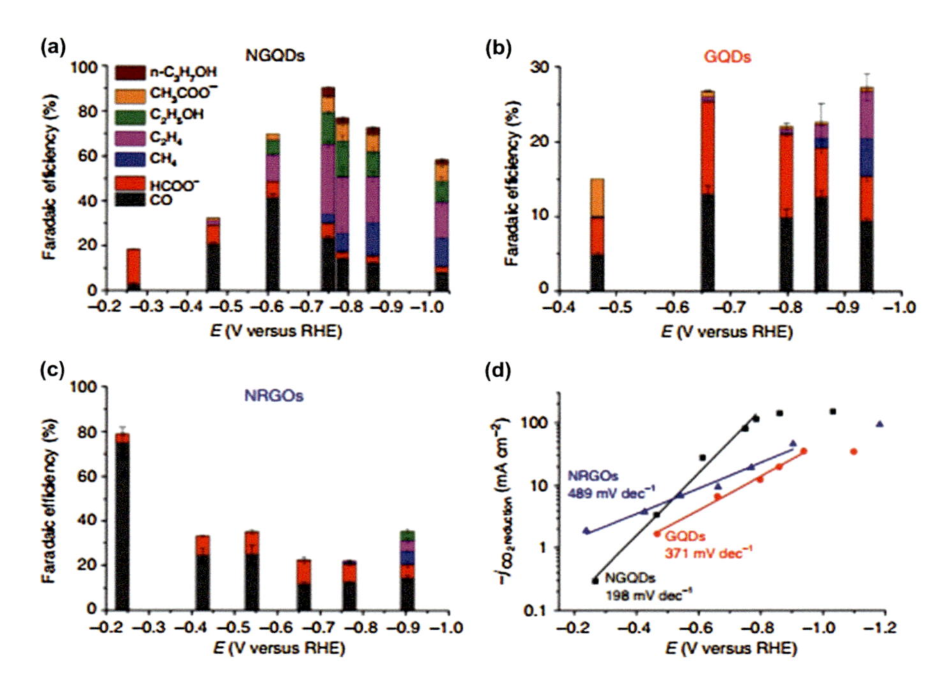

FIGURE 6.7 Electrocatalytic activity of carbon nanostructures towards CO_2 reduction. FE of carbon monoxide (CO), methane (CH_4), ethylene (C_2H_4), formate ($HCOO^-$), ethanol (EtOH), acetate (AcO^-), and n-propanol (n-PrOH) at various applied cathodic potentials for N-GQDs (a). FE of CO_2 reduction products for pristine GQDs (b). Selectivity to CO_2 reduction products for N-RGOs (c). Tafel plots of partial current density of CO_2 reduction *vs.* applied cathodic potential for three nanostructured carbon catalysts (d). The error bars represent the s.d. of three separate measurements for an electrode (Wu et al. (2016). (*This work is licensed under a Creative Commons Attributions 4.0 International License.*)

of micrometre-scale lateral size N-RGOs, the relative content of each N-bonding configuration was found to be similar to that of N-GQDs. Moreover, the N-RGOs also exhibited a similar onset potential to N-GQDs but with the major product as CO. However, the formation of other products occurred at much negative potential (–0.90 V) compared to that of N-GQDs (–0.61 V). The significant difference in the electrocatalytic activity clearly revealed the role of morphology in addition to that of nitrogen doping. Although in both the cases the presence of N-doping active sites (pyridinic-N) was found to be similar, the different location made the difference in the electrocatalytic activity. In the case of nanometre-sized N-GQDs, most of the N-doping active sites (pyridinic-N) were located at the edge plane instead of the basal plane, which are different from those of N-RGOs where most of the N-dopants were located at the basal planes. It was believed that the pyridinic-N located at the edge sites was more active in catalysing C–C bond formation than that located at the basal plane, which results in the formation of C2 and C3 products on N-GQDs.

As we have discussed (*vide supra*), most of the previous researches on NDCs adopted the change in carbonization temperature to adjust the N-doping level and active sites. In contrast to those reports, Cui et al. (2017) adopted a unique steam etching strategy in order to tune the N-doping level and different N-bonding configuration in the as-synthesized N-doped carbon-wrapped CNTs (CN-CNT). It was ascertained that the binding site of water molecules in the NDC framework was around graphitic-N and pyrrolic-N. After selective steam etching, pyrrolic-N was retained with an increased relative concentration from 22.1% to 55.9%, which was highest among other N-functionalities. The as-prepared steam-etched catalyst (CN-H-CNT) was selective for the conversion of CO_2 to CO with the maximum FE of ~88% at −0.5 V *vs.* RHE. In a control experiment, the steam-treated carbon catalyst (H-CNT) was also prepared, which exhibited a negligible electrocatalytic activity towards CO_2RR, clearly suggesting the role of N-doping towards CO_2 RR. Moreover, the onset potential of CN-H-CNTs was found to be more negative compared to that of CN-CNTs, which clearly revealed that the presence of highest pyrrolic-N content in CN-H-CNTs resulted in higher catalytic activity for CO_2 reduction with a maximum selectivity for CO. Moreover, the faster reaction kinetics for CO formation in CN-H-CNTs was revealed by determining the Tafel slope values, which were found to be in the order of H-CNTs (534 mV dec^{-1}) > CN-CNTs (230 mV dec^{-1}) > CN-H-CNTs (124 mV dec^{-1}).

Wu et al. (2016) demonstrated that the change in morphology (N-GQDs) resulted in the selective generation of C2 and C3 products with high FE. In contrast to that, Liu, Yang, and Huang, et al. (2018) synthesized 3D N-doped graphene nanoribbon network (N-GRW) by employing varied ratios of melamine to L-cysteine. The resulted N-GRW exhibited tunable N-content and N-doping configurations, which were found to be selective for the formation of C1 product (CO). Among all the synthesized N-GRW catalysts, GM2 (melamine-to-L-cysteine ratio (2:1)) exhibited the highest catalytic activity with a maximum FE_{CO} of 87.6% at −0.40 V (*vs.* RHE). GM2 catalyst was measured to be 1 nm thick (~ 3 layers of graphene) by atomic force microscopy. Based on the XPS analysis, it was revealed that among the various N-functionalities, pyridinic-N was the main active site, which was directly responsible for CO_2RR. These results were further supported by the theoretical calculations.

Nevertheless, unlike other authors who suggested the pyridinic-N as the main catalytic site for the electrochemical reduction of CO_2, Wang et al. (2018) demonstrated the role of quaternary-N and pyridinic-N as the main active sites for CO_2 reduction to CO employing NDC, which was synthesized by the pyrolysis of MOF (ZIF-8) followed by the acidic treatment. The maximum FE for CO was found to be dependent on pyrolysis temperature. NDC catalyst pyrolysed at 900°C (NC-900) exhibited highest FE_{CO} (78%) with a stable current density of −1.1 mA cm^{-2} at −0.93 V *vs.* RHE, which is much larger than undoped carbon catalyst (C-900) under similar experimental conditions (FE_{CO} (2%) and current density (−0.6 mA cm^{-2})) (Figure 6.8). The study significantly demonstrated the role of active sites (pyridinic-N and quaternary-N) and BET surface area in electrochemical CO_2 reduction. The presence of active sites lowered the energy barrier of intermediate (COOH*) to CO formation by accelerating the transfer of proton–electron pair to CO_2.

So far, we have discussed the role of pyridinic-N and quaternary-N as the main catalytically active sites for the electrochemical CO_2 reduction. In contrast to that, Li, Fechler et al. (2018) have explored the possibility of another N-bonding configuration

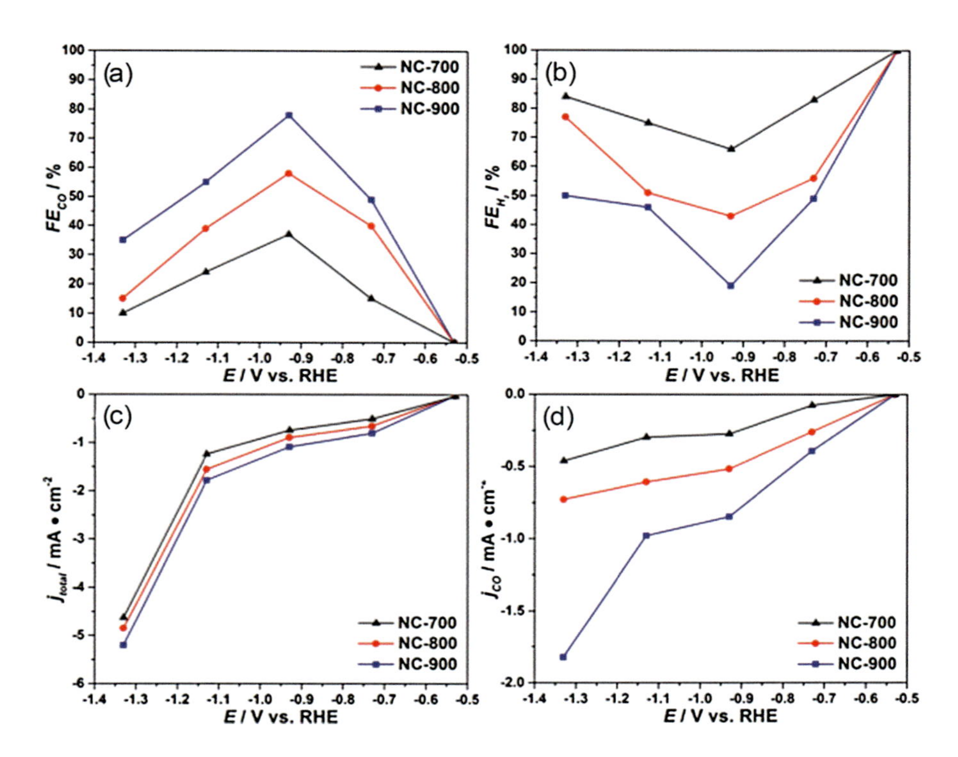

FIGURE 6.8 FE to CO (a), FE to H_2 (b), the total current density (c), and the partial CO current density (d). (Reprinted with permission from Wang et al. (2018). Copyright 2018, American Chemical Society.)

as an active site, i.e. pyrazinic-N towards CO_2RR in the as-synthesized microporous carbons. The pyrazinic-N structure is very much alike to the pyridinic-N structure except it contains two nitrogen atoms in the ring, which may affect the CO_2 reduction mechanism due to the closeness of N-bonds in the ring. However, there has not been much detail in the literature about the role of this nitrogen configuration towards CO_2RR. Moreover, it is very hard to discriminate between pyridinic-N and pyrazinic-N due to the overlapped binding energies. Therefore, more results are needed in order to clearly reveal the role of pyrazinic-N as active site towards CO_2RR.

Recently, Zhu et al. (2019) synthesized a highly efficient NDC electrocatalyst (1D/2D NR/CS-X) by *in situ* aniline polymerization on ultrathin carbon nanosheets (CSs). CS was obtained from the thermal expansion of calcium gluconate and acid treatment, which was subsequently pyrolysed at different temperatures (800°C–1000°C). Among all the synthesized catalysts, 1D/2D NR/CS-900 electrocatalyst exhibited relatively large specific surface area (SSA) (859.97 m^2g^{-1}) and N-content (5.43 at.%), which converted CO_2 to CO with the maximum FE of 94.2% at −0.45 V *vs.* RHE, and demonstrated outstanding stability for up to 30 h (Figure 6.9). This catalyst also exhibited a relatively high current density of 3.78 mA cm^{-2} compared to other prepared catalysts at 800°C (2.96 mA cm^{-2}) and 1000°C

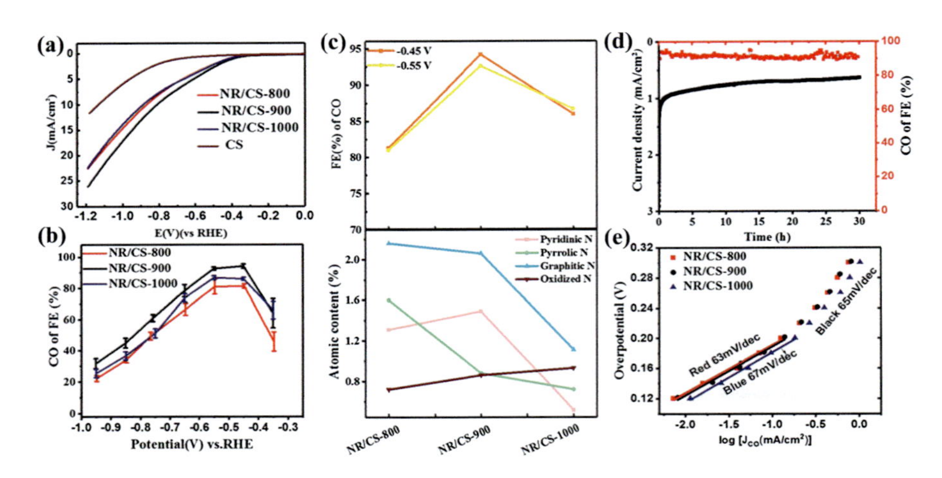

FIGURE 6.9 Linear sweep voltammetry curves of 1D/2D NR/ CS-X and CS at a scanning rate of 1 mV s^{-1} (a). Potential-dependent FE of CO formation for CO_2 RR on 1D/2D NR/ CS-X (b). Maximum FE for CO formation at −0.45 and −0.55 V *vs*. N-species content (c). Stability test on 1D/2D/NR/CS-900 electrode at −0.45 V *vs*. RHE (d). Tafel plots of the 1D/2D NR/ CS-X (e). All electrochemical measurements were taken in CO_2^--saturated aqueous 0.5 M $KHCO_3$ solution. (Reprinted with permission from Zhu et al. (2019). Copyright 2019, Royal society of chemistry.)

(3.08 mA cm^{-2}) and CS (0.59 mA cm^{-2}) at −0.55 V *vs*. RHE. Further, the faster kinetics of the as-synthesized catalyst at 900°C was manifested by the low Tafel slope value (65 mV dec^{-1}). The improved electrocatalytic performance of 1D/2D NR/ CS-900 was ascribed to the N-doping, large SSA, and unique nanostructure. More specifically, the FE for CO formation was related to pyridinic-N content, whereas other N-bonding configurations did not show any correlation. From the experimental outcomes and DFT simulations, it was confirmed that the pyridinic-N was the main active site for the CO_2RR, similar to other reports (Sharma et al. 2015, Liu, Yang, and Huang, et al. 2018).

Hursán et al. (2019) recently explored the role of morphological attributes towards electrochemical CO_2 reduction performance on NDC, unlike other authors who demonstrated the role of N-doping. NDC structures were obtained from hard templating method with uniform meso- and macropores having different pore sizes. The as-synthesized different NDC electrocatalysts exhibited different pore sizes of 13, 27, and 90 nm. NDC catalyst having a pore size of 27 nm (NC-27) demonstrated the best electrochemical performance for CO_2 reduction in terms of catalytic activity and product selectivity with the maximum FE of 76% for the formation of CO, which was achieved at −0.6 V *vs*. RHE. However, NC-27 catalyst did not show the highest content of pyridinic-N and graphitic-N, which was demonstrated to be the main active sites for the CO_2RR by other authors (Sharma et al. 2015, Liu, Yang, and Huang, et al. 2018, Zhu et al. 2019). Therefore, it was suggested that the electrocatalytic activity of NC-27 catalyst was found to be originated due to different morphological attributes.

Recently, Yao et al. (2019) reported N-doped nanoporous carbon sheets (NC-900) from low cost and renewable biomass (Typha) with SSA (910 $m^2 g^{-1}$), pore volume (0.615 $cm^3 g^{-1}$), and pyridinic-N content (1.37%). Calcination temperature was found to be significant in modulating the N-species and porosity of NDCs. Among all the synthesized catalysts, NC-900 exhibited the maximum FE_{CO} of 82% at −0.50 V *vs.* RHE. The improved electrocatalytic activity was attributable to the large pyridinic-N and porous structure of NC-900. Porous structure was instrumental for improving the CO_2 concentration at the reaction interface by accelerating the CO_2 capture. However, the high pyridinic-N content lowered the free energy barrier of intermediate (COOH*), which consequently resulted in improved activity and selectivity.

We have reviewed a variety of NDC materials that have been investigated for the electrochemical reduction of CO_2. The significant effects of total N-content, different N-bonding configurations, surface area, and pore volume have been discussed, which are conducive for modulating electrocatalytic activity and controlling product selectivity. Nonetheless, the morphology of NDC materials is also of great importance, which can result in the conversion of CO_2 to C2 products and even higher hydrocarbons and oxygenates. The pursuit of such kind of metal-free carbon electrocatalysts with high selectivity and activity is highly desirable for future research directions.

6.4 CONCLUSIONS AND OUTLOOK

The electrochemical conversion of CO_2 into low-cost value-added carbon products is a promising way to bring down CO_2 level released into the atmosphere due to excessive combustion of fossil fuels. Regarding this, heteroatom-doped carbonaceous materials have drawn significant attention as electrocatalysts for CO_2RR. Heteroatom dopants in the carbon network break the electroneutrality of the pristine carbon and introduce the defects and catalytically active sites for CO_2 molecule adsorption and key intermediate generation. Specifically, N-doping in the carbon framework is well received due to comparable size with carbon atoms and larger electronegativity. Generally, NDCs have been synthesized by employing *in situ* and post-synthesis treatment for CO_2RR. NDC materials exhibit excellent electrochemical performances such as enhanced current density, excellent selectivity with high FEs, and remarkable stability owing to the presence of different N-bonding configurations. Besides this, the electrochemical performance of NDC electrocatalysts for CO_2RR significantly depends on the type of the N-bonding configuration in carbon structure as different N-bonding configurations have different ability to bind with CO_2 molecule. In most of the researches, pyridinic-N has been considered as the main catalytically active site for CO_2 adsorption and the generation of key intermediates. However, the exact identification/role of active centre in NDC materials, which is involved in enhancing the electrocatalytic activity towards CO_2RR, is still a matter of debate. Additionally, optimum N-content, choice of N-doping precursor, carbon precursor, and thermal annealing conditions are the key considerations for achieving superior NDC electrocatalysts for CO_2RR.

The electrochemical CO_2 reduction on NDCs results in several different reduction products such as carbon monoxide, formate, ethanol, methanol, ethylene, acetate, and C2 and C3 oxygenates. The selectivity of the products significantly relies on the physicochemical properties of NDC catalyst, such as morphology, SSA, pore size, elemental composition, and the presence of defects/active sites. The formation of several different products involves complicated reaction steps and single electron transfer for intermediate generation (CO_2^-), which is a rate-limiting step. However, the detailed reaction mechanism for CO_2RR has not been understood completely yet and needs to be explored further.

Nonetheless, much efforts have been made, and therefore, significant results have been obtained for electrochemical CO_2RR by employing NDCs as electrocatalysts. Interestingly, the electrocatalytic activity of NDC materials is significantly higher than that of metal-based electrocatalysts and undoped carbon materials. However, there is a lot more which is yet to be achieved in view of electrochemical CO_2RR. In view of this, it is needed to synthesize low-cost novel NDC materials with optimum N-content having a large number of active sites. Moreover, the focus should also be made on modulating the surface properties of NDC electrocatalysts, which is beneficial for enhancing the adsorption of CO_2 and the generation of a key intermediate for CO_2RR. Another important aspect is clarification about the reaction mechanism for CO_2RR by employing NDCs which we may get by characterizing the material employing *in situ*/operando techniques. Furthermore, the experimental results should be supported by the theoretical calculations in order to provide in-depth information on electronic structures, active sites, energy barrier, and reaction pathway of CO_2RR. Last but not the least, co-doped carbon materials like the combination of N and S or N and B should also be investigated to great extent, which are expected to show an enhanced electrochemical performance towards CO_2RR due to their synergistic effect between each heteroatom.

REFERENCES

Benson, E. E., Kubaik, C. P., Sathuram, A. J., and Smieja, J. M. 2009. Electrocatalytic and homogeneous approaches to conversion of CO_2 to liquid fuels. *Chem. Soc. Rev.* 38: 89–99.

Canadell, J. G., Quéré C. L., and Raupach, M. R., et al. 2007. Contributions to accelerating atmospheric CO_2 growth from economic activity, carbon intensity, and efficiency of natural skins. *PNAS* 104:18866–70.

Chen, Z., Mou, K., Yao, S., and Liu, L. 2018. Highly selective electrochemical reduction of CO_2 to formate on metal-free nitrogen-doped PC61BM. *J. Mater. Chem. A* 6:11236–43.

Chu, S., and Majumdar, A. 2012. Opportunities and challenges for a sustainable energy future. *Nature* 488: 294–303.

Costentin, C., Robert. M., and Savéant, J.-M. 2013. Catalysis of the electrochemical reduction of carbon dioxide. *Chem. Soc. Rev.* 42:2423:36.

Cui, X., Pan, Z., Zhang, L., Peng, H., and Zheng, G. 2017. Selective etching of nitrogen-doped carbon by steam for enhanced electrochemical CO_2 reduction. *Adv. Energy Mater.* 7:1701456.

Duan, X., Xu, J. and Wei, Z., et al. 2017. Metal-free carbon materials for CO_2 electrochemical reduction. *Adv. Mater.* 29:1701784.

Feng, D.-M., Zhu, Y.-P., Chen, P., and Ma, T.-Y. 2017. Recent advances in transition-metal-mediated electrocatalytic CO_2 reduction: From homogeneous to heterogeneous systems. *Catalysts* 7:373.

Fernandes, D. M., Peixoto, A.F., and Freire, C. 2019. Nitrogen-doped metal-free carbon catalysts for (electro)chemical CO_2 conversion and valorization. *Dalton Trans*. 48: 13508–28.

Gao, T., Xie, T., and Han, N., et al. 2019. Electronic structure engineering of 2D carbon nanosheets by evolutionary nitrogen modulation for synergizing CO_2 electroreduction. *ACS Appl. Energy Mater.* 2:3151–59.

Gao, K., Wang, B., and Tao, L., et al. 2019. Efficient metal-free electrocatalysts from N-doped carbon nanomaterials: Mono-doping and co-doping. *Adv. Mater.* 31:1805121.

Hori, Y., Wakebe, H., Tsukamoto, T., and Koga, O. 1994. Electrocatalytic process of CO selectivity in electrochemical reduction of CO_2 at metal electrodes in aqueous media. *Electrochim. Acta*. 39:1833–39.

Hou, L., Yan, J., Takele, L., Wang, Y., Yan, X., and Gao, Y. 2019. Current progress of metallic and carbon-based nanostructure catalysts towards the electrochemical reduction of CO_2. *Inorg. Chem. Front*. 6:3363–80.

Hursán, D., Samu, A. A., and Janovák, L., et al. 2019. Morphological attributes govern carbon dioxide reduction on N-doped carbon electrodes. *Joule* 3:1719–33.

Jia, C., Dastafkan, K., Ren, W., Yang, W., and Zhao, C. 2019. Carbon-based catalysts for electrochemical CO_2 reduction. *Sustain Energy Fuels* 3:2890–906.

Jiao, Y., Zheng, Y., Jaroniec, M., and Qiao, S. Z. 2015. Design of electrocatalysts for oxygen- and hydrogen-involving energy conversion reactions. *Chem. Soc. Rev.* 44, 2060–86.

Jurewicz, K., Babeł, K., Źiółkowski, A., and Wachowska, H. 2003. Ammoxidation of active carbons for improvement of supercapacitor characteristics. *Electrochim. Acta* 48: 1491–98.

Khandelwal, M., Chandrasekaran, S., Hur, S. H., and Chung, J. S. 2018. Chemically controlled *in-situ* growth of cobalt oxide microspheres on N, S-co-doped reduced graphene oxide as an efficient electrocatalyst for oxygen reduction reaction. *J. Power Sources* 407:70–83.

Khandelwal, M., Li, Y., Hur, S. H., and Chung, J. S. 2018. Surface modification of co-doped reduced graphene oxide through alkanolamine functionalization for enhanced electrochemical performance. *New J. Chem*. 42:1105–114.

Khandelwal, M., Li, Y., Molla, A., Hur, S. H., and Chung, J. S. 2017. Single precursor mediated one-step synthesis of ternary-doped and functionalized reduced graphene oxide by pH tuning for energy storage applications. *Chem. Eng. J.* 330:965–78.

Kondratenko, E. V., Mul, G., Baltrusaitis, J., Larrazábal, G.O., and Pérez-Ramírez, J. 2013. Status and perspectives of CO_2 conversion into fuels and chemicals by catalytic, photocatalytic and electrocatalytic processes. *Energy Environ. Sci*. 6:3112–35.

Kuang, M., Guan, A., Gu, Z., Han, P., Qian, L., and Zheng, G. 2019. Enhanced N-doping in mesoporous carbon for efficient electrocatalytic CO_2 conversion. *Nano Res*. 12:2324–29.

Kuang, M., and Zheng, G. 2016. Nanostructured bifunctional redox electrocatalysts. *Small* 12:5656–75.

Kumar, B., Asadi, M., and Pisasale, D., et al. 2013. Renewable and metal-free carbon nanofibre catalysts for carbon dioxide reduction. *Nat. Commun*. 4:2819.

Li, W., Fechler, N., and Bandosz, T. J. 2018. Chemically heterogeneous nitrogen sites of various reactivity in porous carbons provide high stability of CO_2 electroreduction catalysts. *Appl. Catal. B: Environ*. 234:1–9.

Li, H., and Oloman, C. 2005. The electro-reduction of carbon dioxide in a continuous reactor. *J. Appl. Electrochem*. 35:955–65.

Li, C., Wang, Y., and Xiao, N., et al. 2019. Nitrogen-doped porous carbon from coal for highly efficiency CO_2 electrocatalytic reduction. *Carbon* 151:46–52.

Li, H., Xiao, N. and Hao, M., et al. 2018. Efficient CO_2 electroreduction over pyridinic-N active sites highly exposed on wrinkled porous carbon nanosheets. *Chem. Eng. J.* 351:613–21.

Li, H., Xiao, N., and Wang, Y., et al. 2019. Nitrogen-doped tubular carbon foam electrodes for efficient electroreduction of CO_2 to syngas with potential-independent CO/H_2 ratios. *J. Mater. Chem. A* 7:18852–60.

Li, F., Xue, M., Knowles, G. P., Chen, L., MacFarlane, D. R., and Zhang, J. 2017. Porous nitrogen-doped carbon derived from biomass for electrocatalytic reduction of CO_2 to CO. *Electrochim. Acta* 245:561–68.

Lim, R. J., Xie, M., and Sk, M. A. et al. 2014. A review on the electrochemical reduction of CO_2 in fuel cells, metal electrodes and molecular catalysts. *Catal. Today* 233:169–80.

Liu, T., Ali, S., Lian, Z., Li, B., and Su, D. S. 2017. CO_2 electroreduction reaction on heteroatom-doped carbon cathode materials. *J. Mater. Chem. A* 5:21596–603.

Liu, T., Ali, S., Lian, Z., Si, C., Su, D. S., and Li, B. 2018. Phosphorous-doped onion-like carbon for CO_2 electrochemical reduction: The decisive role of the bonding configuration of phosphorous. *J. Mater. Chem. A* 6:19998–04.

Liu, Y., Chen, S., Quan, X., and Yu, H. 2015. Efficient electrochemical reduction of carbon dioxide to acetate on nitrogen-doped nanodiamond. *J. Am. Chem. Soc.* 137:11631–36.

Liu, J., Guo, C., Vasileff, A., and Qiao, S. 2017. Nanostructured 2D materials: Prospective catalysts for electrochemical CO_2 reduction. *Small Methods* 1:1600006

Liu, X., Li, L., Zhou, W., Zhou, Y., Niu, W., and Chen, S. 2015. High-performance electrocatalysts for oxygen reduction based on nitrogen-doped porous carbon from hydrothermal treatment of glucose and dicyandiamde. *ChemElectroChem* 2:803–10.

Liu, S., Yang, H., and Huang, X., et al. 2018. Identifying active sites of nitrogen-doped carbon materials for the CO_2 reduction reaction. *Adv. Funct. Mater.* 28:1800499.

Ma, C., Hou, P., Wang, X., Wang, Z., Li, W., and Kang, P. 2019. Carbon nanotubes with rich pyridinic nitrogen for gas phase CO_2 electroreduction. *Appl. Catal. B: Environ.* 250: 347–54.

Mao, X., and Hatton, T. A. 2015. Recent advances in electrocatalytic reduction of carbon dioxide using metal-free catalysts. *Ind. Eng. Chem. Res.* 54:4033–42.

Schneider. J., Jia, H., Muckerman, J.T., and Fujita, E. 2012. Thermodynamics and kinetics of CO_2, CO, and H^+ binding to the metal centre of CO_2 reduction catalysts. *Chem. Soc. Rev.* 41:2036–51.

Sharma, T., Sharma, S., Kamyab, H., and Kumar, A. 2020. Energizing the CO_2 utilization by chemo-enzymatic approaches and potentiality of carbonic anhydrases: A review. *J. Clean. Prod.* 247: 119138.

Sharma, P. P., Wu, J., and Yadav, R. M., et al. 2015. Nitrogen-doped carbon nanotube arrays for high-efficiency electrochemical reduction of CO_2: On the understanding of defects, defect density, and selectivity. *Angew. Chem. Int. Ed.* 54:13701–05.

Sreekanth, N., Nazrulla, M. A., Vineesh, T. V., Sailaja, K., and Phani, K. L. 2015. Metal-free boron-doped graphene for selective electroreduction of carbon dioxide to formic acid/ formate. *Chem. Commun.* 51:16061–64.

Sun, Z., Ma, T., Tao, H., Fan, Q., and Han, B. 2017. Fundamentals and Challenges of electrochemical CO_2 reduction using two-dimensional materials. *Chem* 3:560–87.

Sun, X., Kang, X., and Zhu, Q., et al. 2016. Very highly efficient reduction of CO_2 to CH_4 using metal-free N-doped carbon electrodes. *Chem. Sci.* 7:2883–87.

Vasileff, A., Zheng, Y., and Qiao, S. Z. 2017. Carbon solving carbon's problems: Recent progress of nanostructured carbon-based catalysts for the electrochemical reduction of CO_2. *Adv. Energy Mater.* 7:1700759.

Wang, H, Chen, Y., Hou, X., Ma, C., and Tan, T., 2016. Nitrogen-doped graphenes as efficient electrocatalysts for the selective reduction of carbon dioxide to formate in aqueous solution. *Green Chem.* 18:3250–56.

Wang, H., Maiyalagan, T., and Wang, X. 2012. Review on recent progress in nitrogen-doped graphene: Synthesis, characterization and its potential applications. *ACS Catal.* 2: 781–94.

Wang, R., Sun, X., and Ould-Chikh, S., et al. 2018. Metal-organic-framework-mediated nitrogen-doped carbon for CO_2 electrochemical reduction. *ACS Appl. Mater. Interfaces* 10:14751–58.

Wang, X., Zhao, Q., and Yang, B., et al. 2019. Emerging nanostructured carbon-based non-precious metal electrocatalysts for selective electrochemical CO_2 reduction to CO. *J. Mater. Chem. A* 7:25191–202.

Wanninayake, N., Ai, Q., and Zhou, R., et al. 2020. Understanding the effect of host structure of nitrogen doped ultrananocrystalline diamond electrode on electrochemical carbon dioxide reduction. *Carbon* 157: 408–19.

Wu, J., Huang, Y., Ye, W., and Li, Y. 2017. CO_2 reduction: From the electrochemical to photochemical approach. *Adv. Sci.* 4:1700194.

Wu, J., Ma, S., and Sun, J., et al. 2016. A metal-free electrocatalyst for carbon dioxide reduction to multi-carbon hydrocarbons and oxygenates. *Nat. Commun.* 7:13869.

Wu, J., Yadav, R. M., and Liu, M., et al. 2015. Achieving highly efficient, selective, and stable CO_2 reduction on nitrogen-doped carbon nanotubes. *ACS Nano* 9:5364–71.

Xie, J., Zhao, X., Wu, M., Li, Q., Wang, Y., and Yao, J. 2018. Metal-free fluorine-doped carbon electrocatalyst for CO_2 reduction outcompeting hydrogen evolution. *Angew. Chem. Int. Ed.* 57:9640–44.

Yao, P., Qiu, Y., Zhang, T., Su, P., Li, X., and Zhang, H. 2019. N-doped nanoporous carbon from biomass as a highly efficient electrocatalyst for the CO_2 reduction reaction. *ACS sustainable Chem. Eng.* 7:5249–55.

Zhang, S., Kang, P., and Ubnoske, S., et al. 2014. Polyethylenimine-enhanced electrocatalytic reduction of CO_2 to formate at nitrogen-doped carbon nanomaterials. *J. Am. Chem. Soc.* 136:7845–48.

Zhu, Y., Lv, K., Wang, X., Yang, H., Xiao, G., and Zhu, Y. 2019. 1D/2D nitrogen-doped carbon nanorod arrays/ultrathin carbon nanosheets: Outstanding catalysts for the highly efficient electroreduction of CO_2 to CO. *J. Mater. Chem. A* 7:14895–903.

7 Role of Nanotechnology in Conversion of CO_2 into Industrial Products

Ramya Thangamani, Lakshmanaperumal Vidhya, and Sunita Varjani

CONTENTS

7.1 INTRODUCTION

Conversion of waste CO_2 into useful products with minimum non-renewable energy is a heavy challenge since the process needs to be economically competitive and reduces the maximum amount of greenhouse gas (GHG) emissions by making use of the existing technology. Carbon dioxide contains very strong bonds that are not reactive, and to break the bonds, a large amount of input energy is required. So, it is not a simple process to convert the carbon dioxide into other fuels as well as chemicals. The technological limitations are found and emphasized to have a better understanding of the existing state of research in the synthesis of fuels and new chemicals using CO_2 as a main source of materials. Patricio et al. (2017) mentioned that the commercial conversion of CO_2 into chemicals is a limited process (Quadrelli et al. 2011, Topham et al. 2014). The exclusive example of using CO_2 in the synthesis of chemicals is urea, produced in 1922 (Artz et al. 2018). The methane gas is released when ammonia is combined with CO_2 during the production of urea. To synthesize ammonia, hydrogen gas is generally required. In order to balance this process, electrolysis technique is used.

In the 1950s, various stakeholders strategized their business unions to produce cyclic carbonates from carbon dioxide. To be specific, when the carbon dioxide is treated with propylene oxide or ethylene in the presence of essential impetus, it produces propylene carbonate or ethylene carbonate, respectively. Chemicals such as styrene oxide, cyclohexene oxide, and 1,3-propylene oxide can also be utilized along with carbon dioxide in the production of cyclic carbonates, while the produce will be only less in volume. In 2010, the carbon dioxide was considered as a base and was used to produce 80,000 tons of cyclic carbonates (Alper and Yuksel-Orhan, 2017). The carbon dioxide is also used as a feedstock to amalgamate the aliphatic and fragrant polycarbonates. In Asahi Kasei process, the yearly production of polycarbonate is 600,000 tons, which is used as a feedstock in addition to carbon dioxide, bisphenol, and ethylene oxide (Fukuoka et al. 2007). According to Langanke et al. (2014), Covestro constructed a plant to copolymerize the carbon dioxide and propylene oxide in order to yield polymeric polyols (i.e. polyether carbonates), which are commercially called as cardyon. Having been found in froth beddings, these polyols can be used in the production of polyurethanes. The alkali generation produces the carbon dioxide. Approximately 5,000 tons of polymeric polyols is manufactured from this plant every year.

The major advantage is the achievement of 99% efficiency in the conversion of CO_2 and water into chemical compounds that have one, two, three, or four carbon atoms. Being a highly efficient process, there is no need of much electricity too. The carbon compounds can otherwise be utilized as building blocks in the process of making useful materials.

7.2 DEVELOPMENT OF TECHNOLOGIES IN PRODUCT-BASED CONVERSION OF CO_2 TO FUELS

7.2.1 Production of Dimethyl Ether

The conversion and usage of natural gas like carbon dioxide are important alternatives for the new energy production source. In recent times, the methane retrieved from natural gas is converted into high-valued hydrocarbon products. In general, three procedures are considered for the production of natural gas, namely steam reforming, carbon dioxide reforming, and semi-oxidation of methane production (Cho, Baek, Kim et al. 2002, Bradford et al. 1999, Rostrup-Nielsen et al. 1993). Figure 7.1 represents the conversion of carbon dioxide into fuel. Methanol is directly derived from the hydrogenation of carbon dioxide because it can be directly used as a fuel. However, the other intermediates obtained during the process can be used to produce advanced fuels (Arena et al. 2014, Olah et al. 2009, Quadrelli et al. 2011). Figure 7.1 shows the bioenergy conversion of fuels.

During the drying process of methanol, dimethyl ether (DME) can be obtained. Being a stage compound, DME can be used in place of LPG as a spotless fuel. At present, DME is merged either through the drying process of methanol or via syngas.

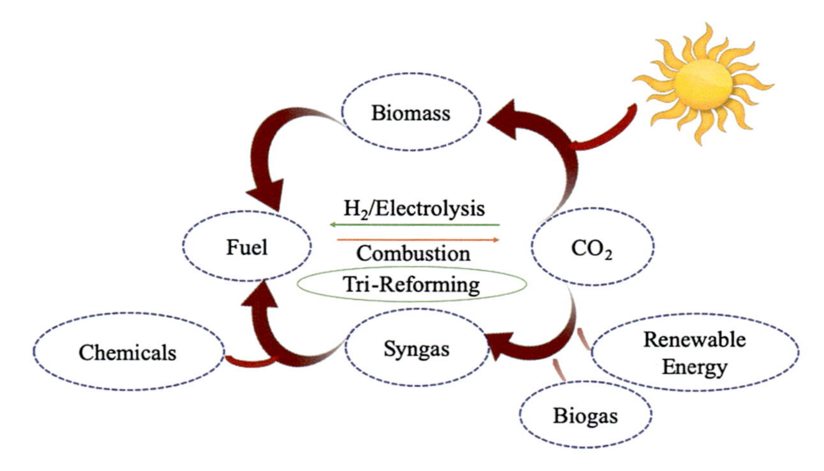

FIGURE 7.1 Bioenergy conversion into fuels.

It is interesting to note that the CO_2 gets immediately converted to DME for comparable motivations. An aberrant technique is deployed by the Korea Gas Corporation to combine DME and CO_2 in large quantities, i.e. 100 tons on a daily basis. This process is inclusive of tri-transforming of the methane, CO_2, and water into syngas, which is then transformed into DME. Although the course can be adapted, it remains financially unfeasible and also not inclusive of sudden changes that may occur, i.e. the conversion of CO_2 feedstock into DME. It is evident that when the hydrogenation of CO_2 to methanol is investigated, it provides information about the arrangement of DME. Further, when there is a pragmatic impetus available for the direct methanol development, it is easy to construct a framework for DME too.

7.2.2 Production of Formic Acid

Formic acid can be produced when hydrogen reacts with carbon dioxide in the acidic medium, as per the study conducted by researchers (Zabranska et al. 2017, De Luna et al. 2018). It is important to note the conversion of carbon dioxide into other useful feedstock chemicals as well as fuels since it has dual advantages: removal of carbon dioxide from the atmosphere and getting independent from using petrochemicals. There has been a significant progress achieved in the recent years in this area; however, various reactions are also commercialized (Oldenburg et al. 2001).

Based on the overall amount of non-emitted carbon dioxide due to its usage in the carbon dioxide utilization (CDU) process, plant feasibility, availability of captured as well as anthropogenic CO_2, and the future demand for the products synthesized using CO_2, the CDU processes may flourish in future. In this research work, the above-discussed scenarios are analysed via different technological, environmental, and economic key performance indicators in the production of formic acid utilizing carbon dioxide. Further, this study also assesses the product's potential usage and penetration in the context of European nations. Being a well-established chemical, formic acid has the capability to be used as a hydrogen carrier as well as a fuel in fuel cells (Licht et al. 2016).

The current study makes use of process flow modelling with simulations that are developed in CHEMCAD software so as to attain the energy and mass balances and the purchase equipment cost of the formic acid plant. Once the financial analysis is performed with the net present value as the chosen metric, the cost involved per ton of formic acid and CO_2 is differentiated in order to make the CDU project financially feasible. The current study ensures that the carbon dioxide is not emitted in CDU process in comparison with other traditional processes under specific conditions. The outcome and effectiveness of CDU process primarily rely on other technologies and/or developments, for instance the presence of renewable electricity and steam.

In 2013, the production capacity of formic corrosive across the globe was 620 kilotons. A two-stage advanced procedure is the well-established technique to produce formic corrosive, i.e. the underlying response of CO and methanol to shape up the methyl formate. After this step, the methyl formate is then transformed into formic corrosive through two different methods: in the first method, the methyl formate reacts with smelling salts and results in the production of formamide, eventually leading to fermentation with sulphuric acid (H_2SO_4); and in the second method, the transformation process occurs via direct hydrolysis using water. If the second method is to be made possible, then heavyweight and abundant quantity of water are required, followed by a quick decrease in the weight and intense cooling in order to produce formic corrosive. The transformation of CO_2 to formic corrosive by leveraging protons or hydrogens and electrons is one of the easiest ways to produce such a significant product. According to Pérez-Fortes et al. (2016), when the yielded formic corrosive is utilized as a hydrogen source, it can be considered as a progressive improvement (Sordakis et al. 2018).

When the carbon dioxide is made to undergo warm hydrogenation, it produces formic corrosive, which is thermodynamically negative at the beginning of the gas-stage reactants. But opined that it turns to be exogenic when the above-said process is conducted in the presence of water (Sordakis et al. 2018). In general scholarly research, a compendium of systems is widely utilized to achieve the results, which include the utilization of a base to deprotonate the formic corrosive. The ester is produced by making the formic corrosive respond to methanol or otherwise a higher-request liquor *in situ*, or through the removal of formic corrosive when it is developed. In research-scale facilities, there are numerous heterogeneous and homogeneous stimuli created for the hydrogenation of CO_2 to formic corrosive. Although there are numerous stimuli with more frequencies, they lie only within the sight of the base, hence resulting in less production of molecules at a high cost.

7.2.3 Production of Methane

In general, methane is used as a fuel as well as in the production of syngas. According to the reports published in 2017 by Environmental Impact Assessment (EIA), 3,500 billion cubic metres of gaseous petrol was used in 2014 alone. In modern days, methane is legitimately collected from the flammable gas and through Sabatier response, which is an established process of hydrogenating CO_2 in the production of methane using nickel impetus, which is expressed in the following equation (Su et al. 2016):

$$CO_2 + 4\ H_2 \rightarrow CH_4 + 2\ H_2O. \quad (7.1)$$

It is not simple to hydrogenate the carbon dioxide to methane on a small scale, and this is not going to happen at the earliest due to low cost and bounteous accessibility of methane from petroleum gas. Moreover, there will be an altogether more significant financial incentive in changing the CO_2 to a number of different synthetic substances and methane. Manthiram et al. (2014) cited the process involved in the reduction of carbon dioxide to methane, i.e. electrochemical process as a common procedure with 80%–94% Faradaic efficiencies by making use of N-doped carbon or at some times, copper-on-carbon impetuses in standard 3-terminal or H cells (Sun et al. 2016, Qiu et al. 2017). The halfway present densities in case of methane arrangement with 38 mA cm^{-2} were responsible for the copper impetus-based electrodeposition in carbon gas dispersion anode. As per the existing thermochemical CO_2-to-methane forms discussed earlier, the electrocatalytic transformation of CO_2 to methane, though, was conducted with an improvement in particular impetuses; there seems to be no valuable outcome on a large scale, given the worldwide accessibility of methane produced from gaseous petrol with minimal effort. Figure 7.2 depicts the production of methane.

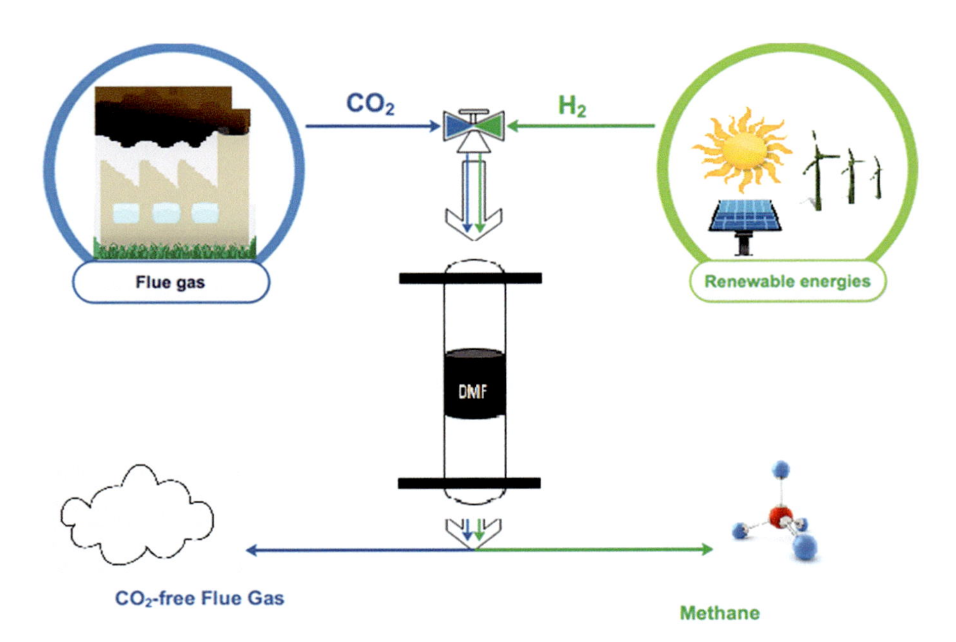

FIGURE 7.2 Production of methane.

7.2.4 Ethylene and Ethanol Production

In plastic production, ethylene remains an important commodity chemical. Ethylene is considered the tractable target in the renewable production of carbon dioxide. It is challenging to prevent the deteriorating performance of the copper electrocatalysts, which are utilized in this conversion under the required basic reaction conditions, from the parallel reaction that happens between carbon dioxide and base to produce bicarbonate. In the electrode designed by Dinh et al. (2018), the base is tolerated through the optimization of carbon dioxide diffusion to the catalytic sites. This catalyst design results in 70% efficiency for about 150 hours.

Being a commodity chemical and an excellent monomer, ethylene is used to generate various chemicals, especially polymers. More than 150 million tons of ethylene was synthesized in 2016, solely from non-renewable energy source derived from its precursors. Even though the conversion of carbon dioxide to ethylene needs huge volumes of input energy, various research investigations are motivated by the possibility of using carbon dioxide as the carbon source for the high-volume generation of commodity chemical.

Recent research studies discovered that the iron catalysts' selectivity for the hydrogenation of CO_2 to ethylene with other little olefins up to 65% is one of the potential strategies with a minimum amount of carbon source in H_2 (Satthawong et al. 2015, Wang et al. 2013). But this method only ends up in the conversion of the minuscule amount of CO_2 and also generates CO and methane along with alkanes. In order to develop a system that produces ethylene at higher volumes, a substantial amount of improvement has to be made in catalyst so as to produce a variety of hydrocarbons from CO_2.

Another method of producing ethylene from CO_2 is through the electroreduction of copper with the help of silver or gold catalysts. Various investigations are conducted with regard to different copper nanostructures as well as their working conditions. This phenomenon enhances the generation capacity of ethylene, i.e. more than 40%, although it is still a challenging process. Few procedures increase the Faradaic efficiency of the ethylene up to 60%–70% with 160–250 mA cm^{-2} (Dinh et al. 2018). One such procedure is the silver alloying with copper, in which the silver production increases the required CO intermediate (Hoang et al. 2018) and the Cu catalyst layer achieves the perfect design within the sandwich-type gas diffusion of the electrode. In Faradaic efficiencies of 10%–30%, a sizeable volume of ethanol is co-produced frequently along with ethylene.

Ethylene and ethanol are the important commodity chemicals by itself and were produced for about 80 million tons in 2016. When ethylene is produced, the ethanol is also produced, thus helping the producers financially and making the process feasible in this regard. Further, ethylene is a gas, whereas ethanol is a fluid, which makes the separation process easier, although the removal of ethanol from the electrolyte is a challenging process. According to Yanming et al. (2017), the Faradaic efficiencies can be achieved up to 93% in ethanol production when boron and nitrogen-co-doped nanodiamond catalysts are used in the experiment. However, it still remains an unsolved issue of whether to use nanodiamond crystals in a gas diffusion electrode and in a fluid electrolyte or membrane-based electrolyser for a successful economic accomplishment. Although there are advancements brought in electroreducing the CO_2 to ethylene and ethanol, there are various challenges associated with it. Few studies concluded that the catalyst degradation and remodelling are the notable issues to be taken care when checking the stability of variants of Cu catalysts. At the same time, there is no specific note about the continuous 2^+ hours of operations in the ethylene-delivering electrolyser. Further, as discussed earlier, the loss of CO_2 to the electrolyte can also lead to a significant drawback.

7.2.5 Production of Polymer

In the recent times, researchers shifted their attention towards the adsorption process, which has a strong affinity and high loading capacity for targeted metal ions, since this process resulted in the alteration of the super adsorbent. Although it is possible for the super adsorbents (polymers) to adsorb heavy metals in aqueous solutions, it is important for the composites to adsorb the heavy metals in order to choose the optimum adsorbent in the removal of metal from aqueous solutions. Polymers are materials that are generally light-weighted and possess flexibility, chemical inertness, and fabrication (Kaya et al. 2008).

In the recent years, there has been much interest towards polymers with highly conjugated chains since these academic interest materials are investigated in electronics (Diaz et al. 1999, and Suhl et al. 2000), optoelectronics (Bredas et al. 1990), and photonics (Ledoux-Rak et al. 2001). For a known period of time, the polymers are known for their excellent thermal stability and fascinating optoelectronics characteristics. The solvent-based polymerization process is employed in the preparation of few poly(azomethines), whereas the spin-on technique is employed to deposit their thin films (Suhl et al. 2002).

To synthesize the polymers, CO_2 as a feedstock can be used either directly or indirectly. Carbon dioxide is utilized as a monomer unit in the direct approach method, which is directly integrated into the polymer. In another method, the carbon dioxide is initially converted into alternate monomers such as ethylene, dimethyl carbonate, methanol, natural carbonates, carbon monoxide, and urea. After this conversion, these monomers are made to undergo polymerization. The polycarbonate plastics with carbonate gatherings are always conjoined when phosgene reacts with 1,2-diols (Qin et al. 2015, Poland and Darensbourg, 2017). One of the synthetic methods is the copolymerization of carbon dioxide with epoxides in the development of polycarbonates. Few researchers mentioned that a number of heterogeneous and homogeneous conversion metal catalysts generated precisely produce polycarbonates rather than the cyclic carbonates from a scope of common monomers, which are inclusive of styrene oxide, ethylene oxide, propylene oxide, cyclohexene oxide, and vinyl oxide among others.

In addition, when the catalyst nature is changed, one can produce either rotating polymers or polymers that have carbonate groups with one carbon dioxide atom in addition to one epoxide atom, or otherwise the statistical polymers, which are always called as polyether carbonates, that may possess ether linkages. In the presence of two open-ringed epoxides, these ether linkages are created. Generally, when the copolymers are altered, it may produce a low glass transition point. This denotes the fact that they can be used in selected applications; for instance, they can be used as binders in glues and ceramics as per Qin et al. (2015). At present, the Empower Materials sell poly(ethylene carbonate), which is produced from carbon dioxide and ethylene. But Econic Technologies produces polycarbonates with 50% of carbon dioxide by weight (Quadrelli et al. 2011).

By and large, rotating copolymers possess low glass change focus, implying that they may be utilized in featured applications, e.g. covers in earthenware production and glues. The organizations like NOVAMER are trying to get rid of the decaying nature of polycarbonates that produces cyclic carbonate, especially by making use of electron-insufficient epoxides. This issue, as well as the inclination issue, i.e. polycarbonates inclined with high CO_2 substance in order to respond to H_2O, can be unravelled on a fundamental level through the sensible utilization of added substances, yet this should be highly assessed. There is a need to create impetuses that are exceptionally dynamic with a wide scope of epoxides that may incite the substitution of polymers with higher glass change temperatures. Further, when the systems are advanced to fuse with some other monomers, for instance cyclohexene oxide into copolymer, the features of the polymer get improved and become the most sought-after ones. Since the CO_2 response compared with that of the epoxides is completely exothermic, it is important to discover the thermally steady impetuses.

As opposed to the restricted applications for the exchange of copolymers, the polyether carbonates obtained from the epoxides and CO_2 are valuable for a lot more extensive scope of uses. Specifically, they can be utilized as a part in polyurethanes, and aside from the business application discussed in the introduction section, both the companies Covestro and NOVOMER are selling polyether carbonates so that it can be used in the conjoining of polyurethane. All things considered, notwithstanding the overall development of this field, there is still space for development. When there is research conducted in understanding the impact of impetus structure upon polymer properties, it will be useful since this can induce the enhancement of polymers with custom-made structures.

The epoxide beginning materials that are derived from sustainable feedstocks, for instance cyclohexadiene oxide, α-pinene oxide, and limonene oxide, are used in the carbon dioxide copolymerization with some substrates. This is considered as a promising research area, which may invite future researchers to enlarge the research space with more number of motivations in this regard. In addition, the investigation about carbon dioxide getting copolymerized with aziridines resulting in the production of polycarbamates and oxetanes is additionally just at a beginning period and ought to likewise be energized. This may prompt the polymers with unexpected properties in comparison with those as of now accessible.

7.2.6 Production of Oxalate and Oxalic Acid

The oxalic acid is generally produced from oxalate salt; however, it can also be produced under laboratory conditions by the cationic reduction of CO_2 in an electrolytic cell. The porous membrane separates the anode and cathode compartments, and the catholyte acts as an organic solvent. The preferred solutes for the catholyte are tetraethylammonium bromide, tetrabutylammonium iodide, tetrabutylammonium perchlorate, tetraethylammonium perchlorate, and tetraethylammonium p-toluenesulfonate. When the anolyte remains the same electrolyte and the solvent as the catholyte, the coulombic yields are as high as 75%. However, close to 97% of sodium oxalate is also obtained when the aqueous solution of sodium salt is used as the anolyte. This process involved in the synthesis of alkali metal salts of glycolic acid, ethylene glycol, alkali metal salts of nitrilotriacetic acid by the hydrogenation of oxalic acid or an alkali metal hydrogen oxalate, alkali metal salts of diglycolic acid and alkali metal salts of glycine may sometimes contain ammonia or otherwise oxalic acid or alkali metal hydrogen oxalate that contains less than two moles of water.

According to Qiao et al. (2014), although the oxalates and oxalic acid cannot be sold as chemicals, they are still produced in large amounts, i.e. 120,000 tons every year. In general, the synthesis of these two chemicals is inclusive of the reactions in which alcohols react with CO and O_2 so as to produce oxalic acid diesters.

This is a real burdening process since this can be handled efficiently and cost-effectively through converting CO_2 into oxalic acid. There may be an increase in the amount of oxalic acid produced in the due time since the oxalic acid is required to act as a suitable feedstock, when chemicals are prepared. There exist different processes to convert dimethyl oxalate to monoethylene glycol. In the recent times, the dimethyl oxalate is prepared from CO and hydrogen; however, the oxalic acid is prepared from carbon dioxide by esterification process, which is an important and sustainable process. Avantium, commonly known as liquid light, is used in the production of ethylene glycol via the traditional process, i.e. electrochemical conversion of carbon dioxide to oxalic acid. In addition, oxalic acid then undergoes conversion into ethylene glycol with the help of standard chemicals (Twardowski et al. 2016).

The add-on research analyses conducted in this area focused on the optimum methods in the production of oxalates and oxalic acid from carbon dioxide. The investigations conducted before 2000 concluded that lead and other catalysts were generally utilized in the preparation of oxalic acid in electrochemical cells.

The Faradaic efficiencies of the electrochemical cells are 70%–98%. But the less rate of the Faradaic efficiencies can only be achieved in a normal electrochemical cell. Currently, mononuclear and binuclear copper complexes are produced since these elements show promising outcomes in the electrochemical reduction of CO_2 to oxalic acid. But it is mandatory to have advanced mechanical considerations for this process so as to reduce the potential and simultaneously increase the selectivity.

7.2.7 Fuel (Hydrocarbon) Production

For photosynthesis, sunlight and chlorophyll present in plants recycle the carbon dioxide and water oxygen and energy-rich organic compounds. Fossil fuels are formed during a particular geological time. Methanol and DME are produced when hydrogen is reduced via recycling the CO_2 from natural and industrial resources. Methanol production at the time of carbon recycling is one of the main aspects to be considered. Various energy sources are available such as atomic energy, geothermal energy, wind energy, and solar energy, which can be utilized in the preparation of required hydrogen and chemical conversion of CO_2. The bi-reforming of methane process and various catalytic or electrochemical conversion processes are developed by the researchers in order to efficiently reduce the carbon dioxide to methanol and DME. In order to store and transport energy, the liquid methanol is mostly preferred since hydrogen is highly volatile and the risk of its explosion is high. The liquid methanol and DME are the best transportation fuels that can be used in Internal combustion engines (ICEs) and fuel cells, and they can also act as flexible starting materials in the preparation of synthetic hydrocarbons and their different variants. As discussed in the above sections, it is possible to make use of GHG like carbon dioxide as a renewable, valuable, and inexhaustible carbon source in future. This way, the environment can become neutral from carbon fuels and derived hydrocarbon products.

The combustion of fossil fuels is the highest contributor of carbon dioxide into the atmosphere. The closed carbon cycle may be the result of the development of processes in which the fuel is produced from carbon dioxide. So, when the concentration of carbon dioxide is increased in the atmosphere, it remains negligible. This is possible only in case of the production of electricity or hydrogen, which is used in the reduction of carbon dioxide from carbon-free sources. Both methane and methanol can be utilized as fuels. As per the discussions made earlier, there is a dynamic system available in the conversion of CO_2 to methane and methanol. As a result, this section discusses the state-of-the-art technology used in the transformation of carbon dioxide into hydrocarbon fuels that possess >2 carbons. In the transformation of CO and hydrogen to liquid fuels, the Fischer–Tropsch process is followed, which can also be extended to the commercial processes on a large scale.

In addition, few studies established the need to have smaller units that come packed with waste carbon streams, furthering the large-scale units for the Fischer–Tropsch process. The previous paragraphs explained how carbon dioxide is converted to carbon monoxide. In the first step of fuel production from carbon dioxide, the CO is first synthesized from carbon dioxide electrochemically. The second thermal step has the conjoining of CO and sustainably produced hydrogen in order to generate fuels following the traditional Fischer–Tropsch process. Various researchers attempt to build the Fischer–Tropsch chemistry so that the process that starts from CO_2 can be achieved with the help of a single reactor as well as a single catalyst itself. In this method, reverse water gas–shift reaction is the first stage that produces carbon monoxide from carbon dioxide. In addition to that, carbon monoxide then undergoes a reaction with hydrogen to form liquid fuel via the mechanism of the conventional Fischer–Tropsch reaction.

Sunfire and INERATEC developed the small-scale demonstration plants, despite the huge costs incurred to supply hydrogen as the process depends on water electrolysis, in order to conduct CO_2-based Fischer–Tropsch chemistry. At the time of reaction in the Fischer–Tropsch process, only a low amount of CO is present when using CO_2. This prevents the chain growth, and accordingly, the product circulation reaches high in light hydrocarbons. These hydrocarbons may tend to be inappropriate as liquid fuels. In reverse water gas–shift reaction as well as the Fischer–Tropsch chemistry, the most widely discovered iron-based catalyst seems to be active (Wei et al. 2017).

Different transition metal-based promoters are supplemented with Fe-based catalysis and are provided with highly efficient product distributions. Further investigations are conducted to identify the mechanism that enhances the activation process. In line with this, the performance of secondary catalysts on materials like SiO_2 seems to be better although these effects can only be compared to some extent. There have been other investigations that are also conducted to develop new catalysts. When the reactors possess the optimal design, it can result in excellent performance since this research area is yet to be explored.

Few researchers also worked on electroreduction or photoreduction of CO_2 to chemical fuels. The carbon dioxide is mostly reduced to C1 feedstocks such as methane, methanol, formic acid, and carbon monoxide. Various systems are available that can develop products with new C–C bonds. To be specific, Cu catalysts show efficacy in the development of products with C–C bonds, for instance ethylene, ethane, and other higher-order hydrocarbons (Ren et al. 2015). In addition to the selective production of C2 or higher products rather than C1 products, one of the major drawbacks is the electrochemical reduction of carbon dioxide to fuels, which prevents the hydrogen evolution reaction that has hydrogen as a by-product. A piece of mechanistic information about the elementary processes that produce C–C bond is not available. There may be an occurrence of beneficial effects if this topic is further investigated. There is a large amount of scope available for further customization of catalysts in the electrochemical reduction of carbon dioxide. Although this is a worrisome problem, it still creates an interesting pathway ahead.

7.2.8 Production of Carbon Nanotube

The cost and practical nature of the GHG removal processes are important in the viewpoint of environmental sustainability since the high-value secondary applications are devised on the basis of carbon-capture and conversion techniques. By utilizing the solar thermal electrochemical process (STEP), the carbon nanofibres (CNFs) and carbon nanotubes (CNTs) are produced using the ambient carbon dioxide collected in molten lithiated carbonates. The electrolytic process is carried out to produce the above-mentioned elements by making use of cheap steel electrodes.

These cost-efficient CO_2-derived CNTs and CNFs are believed to show excellent performance as energy storage materials in lithium-ion as well as in sodium-ion batteries. Due to the synthetic control of sp3 content in the nanostructures produced, the optimized storage capacities are determined above 370 mAh g^{-1} (lithium) and 130 mAh g^{-1} (sodium). There seems to be no loss in the capacity even when they are subjected to the durability tests up to 200–600 cycles, respectively. In this work, the researchers showed that the ambient CO_2, an environmental pollutant, can possibly bring the economic value in grid-scale and portable energy storage systems with STEP scale-up practicality in the context of combined-cycle natural gas electric power generation (Licht et al. 2016).

The CNT synthesis can be controlled as per the findings of the recent studies. This already resulted in the commercial application of nanotubes in bicycles, boat hulls, water fillers, and skis. The CNTs can be applied in a wide range of areas, thanks to their unusual thermal conductivity and mechanical and electrical properties.

But the CNT synthesis is a cost-incurring process that prevents its further applications, since there is only a small-sized market available for this. Carbon monoxide is the precursor in the production of CNTs, as carbon monoxide is obtained from fossil feedstocks, which are used as carbon sources. CNTs have been electrochemically produced from carbon dioxide only on a laboratory scale so far. In this process, the molten carbonate electrolysis occurs in the presence of CO_2. To be precise, it is possible to reduce the molten Li_2CO_3 to produce CNTs as well as Li_2O. In addition, there occurs a reaction between Li_2O and CO_2 so as to form Li_2CO_3 with a special notion that CO_2 is the primary source for nanotubes.

In this step, in spite of the value the CNTs offer, their applications are still confined in comparison with carbon fibres. The CNTs that are generated from carbon dioxide are yet to achieve the favourable features so as to be considered as a substitute for carbon fibres which are often used in aerospace and automotive industries (Rahaman et al. 2007). But CNTs may attain the potentials for a future technology, when suitable methods are established to enhance the CNTs or carbon fibres from carbon dioxide.

7.3 COLLABORATION OF RESEARCH CHALLENGES IN THE CONVERSION OF CARBON DIOXIDE

The above section discussed the technology and research with regard to carbon dioxide product type-based conversion and also about the challenges that need to be overcome for the development of next-generation systems. Although few cases are very much product specific, few common challenges can be found for almost all the products. This section discusses the common challenges that are present in the development of next-generation systems in the conversion of carbon dioxide.

7.4 CHALLENGES IN CATALYST DEVELOPMENT

According to the discussions made earlier, a number of catalysts were developed to convert carbon dioxide into worthy chemicals. Although there are few cases in which the products and accordingly the catalysts cannot be found as commercially viable, the much-needed stability and durability are absent in the catalysts used for the conversion of carbon dioxide. The catalyst decomposition remains the big issue even in case of high turnover frequencies. This is the primary challenge that needs to be addressed since most of the small-scale systems make use of purified carbon dioxide. When these catalysts can be operated using non-purified CO_2, the chances of decomposition are high and rapid. If the carbon dioxide needs to be purified, it increases the process cost; however, the improvements in the system are compatible with the conversion of carbon dioxide, and numerous fills present in the flue gas may ease the commercialization process. In line with this, when the interface between carbon dioxide capture and conversion is improved, it can support the application of carbon dioxide from gaseous waste streams. When compared, the number of catalysts available for CO_2 conversion is low, yet it can be made use of, in association with CO_2-capture systems. It can play an important role in the development of efficient tactics to make use of carbon dioxide from waste streams.

7.5 CONCLUSION

The excellent test is to convert CO_2 waste streams into useful products, which are insignificant measures of non-renewable vitality. The products are again of high cost, and this conversion can result in a decreased depletion of ozone in contrast with existing innovation. Currently, there are a few mechanical procedures available that convert CO_2 into fills or synthetic substances. The foremost innovative work of CO_2-use forms is focused on C1 mixes (methane, CO, methanol, and formic corrosive) with a couple of conspicuous special cases, e.g. natural carbonates and polymers. The procedures for the chemical conversion of carbon dioxide into chemicals and fuels need to be improved. In numerous carbon dioxide change forms, the catalyst execution is still a challenging one. The heterogeneous, homogeneous, and (photo) electrochemical impetuses need to be generally improved. Substantial difficulties present in these processes confine the vitality input, which is needed for the transformation of CO_2 and the improvement of impetus selectivity, solidness, and resilience to basic debasements in waste CO_2 streams.

REFERENCES

Alper E., Yuksel-Orhan O. (2017) CO_2 utilization: Developments in conversion processes, *Petroleum*, 3: 109–126.

Arena F., Mezzatesta G, Spadaro L, Trunfio G. (2014) Latest advances in the catalytic hydrogenation of carbon dioxide to methanol/dimethylether, in: Transformation and utilization of carbondioxide, *Green Chemistry and Sustainable Technology*, 303–334, doi: 10.1007/978-3-642-44988-8.

Artz J., Muller T.E., Thenert K., Kleinekorte J., Meys R., Sternberg A., Bardow A., Leitner W. (2018) Sustainable conversion of carbon dioxide: An integrated review of catalysis and life cycle assessment, *Chemical Reviews*, 118: 434–504.

Bradford M.C.J, Vannice M.A. (1999) CO_2 reforming of CH_4, *Catalysis Reviews*, 41: 1–42.

Bredas J.L., Chance. R.R. (1990) Conjugated polymeric materials: opportunities in electronics, optoelectronics and molecular electronics, Kluwer Academic, Dordrecht, *Journal of Chemistry*, 92:711–28.

Cho W, Baek Y, Kim Y.C, Anpo M. (2002) Plasma catalytic reaction of natural gas to C_2 product over Pd-NiO/Al_2O_3 and Pt-Sn/Al_2O_3 catalysts, *Research on Chemical Intermediates*, 28: 343–357.

Cho W, Baek Y, Moon S.K, Kim Y.C. (2002) Oxidative coupling of methane with microwave and RF plasma catalytic reaction over transitional metals loaded on ZSM-5, *Catalysis Today*, 74: 207–223.

De Luna P., Quintero-Bermudez R., Dinh C.-T., Ross M.B., Bushuyev O.S., Todorović P., Regier T., Kelley S.O., Yang P., Sargent E.H. (2018) Catalyst electro-redeposition controls morphology and oxidation state for selective carbon dioxide reduction, *Nature Catalysis*, 1: 103–110.

Diaz F.R., Godoy A., Moreno J., Bernede J. Del Valle C, Tagle M.A. (2007) East Synthesis and characterization of poly(pyridylurea), potentially semiconducting polymers, *Journal of Applied Polymer Science*, 105: 1344–1350.

Dinh C.T., Burdyny T., Kibria M.G., Seifitokaldani A., Gabardo C.M., de Arquer F.P.G., Kiani A., Edwards J.P., Luna P.D., Bushuyev O.S., Zou C., Quintero-Bermudez R., Pang Y., Sinton D., Sargent E.H., (2018) CO_2 electroreduction to ethylene via hydroxide-mediated copper catalysis at an abrupt interface, *Science*, 360: 783–787.

Fukuoka S., Tojo M., Hachiya H., Aminaka M., Hasegawa K. (2007) Green and sustainable chemistry in practice: Development and industrialization of a novel process for polycarbonate production from CO_2 without using phosgene, *Polymer Journal*, 39: 91–114.

Hoang T.T.H., Verma S., Ma S., Fister T.T., Timoshenko J., Frenkel A.I., Kenis P.J.A., Gewirth A.A., (2018) Nanoporous copper–silver alloys by additive-controlled electrodeposition for the selective electroreduction of CO_2 to ethylene and ethanol, *Journal of the American Chemical Society*, 140: 5791–5797.

Kaya İ., Yıldırım M., Avcı A., Kamacı M., (2009) Synthesis and thermal characterization of novel poly(azomethine-urethane) s derived from azomethine containing phenol and polyphenol species, *Journal Title Macromolecular Research*, 19: 286–293.

Langanke J., Wolf A., Hofmann J., Bohm K., Subhani M.A., Muller T.E., Leitner W., Gurtler C. (2014) Carbon dioxide (CO_2) as sustainable feedstock for polyurethane production, *Green Chemistry*, 16: 1865–1870.

Ledoux Rak I., Dodabalapur A., Blom P. (2001) Electronic olfaction with organic and polymer transistors, *Applied Physics Letter*, 78:22–29.

Licht S., Douglas A., Ren J., Carter R., Lefler M., Pint C.L. (2016) Carbon nanotubes produced from ambient carbon dioxide for environmentally sustainable lithium-ion and sodium-ion battery anodes, *ACS Central Science*, 2: 162–168.

Manthiram K., Beberwyck B.J., Alivisatos A.P. (2014) Enhanced electrochemical methanation of carbon dioxide with a dispersible nanoscale copper catalyst, *Journal of the American Chemical Society*, 136: 13319–13325.

Olah G.A., Goeppert A., Prakash G.K.S. (2009) Chemical recycling of carbon dioxide to methanol and dimethyl ether: From greenhouse gas to renewable, environmentally carbon neutral fuels and synthetic hydrocarbons, *Journal of Organic Chemistry*, 74: 487–498.

Oldenburg C.M., Pruess K., Benson S. M. (2001) Process modeling of CO_2 injection into natural gas reservoirs for carbon sequestration and enhanced gas recovery, *Energy Fuels*, 15: 293–298.

Patricio J., Angelis-Dimakis A., Castillo-Castillo A., Kalmykova Y., Rosado L. (2017) Method to identify opportunities for CCU at regional level-Matching sources and receivers. *Journal of CO_2 Utilization*, 22: 330–345.

Pérez-Fortes M., Schoneberger J.C., Boulamanti A., Harrison G., Tzimas E. (2016) Formic acid synthesis using CO_2 as raw material: Techno-economic and environmental evaluation and market potential, *International Journal of Hydrogen Energy*, 41: 16444–16462.

Poland S.J., Darensbourg D.J. (2017) A quest for polycarbonates provided via sustainable epoxide/CO_2 copolymerization processes, *Green Chemistry*, 19: 4990–5011.

Qiao J., Liu Y., Hong F., Zhang J., (2014) A review of catalysts for the electroreduction of carbon dioxide to produce low-carbon fuels, *Chemical Society Reviews*, 43: 631–675.

Qin Y., Sheng X., Liu S., Ren G., Wang X., Wang F. (2015) Recent advances in carbon dioxide-based copolymers, *Journal of CO_2 Utilization*, 11: 3–9.

Qiu Y.L., Zhong H.X., Zhang T.T., Xu W.B., Li X.F., Zhang H.M., (2017) Copper electrode fabricated via pulse electrodeposition: Toward high methane selectivity and activity for CO_2 electroreduction, *ACS Catalysis*, 7:6302–6310.

Quadrelli E.A., Centi G., Duplan J.L., Perathoner S. (2011) Carbon dioxide recycling: Emerging large-scale technologies with industrial potential, *ChemSusChem*, 4: 1194–1215.

Rahaman M. S.A., Ismail A.F., Mustafa A. (2007) A review of heat treatment on polyacrylonitrile fiber, *Polymer Degradation and Stability*, 92: 1421–1432.

Ren D., Deng Y, Handoko A.D, Chen C.S, Malkhandi S, Yeo B.S, (2015) Selective electrochemical reduction of carbon dioxide to ethylene and ethanol on copper(I) oxide catalysts, *ACS Catalysis*, 5: 2814–2821.

Rostrup Nielsen J.R, Bak Hansen J.H. (1993) CO_2-reforming of methane over transition metals, *Journal of Catalysis*, 144: 38–49.

Satthawong R., Koizumi N., Song C., Prasassarakich P., (2015) Light olefin synthesis from CO_2 hydrogenation over K-promoted Fe–Co bimetallic catalysts, *Catalysis Today*, 251: 34–40.

Sordakis K., Tang C., Vogt L.K., Junge H., Dyson P.J., Beller M., Laurenczy G. (2018) Homogeneous catalysis for sustainable hydrogen storage in formic acid and alcohols, *Chemical Reviews*, 118: 372–433.

Su X., Xu J., Liang B., Duan H., Hou B., Huang Y. (2016) Catalytic carbon dioxide hydrogenation to methane: A review of recent studies, *Journal of Energy Chemistry*, 25:553–565.

Suhl, S.C., Shim S.C. (2000) Synthesis and Properties of novel poly (azomehine) the polymer with high photo conductivity and second-order non linearity, *Journal of Synthetic Metals*, 114:91–95.

Sun X., Kang X., Zhu Q., Ma J., Yang G., Liu Z., Han B., (2016) Very highly efficient reduction of CO_2 to CH_4 using metal-free N-doped carbon electrodes, *Chemical Science*, 7:2883–2887.

Topham S., Bazzanella A., Schiebahn S., Luhr S., Zhao L., Otto A., Stolten D. (2014) *Carbon Dioxide in Ullmann's Encyclopedia of Industrial Chemistry*, Weinheim: Wiley-VCH, doi: 10.1002/14356007.a05_165. pub2.

Twardowski Z., Cole E.B., Kaczur J.J., Teamey K., Keets K.A., Parajuli R., Bauer A., Sivasankar N., Leonard G., Kramer T.J., (2016) Method and system for production of oxalic acid and oxalic acid reduction products, Google Patents. EP2935654A4.

Wang J., You Z., Zhang Q., Deng W., Wang Y. (2013) Synthesis of lower olefins by hydrogenation of carbon dioxide over supported iron catalysts, *Catalysis Today*, 215: 186–193.

Wei J., Ge Q., Yao R., Wen Z., Fang C., Guo L., Xu H., Sun J., (2017) Directly converting CO_2 into a gasoline fuel, *Nature Communications*, 8: 151–74.

Yanming L., Yujing Z., Kai C., Xie Q., Xinfei F., Yan S., Shuo C., Huimin Z., Yaobin .Z, Hongtao Y., Hoffmann M.R. (2017) Selective electrochemical reduction of carbon dioxide to ethanol on a boron- and nitrogen-co-doped nanodiamond, *AngewandteChemie International Edition*, 56:15607–15611.

Zabranska, D.P. (2017) Bioconversion of carbon dioxide to methane using hydrogen and hydrogenotrophic methanogens, *Biotechnology Advances* doi: 10.1016/j.biotechadv.2017.12.003.

8 Application of Nanomaterials in CO_2 Sequestration

Anirban Biswas, Suvendu Manna, and Papita Das

CONTENTS

8.1 INTRODUCTION

The most commonly produced greenhouse gas is carbon dioxide. Atmospheric carbon dioxide is captured through the mechanism called the carbon sequestration. So, the carbon capture and sequestration is the process of reducing atmospheric carbon dioxide, thus getting control over the global climate change. It also enables the low carbon economy and maintains the global carbon budget.

Greenhouse gas emission sources are different. Even the sources vary country-wise and are directly proportional to economic development and population demand. As has been described in the report of U.S. Inventory of Greenhouse Gas Emissions and Sinks, 40% or more CO_2 emissions in the United States come from the electric power generation sector alone. Source-specific carbon-capture and sequestration mechanisms are being invented to reduce the carbon dioxide potential. Fossil fuel-based power plants using those industry-specific technologies can reduce their carbon dioxide emission from 80% to 90% efficiently, which is comparable to planting more than 62 million trees and to wait for their 10 years' carbon capture.

The carbon sequestration practices got new outlook considering nanomaterials' uses in the carbon sequestration process. Scientific methods have been developed using nanotechnology. More insights are needed in this process. This chapter highlights the recent trend of researches in carbon sequestration with the application of nanomaterials and nanocomposites and their pros and cons in comparison with other available technologies.

8.1.1 Terrestrial Carbon Storage and Sequestration

Terrestrial (or biologic) sequestrations imply the atmospheric CO_2 sequestration and store the same as carbon in the plant parts and into the soil biome. During photosynthesis, plants use carbon dioxide and produce oxygen (O_2), which dissipates it into the atmosphere. So, plants actually hold on to and use the carbon to sustain and grow. After the death of any plant, part of the carbon from the plant gets to be preserved (stored) in the soil environment. Terrestrial sequestration is one of the automatic management practices to maximize the carbon storage in the soil and plant materials. Reforestation, rangeland management, and wetland management are prominent examples of terrestrial carbon sequestration practices (MIT 2007). Terrestrial carbon dioxide sequestration is technically low-cost and easy process compared with resource-intensive and complex technology-based geologic sequestration activities, which in turn also improve the habitat and water quality, and overall, it preserves the forests. Terrestrial sequestration alone offsets one-third of the global anthropogenic carbon dioxide emissions. Soil can store the carbon up to a certain level, and over the next 50–100 years, the soil will be saturated with no more space for excess carbon (Paustian and Cole 1998).

8.1.2 Mechanisms for Terrestrial Storage

Various water and land management practices increase the terrestrial carbon storage; the current processes are conservation tillage, soil erosion and distribution management practices, maintaining the buffer strips all along waterway, consideration of land resources in conservation methods, wetland restoration and management, limiting the summer crops, using winter cover crops and perennial grasses, and increase in afforestation (de Silva et al. 2005).

8.1.3 Geologic Sequestration of Carbon Dioxide

It is a one-step process to sequester the atmospheric carbon dioxide. In this process, carbon dioxide is injected deep into the underground to hold it back permanently. Due to the population growth and rapid industrialization, there is an ever-increasing trend of carbon dioxide production in the environment causing greenhouse condition for the mother Earth. Although working as the carbon dioxide sinks, plants seem not enough to hold all those produced. So scientists are continuously trying new methods to accelerate the carbon sequestration mechanisms. Here lies the importance of the geological carbon dioxide sequestration. In many cases, injection of carbon dioxide into a geological formation increases the hydrocarbon recovery, providing economic values that can offset the carbon dioxide sequestration costs. Geological carbon sequestration involves the separation and capture of carbon dioxide at the point of emissions followed by storage in deep underground geologic formations. The method may be of two types, which are discussed below.

8.1.4 Physical Process of Geological Carbon Sequestration

This process involves carbon dioxide trapping within a cavity of any underground rock. The cavities may be either man-made large cavities (e.g. caverns and mines) or the natural pore space within rock formations (e.g. structural traps in depleted oil and gas reservoirs, aquifers). Precise conditions are needed within the deep subsurface geological environment to store the injected carbon dioxide effectively. The selected space should be with some shield so that it cannot migrate out of the storage to cause some environmental hazards. If the reservoirs are porous and permeable, there must be a confining unit overlying as the shield beneath which carbon dioxide should be stored in its supercritical state. Characteristically, there are two types of reservoirs to support the geological carbon dioxide sequestration:

a. Sandstone or other reservoirs contain salt water,
b. Injection into any hydrocarbon-bearing strata such as oil reservoir, gas reservoir, and coal seam. In these cases, actually replacement of the space by carbon dioxide happens when those economically important materials are taken out of the soil core. The CO_2 has a higher affinity towards coal than does the methane that is usually found in coal beds.

8.1.5 Chemical Process of Geological Carbon Sequestration

Chemical mechanisms of trapping the atmospheric carbon dioxide involve transforming the gas or binding it chemically to another substance in the ground. The chemical binding may be done as follows:

a. Dissolving the carbon dioxide directly into underground water or reservoir oil,
b. Decomposing the carbon dioxide into its ionic parts,
c. Locking the carbon dioxide into a steady mineral precipitate,
d. Trapping by adsorption.

8.2 CO_2 SEQUESTRATION: NON-NANOMATERIAL-BASED PROCESSES

8.2.1 CONVENTIONAL ADSORBENTS

Adsorbents such as activated charcoal, natural zeolite, and alkali-based materials have frequently been used for CO_2 capture from air. NaX zeolite (Li et al. 2013), zeolite 5A (Saha et al. 2010), and zeolite 13X have been tested in the laboratory for CO_2 sequestration, and the aforementioned studies indicated that the zeolite-based materials have a very high capacity for CO_2 capture. Although zeolite-based materials show high sorption capacity, they need high temperature for regeneration.

Another emerging conventional adsorbent is activated charcoal. As it is economically feasible and thermally stable, its efficacy in CO_2 sequestration has been investigated. It was noticed that activated charcoal has very high CO_2-capturing capacity at increased pressure (Cheng-Hsiu, Chih-Hung, and Tan 2012; Himeno, Komatsu, and Fujita 2005). Also, its regeneration is easy. However, the use of activated charcoal has drawbacks such as low sorption at low pressure, low selectivity, and low efficacy, which are attributable to the presence of NO_x and SO_x.

Alkali-based materials are being used for CO_2 sequestration for a very long time. Alkali-based K_2CO_3/TiO_2 (Lee et al. 2006) and Li_2ZrO_3 (Ochoa-Fernández et al. 2005) have been used for CO_2 sequestration and found to be workable. These types of materials are unique as they can be easily regenerated for more than 20 regeneration–reuse cycles. Also, the volume change due to recycling is found to be very low. However, these types of sorbents have also some drawbacks such as low sorption efficacy and slow sorption kinetics compared with the other conventional materials.

8.2.2 IONIC LIQUIDS

Ionic liquids are very emerging groups of adsorbents that are being used for CO_2 sequestration. Ionic liquids are composed of organic or inorganic anions and bulky asymmetric cations. These types of materials are thermally stable, have very minimum vapour pressure, and have tunable physicochemical characteristics with very high CO_2 solubility. Thus, they have been used for CO_2 sequestration (Park et al. 2015). As these materials are easy to modify using different functional groups for enhancing their CO_2 sorption capacity, many researchers modified the ionic liquids with amine (Galán Sánchez, Meindersma, and de Haan 2007) or hydroxyl (Lee et al. 2012) groups and found that the CO_2-capturing capacity of the modified ionic liquids is enhanced compared with that for the unmodified ionic liquids. As CO_2 gas is highly soluble in ionic liquid, the ionic liquid is an ideal choice for CO_2 capture from the natural environment (Ramdin et al. 2015).

Chemicals used for CO_2 sequestration along with their CO_2-capturing capacities are given in Table 8.1 (Sharma et al. 2019).

8.2.3 MODIFIED POROUS SUPPORT

Porous support is another emerging CO_2 adsorbent. Recently, Xia et al. (2011) developed zeolite template N-doped carbons and found that these porous supports

TABLE 8.1
Chemical Agents Used for CO_2 Sequestration (Sharma et al. 2019)

Name of the chemical	CO_2 absorption Capacity (mol CO_2/mol solvent)
DETA-AMP-PMDETA	3.17
Triethylenetetramine (TETA)-polyethylene glycol (PEG200)	1.63
DETA/sulfolane	1.78
Dibutylamine (DBA)/water/ethanol	0.82
Diglycolamine-PEG200	0.438
[C_2OHmim][Lys]/MDEA/water	0.75
Triethylenetetramine lysine	2.59
TETA, and fluoroboric acid (HBF4) [TETAH]+[BF4]	0.96
1-Aminoethyl-2,3-dimethylimidazolium cation and amino acid taurine anion [aemmim][Tau]	0.9
1-(3-Propyamino)-3-butyllimidazoliumtetrafluoroborate ([apbim]BF4)	0.5

TABLE 8.2
Porous Carbon-Based Materials (Sharma et al. 2019)

Name of the materials	CO_2-capture capacity (mmol g^{-1})
Hierarchically porous carbon	3.7
Zeolite–chitosan	1.7
N-doped porous carbon	3.88
N-doped phenolic resin	7.13
ZIF-100 MOF	0.7

have very high CO_2-capturing capacity than carbonaceous or other porous materials. Various researchers (Przepiórski, Skrodzewicz, and Morawski 2004; Kim et al. 2008; Son, Choi, and Ahn 2008) also reported similar observations. The CO_2-capturing capacity of different porous materials is given in Table 8.2 (Sharma et al. 2019).

8.3 CO_2 SEQUESTRATION: NANOMATERIAL-BASED PROCESSES

The nanomaterial is defined as a natural, manufactured, or incidental material that has one nanoscale dimension – the dimension range is in between 1 and 100 nm. Nanodimension of any materials can improve their catalytic, electrical, mechanical, magnetic, thermal, and/or imaging properties, which are highly desirable in the military, medical, commercial, and environmental sectors. Till date, many nanomaterials are synthesized and can be categorized as carbonaceous (carbon based) or non-carbonaceous. Often non-carbonaceous nanomaterials mixed with other nanomaterials or macroscopic carbonaceous or non-carbonaceous materials are used for the synthesis of nanocomposites. In these subsections, all of these materials will be briefly described.

8.3.1 Carbonaceous

Carbon is one of the most versatile types of elements found in Earth. Due to its allotropic nature, carbon forms two completely different compounds by rearranging the adjacent carbon atoms. Since the last few decades, carbon compounds in nanodimensions are being synthesized and being used for many advanced and innovative applications. Carbon-based nanomaterials with different forms, e.g. tube, spherical, ellipsoids, or sheets, are frequently being synthesized. In this subsection, each type of carbonaceous nanomaterials along with their application potential will be briefly discussed.

8.3.1.1 Carbon Nanotube (CNT)

Carbon nanotube (CNT) is a cylindrical carbon nanostructure that can be compared with the rolled graphene sheets. CNT has been considered as one of the most studied carbon nanostructures in the last few decades. This nanomaterial has a diameter of only a few nanometres with a length of very higher magnitude compared to the diameter that makes them nearly one dimensional in its structure. CNT has very unique mechanical and electrical properties when compared with other carbon allotrophs such as graphite and diamond. In 1991, Sumio Lijima first reported a pathway of CNT synthesis. Thereafter, numerous applications of CNTs make them to be studied rigorously. CNTs have many outstanding properties such as high reflex strength and elasticity; excellent structural, chemical, and thermal stability; and the high conductivity (Shenderova, Zhirnov, and Brenner 2002; Katz and Willner 2004; Gohardani, Elola, and Elizetxea 2014). Single-walled carbon nanotube (SWNT) and multiwalled carbon nanotube (MWNT) are the two main structural configurations. SWNT is a simple cylindrical structure, whereas MWNT contains several concentric SWNTs. Thus, the properties of these two forms of CNTs are different from one another. All the properties of CNTs mainly depend on diameter, length, and the morphology of the nanotubes. Till date, many methods have been proposed for CNT synthesis. Among them, three different procedures, namely pulsed laser, electric arc, and chemical vapour deposition (CVD), are mainly used to synthesize CNT from graphite (Koziol, Boskovic, and Yahya 2010; Dresselhaus, Dresselhaus, and Eklund 1996; Gohardani, Elola, and Elizetxea 2014).

Due to its unique properties, CNT has been utilized in different applications. For the preparation of sensors and actuators, CNT has been frequently used in electronic nanodevices. Biomolecules are often integrated with CNTs for their application in diagnosing disease, drug delivery, and processing of biomaterials (Katz and Willner 2004; Willner and Willner 2010; Gooding 2005; Wang 2005; Allen, Kichambare, and Star 2007; Balasubramanian and Burghard 2006). SWNT-based field-effect transistor is also prepared for detecting biological species in the control environment (Allen, Kichambare, and Star 2007; Balasubramanian and Burghard 2006). Due to its large surface area, high electrical conductivity, mechanical strength, and lightweight, CNT is being used readily as supercapacitor for energy storage applications. Optical transparency and flexibility of CNT are also advantageous as it allows developing flexible electronic devices such as pliable cellphones and computers. Metal oxides are often incorporated into CNT to reduce its contact resistance and improve its energy storage capacity (Li and Wei 2013; Li et al. 2013). CNT has five times higher elasticity and 50–200 times higher tensile strength than steel. In addition, CNT has high lightweight than any metal. These properties make it ideal for use in the preparation of aircraft materials (Gohardani, Elola, and Elizetxea 2014). Threads made of CNT are being used to prepare the bullet-proof jacket (Obradovic et al. 2015). CNT has also been used for the improvement of electrostatic shielding in plastic composites (Pande et al. 2014). CNT has been used for improving the mechanical strength of composites, including biocomposites, and has been applied in tissue engineering, bone, and cartilage replacement (Gupta et al. 2014; Wang et al. 2015).

The use of CNTs, before or after their modification, for CO_2 capture is a new trend since the last decades. Esteves et al. (2008) indicated that the CO_2 is adsorbed onto CNTs preferentially compared with CH_4 and H_2S. CNTs have often been utilized as fillers for the preparation of nanocomposites that are useful for gas separation applications. Yoo et al. (2014) indicated that the incorporation of CNTs dramatically improves the kinetics of gas storage nanomaterials. Recently, Kang et al. (2015) showed that MWCNTs can improve the CO_2 storage capacity of the nanocomposites and the CNTs. Also, they monitored around 70% increase in CO_2 captured for the nanocomposites compared with that for CNTs itself. Lu et al. (2008) also modified CNTs with silane reagents and found that the capacity of CO_2 adsorption improved by 40% compared with unmodified CNTs. Although the use of CNTs for the large-scale storage of CO_2 is found to be promising, it still needs further researches to be implemented. All the current studies related to gas-capturing properties of CNTs are mainly laboratory based and need to be validated by large-scale gas storage applications. CO_2-capturing capacities of different CNT-based materials are given in Table 8.3.

8.3.1.2 Graphene and Its Derivatives

Graphene is one of the exceptional carbon-based nanomaterials with extraordinary chemical, electrical, and mechanical properties. It is a single-layered two-dimensional

TABLE 8.3
CNT-Based CO_2 Capture (Kang et al. 2015)

Name of the nanomaterials	Capacity ($cm^3 g^{-1}$)
MWCNTs	7.4
JUC-32	41.7
Physical mixture	42.8
MWCNTs@JUC32-1	34.5
MWCNTs@JUC32-2	31.0

sheet where carbon atoms are arranged in a hexagonal lattice structure. Graphene is being used in batteries, fuel cells, sensors, supercapacitors, solar cells, and biotechnological processes. Graphene has $2.5 \times 10^5 cm^2\ Vs^{-1}$ electron mobility at ambient temperature, more than 1 TPa Young's modulus, and more than 3000 W mK^{-1} thermal conductivity. In addition, graphene is impermeable to any gas and sustains a current density far more than that of copper. Also, it can be easily functionalized. These exceptional properties of graphene make it an obvious choice for many advanced technological innovations.

The synthesis process of graphene can be divided into two processes, namely the top-down process and the bottom-up process. In the bottom-up process, graphene sheets are directly synthesized from molecular organic precursors. The precursors contain a benzene ring with highly reactive functional groups. Another process is the use of catalyst substances for graphene synthesis via CVD, arc discharge, or SiC epitaxial growth. However, these processes are expensive and cannot produce a uniform and large amount of graphene. The top-down method is rather easy, simple, and used for the large-scale graphene production through chemical and mechanical exfoliation methods of graphite. However, this method also suffers from limitation like the preparation of two-dimensional graphene sheets. Graphene oxide (GO) is an intermediate of this process. GO is converted into a graphene-like material through the reduction process. Though, the product will have structural differences compared with the pure graphene.

Graphene and its derivatives are being used in many applications, including energy storage and dye-sensitizing solar cell (Dhand et al. 2013). It has been utilized in lithium ion batteries (Dhand et al. 2013). Being transparent, flexible, highly conductive, mechanically resistant, and lightweight, graphene can be an excellent alternative for supercapacitor applications. Graphene can be used in photonics and as a photodetector, polarization controller, insulator, and solid-state laser (Novoselov et al. 2012). Alike CNT, graphene is also used in the synthesis of sensors and biosensors (Kuila et al. 2011; Huang et al. 2010). In addition, it has a promising future in the field of medicine.

Graphene and its derivatives are relatively inexpensive when compared with the available solid-state gas adsorbents. Graphene nanosheets have been utilized for CO_2 storage and showed very high CO_2-capturing abilities (Mishra and Ramaprabhu 2011). Graphene is often modified with different ways to improve its gas-capturing capacities. Modification of GO with porous MgO (Ning et al. 2012) and double-layered Mg–Al hydroxide (Garcia-Gallastegui et al. 2012) has been reported for the improvement of the storage capacity of the gas. It was noticed that the CO_2 storage capacity of the double-layered metal hydroxide can be improved by around 62% by the addition of 7% GO. Also, the use of GO improves the stability of the double-layered metal hydroxide upon reuse and increases the effective surface area. Recently, Fe_2O_3 nanoparticle-coated graphene has been developed by Mishra and Ramaprabhu (2014). The Fe_2O_3-coated graphene shows very high CO_2-capturing capacities. As the sorption of the gas molecule on GO surface depends on electrostatic and van der Waals interactions, charge transfer, and dispersion interactions, the related modifications of the graphene surface can be beneficial for the improvement in its gas sorption ability.

8.3.1.3 Nanoporous Carbon

Recently, a series of nanoporous carbon materials have been synthesized for their application in the storage of natural gases (Liu et al. 2006). These nanoporous materials have large pore amount, definite surface area, thermal stability, mechanical stability, and organized pore structure. All these properties make them an exceptional choice for use in the storage of gas. Biomasses from different origins have been utilized for the preparation of nanoporous carbon (Lee, Han, and Hyeon 2004; Kumar et al. 2018). For the synthesis of nanoporous carbon structure, different biomasses have been modified either using an activating agent or using some templates like silica.

8.3.2 Non-Carbonaceous

8.3.2.1 Nanosized Zeolites

Zeolites are highly porous and three-dimensional crystalline aluminosilicates containing alkaline earth metals such as Na^+, K^+, and Ca^+. Due to their highly ordered porous structure, zeolites are being used for many applications, including gas separation (Yang and Xu 1997). Although zeolites favour CO_2 adsorption, their affinity decreases with increased temperature and under humid condition. To improve the separation ability, zeolites are modified in the nanoscale materials. Jiang et al. (2013) synthesized a nanosized T-type zeolite with the pore size of 0.36×0.51 nm. The pore size of T-type zeolite is smaller than those of N_2 and methane molecules, however larger than the CO_2 molecule. The researchers found that such modification enhances the separation ability of the nanosized zeolites by 30%. Nanosized zeolite and silica-based composites have also developed for the improvement of gas separation capacity of zeolites (Auerbach, Carrado, and Dutta 2003).

8.3.2.2 Mesoporous Silica Nanoparticles

Mesoporous silica nanoparticles have gained attention due to their exceptional pore size of 30 nm. This silica nanoparticle has ordered structure of the hexagonal array. This nanoparticle shows around 31% higher gas separation ability than the activated carbon in identical condition (Liu et al. 2006). Qi et al. (2011) studied that amine-functionalized mesoporous silica-based gas separator has been prepared and used for CO_2 separation. This process is reported to be inexpensive and easy to recover CO_2 from the reaction. Atriamine-grafted mesoporous silica was also synthesized for CO_2 separation by researchers (Belmabkhout, De Weireld, and Sayari 2009) and found to separate CO_2 more efficiently.

8.3.2.3 Nanoparticle-Embedded Metal–Organic Framework

The organo-metal framework is a new emerging class of nanoporous materials forming a 3D network with a centralized cation surrounded by an organic framework. These materials are getting attention in recent times due to the possibility of using them for catalysis, separation, gas storage, and optics applications. However, Bloch et al. (2013) showed that the metal–organic framework nanoparticle (MOFN) often shows poor gas adsorption capacities in low CO_2 partial pressure. Many reports are available for the enhancement of gas adsorption properties of MOFNs. In some studies, the pore surface was functionalized with an amine to enhance the gas adsorption capacities of MOFN. Recently, zinc-based MOFN has been synthesized that showed very high CO_2 adsorption capacities (Millward and Yaghi 2005). Amino-functionalized zirconium-based MOFN has also been developed that showed good CO_2 storage capacities (Abid et al. 2013). However, it should be mentioned that there are some technical difficulties related to the use of MOFN for packed bed column. CO_2 adsorption capacities of MOFs were further enhanced by the incorporation of CNT-like nanomaterials (Khdary and Ghanem 2012; Xiang et al. 2011).

8.3.2.4 Metal Oxide Nanoparticles

Nanostructured metal and metal oxide nanoparticles are attractive options for use in gas separation. Mesoporous MgO with nanostructure has been produced by using mesoporous carbon as a template, and the materials showed CO_2 adsorption capability (Bhagiyalakshmi, Lee, and Jang 2010). Nano-hollow-structured CaO has also been synthesized and showed significant CO_2 adsorption properties (Yang et al. 2009). These reports also indicated that CaO nanopods showed higher CO_2 adsorption capacity than the commercial CaO.

Other metal nanoparticles such as hydroxylated Fe_2O_3, Al_2O_3, and TiO_2 were also tested for CO_2 adsorption and found to be useful (Baltrusaitis et al. 2011). The reports indicated that hydroxylated Fe_2O_3 and Al_2O_3 capture CO_2 in the form of bicarbonate. However, TiO_2 nanoparticles capture CO_2 in the form of bidentate carbonates. Recently, Fe_2O_3-amended activated charcoal has been developed (Hakim et al. 2015) for enhanced CO_2 adsorption. CO_2 adsorption was enhanced after the nanoparticle amendment.

8.4 CONCLUSIONS

This chapter indicates that different types of non-nanomaterials and nanomaterial-based adsorbents have been developed and utilized for CO_2 sequestration. Nanomaterial-based adsorbents are found to be better or comparable to the non-nanomaterial-based process. However, some emerging non-nanomaterial-based adsorbents like ionic liquids show very high CO_2 sorption capacity. Also, they can be regenerated and recycled relatively easily. Although nanomaterials show high sorption, their regeneration and reuse efficiency are unclear until now. Also, nanomaterial-based CO_2 adsorption processes need high pressure for CO_2 capture. In addition, some nanomaterials show poor CO_2 adsorption capacity compared with the activated charcoal. Thus, it should be concluded that the nanomaterials can be promising alternatives, but more research is needed to establish their efficacy over the conventional alternatives.

8.5 FURTHER RESEARCH NEEDED

Even though CO_2 geologic sequestration appears feasible, further research is needed before this option can be fully implemented. The researchers need to have an idea about the sequestration sites and their capacities along the geological formations to hold back CO_2. They also need technologies for continuous monitoring to assure the public that the safe CO_2 storage is possible – this may be the most important objective of all. In addition, improvements in computer-based simulation models for predicting the CO_2 performance during the geological sequestration are needed. Scientists are also looking for ways to lower the overall cost of geological sequestration by improving the technology for CO_2 capture. If they learn that some less-pure CO_2 waste streams (like those containing H_2S and/or SO_2) can enhance the sequestration process, or at least cannot interfere with it, the cost of separating CO_2 from a coal gasification or power-plant flue gases may drop dramatically.

ACKNOWLEDGEMENTS

A.B. acknowledges the University Grant Commission for providing research opportunity with the Dr. D. S. Kothari Post-Doctoral Fellowship (UGC-DSKPDF, File No: F.4-2/2006 (BSR)/OT/18–19/0009). The authors also acknowledge the findings of the research contributors who worked on these issues for the last several decades and also the acknowledge the anonymous persons responsible for making the web databases available during extensive database search.

TiO_2: titanium dioxide

REFERENCES

Abid, H.R., J. Shang, H.-M. Ang, and S. Wang. 2013. "Amino-functionalized Zr-MOF nanoparticles for adsorption of CO_2 and CH_4." *Int. J. Smart Nano Mater.* 4: 72–82.

Allen, B.L., P.D. Kichambare, and A. Star. 2007. "Carbon nanotube field-effect-transistor-based biosensors." *Adv. Mater.* 19: 1439–1451.

Auerbach, S.M., K.A. Carrado, and P.K. Dutta. 2003. *Handbook of Zeolite Science and Technology.* Boca Raton, FL: CRC Press.

Balasubramanian, K., and M. Burghard. 2006. "Biosensors based on carbon nanotubes." *Anal. Bional. Chem.* 385: 452–468.

Baltrusaitis, J., J. Schuttlefield, E. Zeitler, and V.H. Grassian. 2011. "Carbon dioxide adsorption on oxide nanoparticle surfaces." *Chem. Eng. J.* 170: 471–481.

Belmabkhout, Y., G. De Weireld, and A. Sayari. 2009. "Amine-Bearing Mesoporous Silica for CO_2 and H_2S Removal from Natural Gas and Biogas." *Langmuir* 25: 13275–13278.

Bhagiyalakshmi, M., J.Y. Lee, and H.T. Jang. 2010. "Synthesis of mesoporous magnesium oxide: Its Application to CO_2 Chemisorption." *Int. J. Greenhouse Gas Control* 4: 51–56.

Bloch, W.M., R. Babarao, M.R. Hill, C.J. Doonan, and C.J. Sumby. 2013. "Post-synthetic structural processing in a metal–organic framework material as a mechanism for exceptional CO_2/N_2 selectivity." *J. Am. Chem. Soc.* 135: 10441–10448.

Cheng-Hsiu, Y., H. Chih-Hung, and C.S. Tan. 2012. "Review of CO_2 capture by absorption and adsorption." *Aerosol Air Qual. Res.* 12: 745–769.

Dhand, V., K.Y. Rhee, H.J. Kim, and D.H. Jung. 2013. "A comprehensive review of graphene nanocomposites: Research status and trends." *J. Nanomater*, 2013: 1–14.

Dresselhaus, M.S., G. Dresselhaus, and P.C. Eklund. 1996. Science of Fullerenes and Carbon Nanotubes. Elsevier. 965.

Esteves, I.A.A.C., M.S.S. Lopes, P.M.C. Nunes, and J.P.B. Mota. 2008. "Adsorption of natural gas and biogas components on activated carbon." *Sep. Purif. Technol.* 62: 281–296.

Galán Sánchez L.M.; Meindersma G.W.; de Haan A.B. 2007. "Solvent properties of functionalized ionic liquids for CO_2 absorption." *Chem. Eng. Res. Des.* 85: 31–39.

Garcia-Gallastegui, A., D. Iruretagoyena, V. Gouvea, M. Mokhtar, A.M. Asiri, S.N. Basahel, S.A. Al-Thabaiti, A.O. Alyoubi, D. Chadwick, and M.S.P. Shaffer. 2012. "Graphene oxide as support for layered double hydroxides: Enhancing the CO_2 adsorption capacity." *Chem. Mater.* 24: 4531–4539.

Gohardani, O., M.C. Elola, and C Elizetxea. 2014. "Potential and prospective implementation of carbon nanotubes on next generation aircraft and space vehicles: A review of current and expected applications in aerospace sciences." *Prog. Aerosp. Sci.* 70: 42–68.

Gooding, J.J. 2005. "Nanostructuring Electrodes with Carbon Nanotubes: A Review on Electrochemistry and Applications for Sensing." *Electrochim. Acta* 5: 3049–3060.

Gupta, A., B.J. Main, B.L. Taylor, M. Gupta, C.A. Whitworth, C. Cady, J.W. Freeman, and S.F. El-Amin III. 2014. "In vitro evaluation of three-dimensional single-walled carbon nanotube composites for bone tissue engineering." *J. Biomed. Mater. Res. A* 102: 4118–4126.

Hakim, A., M.N. Abu Tahari, T.S. Marliza, W.N.R. Wan Isahak, M.R. Yusop, M.W. Mohamed Hisham, and M.O. Yarmoa. 2015. "Study of CO_2 adsorption and desorption on activated carbon supported iron oxide by temperature programmed desorption." *Jurnal Teknologi* 77: 75–84.

Himeno, S., T. Komatsu, and S. Fujita. 2005. "High-pressure adsorption equilibria of methane and carbon dioxide on several activated carbons." *J. Chem. Eng. Data* 50: 369–376.

Huang, Y., X. Dong, Y. Shi, C.M. Li, L.J. Li, and P. Chen. 2010. "Nanoelectronic biosensors based on CVD grown graphene." *Nanoscale* 2: 1485–1488.

Jiang, J-Q., C-X. Yang, and X-P. Yan. 2013. "Zeolitic imidazolate framework-8 for fast adsorption and removal of benzotriazoles from aqueous solution." *ACS Applied Mater. Interfaces* 5 (19): 9837–9842.

Kang, Z., M. Xue, D. Zhang, L. Fan, Y. Pan, and S. Qiu. 2015. "Hybrid metal-organic framework nanomaterials with enhanced carbon dioxide and methane adsorption enthalpy by incorporation of carbon nanotubes." *Inorg. Chem. Commun.* 58: 79–83.

Katz, E., and I. Willner. 2004. "Biomolecule-functionalized carbon nanotubes: Applications in Nanobioelectronics." *ChemPhysChem* 5: 1085–1104.

Khdary, N.H., and M.A. Ghanem. 2012. "Metal-organic-silica nanocomposites: Copper, silver nanoparticles-ethylenediamine-silica gel and their CO_2 adsorption behaviour." *J. Mater. Chem.* 22: 12032–12038.

Kim, S.-N., W.-J. Son, J.-S. Choi, and W.-S. Ahn. 2008. "CO_2 adsorption using amine-functionalized mesoporous silica prepared via anionic surfactant-mediated synthesis" *Microporous Mesoporous Mater* 115: 497–503.

Koziol, K., B.O. Boskovic, and N. Yahya. 2010. "Synthesis of carbon nanostructures by CVD method." In *Carbon and Oxide Nanostructures, Advanced Structured Materials.* (Eds) N. Yahy, 23–49.

Kuila, T., S. Bose, P. Khanra, A.K. Mishra, N.H. Kim, and J.H. Lee. 2011. "Recent advances in graphene-based biosensors, biosens." *Bioelectron* 26: 4637–4648.

Kumar, K.T., G.S. Sundari, E. S. Kumar, A. Ashwini, M. Ramya, P. Varsha, R. Kalaivani, et al. 2018. "Synthesis of nanoporous carbon with new activating agent for high-performance supercapacitor." *Mater. Lett.,* 218: 181–84.

Lee, J., S. Han, and T. Hyeon. 2004. "Synthesis of new nanoporous carbon materials using nanostructured silica materials as templates." *J. Mater. Chem.,* 14: 4287–4486.

Lee, S.C., B.Y. Choi, T.J. Lee, C.K. Ryu, Y.S. Ahn, and J.C. Kim. 2006. "CO_2 absorption and regeneration of alkali metal-based solid sorbents." *Catal Today* 111: 385–390.

Lee, Z.H., K.T. Lee, S. Bhatia, and A.R. Mohamed. 2012. "Post-combustion carbon dioxide capture: Evolution towards utilization of nanomaterials." *Renew. Sust. Energ. Rev.* 16: 2599–2609.

Li, X., and B. Wei. 2013. "Supercapacitors based on nanostructured carbon." *Nano Energ.* 2: 159–173.

Li, Y., H. Yi, X. Tang, F. Li, and Q. Yuan. 2013. "Adsorption separation of CO_2/CH_4 gas mixture on the commercial zeolites at atmospheric pressure." *Chem. Eng. J.* 229: 50–56.

Liu, X., L. Zhou, J. Li, Y. Sun, W. Su, and Y. Zhou. 2006. "Methane sorption on ordered mesoporous carbon in the presence of water." *Carbon* 44: 1386–1392.

Lu, C., H. Bai, B. Wu, F. Su, and J.F. Hwang. 2008. "Comparative study of CO_2 capture by carbon nanotubes, activated carbons, and zeolites." *Energy Fuel* 22: 3050–3056.

Millward, A.R., and O.M. Yaghi. 2005. "Metal–organic frameworks with exceptionally high capacity for storage of carbon dioxide at room temperature." *J. Am. Chem. Soc.* 127: 17998–17999.

Mishra, A.K., and S. Ramaprabhu. 2011. "Carbon dioxide adsorption in graphene sheets." *AIP Adv.* 1: 032152.

Mishra, A.K., and S. Ramaprabhu. 2014. "Enhanced CO_2 capture in Fe_3O_4-graphene nanocomposite by physicochemical adsorption." *J. Appl. Phys* 116: 064306.

MIT. 2007. "The Future of Coal: Options for a Carbon-Constrained World." https://web.mit.edu/coal/The_Future_of_Coal.pdf.

Ning, G., C. Xu, L. Mu, G. Chen, G. Wang, J. Gao, Z. Fan, W. Qian, and F. Wei. 2012. "High capacity gas storage in corrugated porous graphene with a specific surface area-lossless tightly stacking manner." *Chem. Commun.* 48: 6815–17.

Novoselov, K.S., V.I. Fal′ko, L. Colombo, P.R. Gellert, M.G. Schwab, and K. Kim. 2012. "A Roadmap for Graphene." *Nature* 412: 192–200.

Obradovic, V., D.B. Stojanovic′, I. Živkovic′, P.S. Radojevic′, R. Uskokovic′, and R. Aleksic′. 2015. "Dynamic mechanical and impact properties of composites reinforced with carbon nanotubes." *Fiber. Polym.* 16: 138–145.

Ochoa-Fernández, E., H.K. Rusten, H.A. Jakobsen, M. Rønning, A. Holmen, and D. Chen. 2005. "Sorption enhanced hydrogen production by steam methane reforming using Li_2ZrO_3 as sorbent: Sorption kinetics and reactor simulation." *Catal. Today* 106: 41–46.

Pande, S., A. Chaudhary, D. Patel, B.P. Singh, and R.B. Mathur. 2014. "Mechanical and electrical properties of multiwall carbon nanotube/polycarbonate composites for electrostatic discharge and electromagnetic interference shielding applications." *RSC Adv.* 4: 13839–13849.

Park, Y., K.Y. Lin, A.H. Park, and C. Petit. 2015. "Recent advances in anhydrous solvents for CO_2 capture: Ionic liquids, switchable solvents, and nanoparticle organic hybrid materials." *Frontiers Energ. Res.* 3: 1–4.

Paustian, K.H., and C.V. Cole. 1998. "CO_2 mitigation by agriculture – an overview: Climatic change." *Clim. Change* 40 (1): 135–162. doi:10.1023/A:1005347017157.

Przepiórski, J., M. Skrodzewicz, and A.W. Morawski. 2004. "High temperature ammonia treatment of activated carbon for enhancement of CO_2 adsorption." *Appl. Surf. Sci.* 225: 235–242.

Qi, G., Y. Wang, L. Estevez, X. Duan, N. Anako, A.-H.A. Park, W. Li, C.W. Jones, and E.P. Giannelis. 2011. "High efficiency nanocomposite sorbents for CO_2 capture based on amine functionalized mesoporous capsules." *Energ. Environ. Sci.* 4: 444–452.

Ramdin, M., S.P. Balaji, A. Torres-Knoop, D. Dubbeldam, T. W. de Loos, and T.J. H. Vlugt. 2015. "Solubility of natural gas species in ionic liquids and commercial solvents: Experiments and monte Carlo simulations." *J. Chem. Eng. Data* 60: 3039–3045.

Saha, D., Z. Bao, F. Jia, and S. Deng. 2010. "Adsorption of CO_2, CH_4, N_2O, and N_2 on MOF-5, MOF-177, and Zeolite 5A." *Environ. Sci. Technol.* 44: 1820–1826.

Sharma, T., S. Sharma, H. Kamyab, and A Kumar. 2019. "Energizing the CO_2 utilization by chemo-enzymatic approaches and potentiality of carbonic 2 anhydrases: A Review." *J. Clean. Prod.* 247: 119–138.

Shenderova, O.A., V.V. Zhirnov, and D.W. Brenner. 2002. "Carbon nanostructures." *Rev. Solid State Mater. Sci.*, 227–356.

de Silva, L.L., L.J. Cihacek, F.L. Leistritz, T.C. Faller, D.A. Bangsund, J.A. Sorensen, E.N. Steadman, and J.A. Harju. 2005. The Contribution of Soils to Carbon Sequestration *In* Plains CO_2 Reduction (PCOR) Partnership Practical, Environmentally Sound CO_2 sequestration, University of North Dacota EERC, 1–23.

Son, W.-J., J.-S. Choi, and W.-S. Ahn. 2008. "Adsorptive removal of carbon dioxide using polyethyleneimine-loaded mesoporous silica materials." *Microporous Mesoporous Mater.* 113: 31–40.

Wang, H., C. Chu, R. Cai, S. Jiang, L. Zhai, J. Lu, X. Li, and S. Jiang. 2015. "Synthesis and bioactivity of gelatin/multiwalled carbon nanotubes/hydroxyapatite nanofibrous scaffolds towards bone tissue engineering." *RSC Adv.* 5: 53550–53558.

Wang, J. 2005. "Carbon-nanotube based electrochemical biosensors: A review." *Electroanalysis* 17: 7–14.

Willner, I., and B. Willner. 2010. "Biomolecule-based nanomaterials and nanostructures." *Nano Lett* 10: 3805–3815.

Xia, Y., R. Mokaya, G.S. Walker, and Y. Zhu. 2011. "Superior CO_2 adsorption capacity on Ndoped, high-surface-area, microporous carbons templated from zeolite." *Adv. Energy Mater.* 1: 678–683.

Xiang, Z., Z. Hu, D. Cao, W. Yang, J. Lu, B. Han, and W. Wang. 2011. "Metal–organic frameworks with incorporated carbon nanotubes: Improving carbon dioxide and methane storage capacities by lithium doping." *Angew. Chem. Int. Ed.* 50: 491–494.

Yang, C., and Q. Xu. 1997. "Aluminated zeolites β and their properties Part 1.-alumination of zeolites β." *J. Chem. Soc. Faraday Trans* 93: 1675–1680.

Yang, Z., M. Zhao, N.H. Florin, and A.T. Harris. 2009. "Synthesis and characterization of CaO nanopods for high temperature CO_2 capture." *Ind. Eng. Chem. Res.* 48: 10765–10770.

Yoo, J., S. Lee, C.K. Lee, C. Kim, T. Fujigaya, H.J. Park, N. Nakashima, and J.K. Shim. 2014. "Homogeneous decoration of zeolitic imidazolate framework-8 (ZIF-8) with core-shell structures on carbon nanotubes." *RSC Adv* 4: 49614–49619.

9 Porous Materials for CO_2 Fixation

Activated Carbon, MOFs, Nanomaterials

Maryam Takht Ravanchi and Mansooreh Soleimani

CONTENTS

9.1 INTRODUCTION

Carbon dioxide capture from flue gas is a subject of interest nowadays, as it reduces the unfavourable influence of CO_2 on global climate change (Noh et al., 2019). It was reported that in 2017, U.S. coal plants produced about 1207 million metric tons of carbon dioxide, which was about 69% of CO_2 emissions by entire U.S. electric power sectors. Energy Information Administration (EIA) recently reported that between 2010 and 2040, energy consumption is increased by 56%. Till 2040, about 80% of the world's energy is supplied by fossil fuels, and more than 50% of the delivered energy is consumed by the industrial section by which the most significant share of emissions is obtained. On the other hand, it is estimated that CO_2 emissions from power plants will be increased by 46% until 2040 (https://www.nanowerk.com/nanotechnology-news2/newsid=51841.php).

Three different technologies can be used for the reduction of carbon dioxide emission from fossil fuel-based power plants (Figure 9.1), namely pre-combustion capture, post-combustion capture, and oxy-fuel capture, by which approximately 90% of CO_2 can be captured. The amount of capture can be increased but with the expense of an increase in separation cost (Takht Ravanchi et al., 2011; Songolzadeh et al., 2014; Takht Ravanchi et al., 2014; Sharma et al., 2019):

- *Pre-combustion capture*: The first step, before combustion, is fuel decarbonization. In the air separation unit of coal-powered plants, pure oxygen is produced by this technology. In the next step, oxygen and pulverized coal are subjected to react and synthesize syngas. In water–gas shift reaction (WGSR), syngas and steam are reacted to produce CO_2 and H_2. Finally, the produced stream is physically washed, and CO_2 is separated and dehydrated. Hydrogen, which is a clean source of energy, is obtained as well (Jin and Zhang, 2011; Yu et al., 2012). The main advantage of this method is the production of a gas stream that has high CO_2 partial pressure and can be easily separated by different solvents; of course, power generation plant needs some modifications and alterations.
- *Post-combustion capture*: In this process, the flue gas stream (which consists of nitrogen, oxygen, and water vapour) is combusted, and CO_2 is

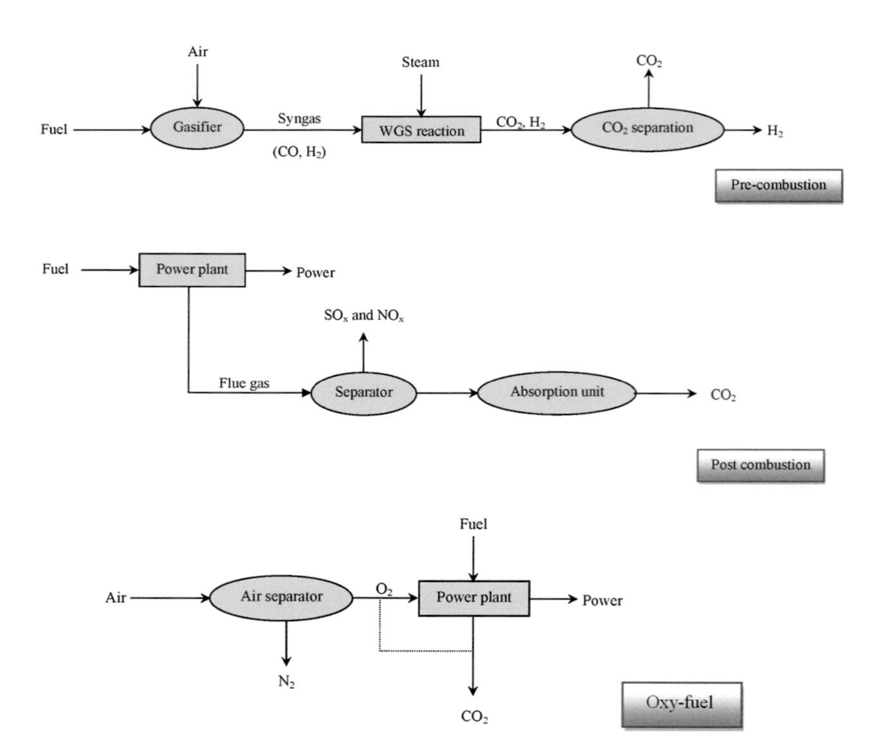

FIGURE 9.1 Different configurations of CO_2 emission reduction in power plants.

separated from it. The effluent stream is sent to another section where 'fly ash' (small particle) and sulphur are omitted. This process is a downstream unit. The drawbacks and limitations of this technique are low CO_2 partial pressure, a large amount of CO_2, and high temperature of the flue gas (Figueroa et al., 2008; Olajire, 2010).

- *Oxy-fuel capture*: At first, oxygen and nitrogen of the airstream must be separated. Then, under a pure O_2 environment, the fuel is burned, and water vapour and pure carbon dioxide are produced that can be recovered by a condensation unit. In this process, pure oxygen and burned fuel are recirculated by which all contents are entirely combusted, and CO_2 concentration of the flue gas is increased.

It is worth mentioning that post-combustion is the most common and advanced one.

9.2 EXISTING TECHNOLOGIES FOR CO_2 CAPTURE

The carbon dioxide-capture cycle consists of two or three main unit operations by which CO_2 is separated from the gas mixture and sent to a final destination. In the first unit, utilizing a medium (such as solvent or adsorbent), CO_2 is separated from the gas phase. In the second unit, the medium might be regenerated (or not), after which CO_2 is released. In the third unit, captured CO_2 is compressed and cooled in order to be handled as a liquid or at high density (Araujo et al., 2014, Songolzadeh et al.; 2015; Smit, 2016).

For post-combustion carbon capture, different technologies can be used, which are depicted in Figure 9.2.

Researchers must take into account several parameters for the commercialization of a separation process. The first important parameter is material selection. Gas molecule size and its electronic behaviour are two parameters that affect the material selection for CO_2 capture. It is worth mentioning that electronic properties are more critical than molecular size; for instance, the molecular size of CH_4, CO_2, and N_2 is 3.76, 3.30, and 3.64 Å, respectively, that has no significant difference but their quadra-polar moment and polarization are significantly different (Li et al., 2011a, b).

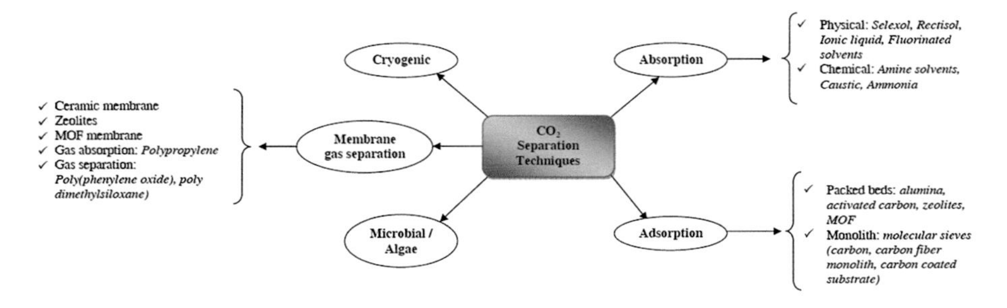

FIGURE 9.2 Various carbon dioxide separation techniques.

Cryogenic distillation is an energy-intensive one, as it has a phase transition. In this method, there is a difference between temperature and pressure of the liquid-state flue gas; carbon dioxide is cooled, condensed, and removed from the flue gas (Song et al., 2012).

In the carbon dioxide absorption process, CO_2 is absorbed from a gas stream to absorbent and the gas stream, which is free from CO_2, is passed through the absorption column. Then, the absorbent is regenerated to be reused for carbon dioxide capture. The main disadvantages of CO_2 absorption are as follows (Alavinasab et al., 2011; Yan et al., 2012; Zhang et al., 2013):

- A need to control equipment corrosion,
- Large energy requirement for the regeneration sector,
- Solvent sensitivity to probable impurities present in flue gas stream.

Chemical absorption of CO_2 by alkanolamines has advantages; at low temperatures and high CO_2 fugacity, it can reversibly bind with alkanolamines; on the other hand, at high temperature and low CO_2 fugacity, stripping of the absorbed CO_2 is done, and the regeneration of the solvent takes place. MEA (monoethanolamine), DEA (diethanolamine), and MDEA (methyl diethanolamine) and their blends are some of the common solvents used in gas absorption, which are given in Table 9.1 (Medeiros et al., 2013).

As amine-based technologies have some limitations, aqueous ammonia is another alternative to be used as a solvent. The main advantage of ammonia is its low heat of absorption. On the other hand, other impurities such as NO_x and SO_x can also be absorbed by ammonia. The main disadvantage of ammonia-based systems is the need to cool the flue gas before entering the absorption column, due to which

TABLE 9.1

A Comparison Between Common Solvents Used for CO_2 Absorption

Amine	Advantages	Disadvantages
MEA	Acceptable capacity for adsorption	High energy requirement for regeneration
	Low cost	Thermal or chemical degradation or corrosion
	Fast kinetics	Negligible evaporation loss, due to its low boiling point
	Water miscibility	Amine stability during regeneration
	Low corrosion	
DEA	Low energy requirement	Probable high corrosion
	Less volatile and corrosive	High pressure (>200 psig) requirement for reaching high purity
	Degradation resistance	
MDEA	Highest CO_2 equilibrium capacity	High cost
	High selectivity	High pressure (>500 psig) requirement for reaching high purity
	Lowest regeneration cost	Very low absorption rate
	Lowest reaction enthalpy	
	Lowest heat of regeneration	
	Degradation resistance	

high energy is required for the treatment of a large volume of flue gas. Moreover, heat exchanger fouling may occur due to the deposition of ammonium bicarbonate (Zhuang et al., 2011).

Another alternative for amine-based solvents are ionic liquids (ILs). Low volatility, excellent thermal stability, the possibility of structure tuning, and high CO_2 solubility are the valuable properties of ILs. The major disadvantage of ILs is their high viscosity after CO_2 absorption. One way to improve its CO_2-capture capability is the incorporation of IL inside the pores of metal–organic framework (MOF) (which will be discussed in Section 9.4.3.1) (Ramdin et al., 2014; Zhou et al., 2014).

The two remaining alternatives, namely membrane separation and adsorption, are good candidates due to their low capital cost, energy efficiency, and large separation capacity (D'Alessandro et al., 2010; Gholami et al., 2013).

For CO_2 separation from the flue gas, recently researchers are interested in using the adsorption process. It is a physical process in which fluid is attached to the solid surface of the adsorbent (Songolzadeh et al., 2019). The main advantage of the adsorption process is their ease of regeneration using pressure or thermal processes, adsorbent durability and selectivity for carbon dioxide, adsorbent stability after regeneration cycles, and CO_2 adsorption capacity (Songolzadeh et al., 2012). Novel adsorbents are capable of reversible CO_2 capture from gas streams with several advantages, the most important of which are reduced regeneration energy, secure handling, larger capacity, and selectivity. Pressure, temperature, and vacuum swing adsorptions (PSA, TSA, and VSA, respectively) are separation techniques with reduced costs, and their required energy for regeneration is lower than amine-based absorption processes. On the other hand, the heat capacity of a solid sorbent is lower than that of amine solvent. It is worth mentioning that a suitable adsorbent is the one that has high CO_2 selectivity and adsorption capacity with a durable and long lifetime, and fast adsorption–desorption kinetics.

In the membrane separation systems, a polymer or ceramic membrane is used for CO_2 separation from the flue gas. Large-scale membrane modules have challenges, due to the specific design of the membrane required to be operable at relatively high temperature.

In the biofixation method, microalgae as a photosynthetic organism is used in order to capture anthropogenic carbon dioxide. Aquatic microalgae are preferred due to their higher carbon fixation rates. Bioprocesses that used microalgae are the expensive ones, but they produce other high-value compounds. In the microalgal photosynthesis process, calcium carbonate is precipitated as well, which will be a long-lasting carbon sink (Nakamura et al., 2003).

In this chapter, the adsorption process is selected to be evaluated for CO_2 separation, and various adsorbents (such as carbon-based materials, zeolites, and MOFs) are studied in detail.

9.3 CRITERIA FOR THE SELECTION OF SOLID SORBENT

The selection of a solid to be as an adsorbent for CO_2 separation is critical as it must have some specific criteria to be economic and operable for CO_2 separation (Yang, 2003; Ho et al., 2008).

Solid sorbent must have specific characteristics that will be suitable for CO_2 capture that can be named as below:

- *High CO_2 selectivity*: An essential parameter of the adsorbent material is its selectivity as it is detrimental to the purity of the captured CO_2. This purity influences its transportation and sequestration and consequently, the economics of the separation process. This parameter is defined as the ratio of CO_2 uptake to the adsorption of other gases present in the feed stream, such as N_2 (for post-combustion) and methane (for natural gas). Typically, the effluent gas of a fossil fuel power plant contains nitrogen and oxygen. Suitable CO_2 adsorbent is the one that has high CO_2 selectivity in the presence of these impurities. Moreover, it must have high CO_2 capacity in the presence of water vapour as well.
- *High CO_2 adsorption uptake*: Any reliable sorbent performance is determined by its adsorption capacity. It can be reported in terms of the amount of adsorbed CO_2 per unit mass of sorbent ($cm^3_{CO_2}/g_{sorbent}$ or $g_{CO_2}/g_{sorbent}$). This form of representation is called the gravimetric uptake. Another form of capacity representation is the volumetric uptake, which is typically reported as CO_2 uptake per volume of sorbent ($cm^3_{CO_2}/cm^3_{sorbent}$ or $g_{CO_2}/cm^3_{sorbent}$). The size of the adsorption column and the required energy for regeneration are determined by the adsorption capacity of the sorbent. CO_2 adsorption isotherm determines its equilibrium adsorption capacity, which is an essential parameter for the evaluation of the capital cost of the capture system (equilibrium adsorption capacity determines the amount of the required adsorbent by which the volume of adsorption vessel is determined). The required amount of adsorbent and the size of adsorption equipment can be reduced when the CO_2 adsorption capacity of the adsorbent is high. The adsorbent that has a minimum of 2–4 mmol CO_2 g^{-1} adsorption capacity is a suitable one for CO_2 separation.
- *Adequate kinetics*: The adsorbent that has a high adsorption rate and fast kinetics is a suitable one. Kinetics of CO_2 adsorption–desorption profile, which is determined by breakthrough experiments, influences the time of the adsorption–regeneration cycle. The shorter this time, the better is the sorbent as its required amount is lower, and CO_2 separation cost is low. Gas-phase mass transfer through adsorbent structure and intrinsic kinetics of CO_2 reaction with adsorbent functional groups have an influence on the overall kinetics of CO_2 adsorption on a functionalized adsorbent. In order to minimize the resistance to diffusion, the structure of the functionalized porous adsorbent can be tailored. In order to reduce the amount of the required adsorbent for a definite volume of flue gas, the adsorbent must have fast CO_2 adsorption/desorption.
- *Adsorption enthalpy*: This parameter determines the energy required for sorbent regeneration by which the cost of the regeneration process can be determined. Usually, for physisorption, the heat of adsorption is 25–50 kJ mol^{-1}, and that of chemisorption is 60–90 kJ mol^{-1} (Adams, 2010). Adsorbent

must be regenerable by a suitable protocol and must keep its sorption capacity after several regeneration cycles. The smoother the regeneration, the lower the CO_2-capture cost.

- *Mechanical strength*: During the cycles of adsorption and regeneration, the adsorbent must have structural and morphological stabilities and keep its CO_2-capture capacity, by which the adsorbent can maintain its high kinetics. On different operating conditions such as high temperature or vibration of high flow rate, a suitable adsorbent must have stable morphology and microstructure and should tolerate the presence of any impurity (such as SO_2, water vapour, and O_2) in the feed. Otherwise, a great make-up rate of adsorbent is required. Obviously, mechanical strength has a direct impact on the separation process economics.
- *Chemical stability to impurities*: Adsorbents suitable for CO_2 capture, especially amine-functionalized ones, should be tolerant to the oxidizing environment and resistant to common contaminants such as NO_x and SO_x. Of course, these impurities should be separated from the gas stream.
- *Stability during adsorption–regeneration cycles*: The lifetime of adsorbent and the frequency of its replacement are determined by its stability. If the solid sorbent has stability under various conditions (such as flue gas condition, operating condition, and different cycles of adsorption–regeneration), operating costs of the adsorption process will be reduced. Moreover, the sorbent must be stable in the presence of water vapour as well.
- *Sorbent cost*: Adsorbent cost has a direct impact on separation process economics. The suitable adsorbent is the one that has excellent adsorption attributes and has a low cost. Tarka et al. (2006) reported that the adsorbent cost of \$5/kg is an economical choice, but that of \$15/kg is not. For an ideal adsorbent, it is desirable and economical that CO_2-capture cost be approximately \$10/kg. Of course, in their synthesis protocol, environmental concerns must be considered as well.

The suitable and useful adsorbent is the one that can *economically* and *effectively* capture CO_2 from the gas stream.

9.4 POROUS MATERIALS FOR CO_2 FIXATION

During the years, researchers reported various materials to be used for CO_2 capture. General CO_2 adsorbents are the physical and chemical ones. From the viewpoint of energy efficiency, physical adsorbents are preferred to the chemical ones. Moreover, in physical adsorption, there is a balance between adsorbent affinity for CO_2 removal from a gas mixture and the required energy for regeneration. Three mechanisms were proposed for the adsorptive separation (Li et al., 2009):

- *Kinetic effect*, which is based on diffusion rate difference between gas stream components;

- *Molecular sieving effect*, which is based on shape or size difference between gas stream components;
- *Thermodynamic effect*, which is based on the interaction between the surface and the adsorbate.

Metal oxides, activated carbon (AC), zeolites, carbon molecular sieves (CMSs), hydrotalcite-like compounds, carbon nanotubes (CNTs), and chemically modified mesoporous materials are various adsorbents reported in the literature that can be used for the physical CO_2 separation (Choi et al., 2009; Zhao et al., 2010; Akhtar et al., 2012). The most important point is that they are hardly disturbed during adsorption. In physical adsorption, the pore size plays a crucial role in useful separation. For instance, microporous (pore size <2 nm) adsorbents have better selectivity for CO_2 separation from a CO_2–CH_4 mixture. AC is a suitable adsorbent as it has high CO_2 adsorption capacity, low cost, high hydrophobicity, and the low energy requirement for regeneration, and it is insensitive to moisture. For CO_2 separation from the CO_2–N_2 mixture, zeolite is a better carbonaceous material.

These adsorbents and other examples are discussed in detail in the below sections.

9.4.1 Carbon-Based Adsorbents for CO_2 Capture

Adsorbents that are based on carbon are valuable materials for CO_2 capture, as they are chemically inert and cheap, and have a high surface area (Yong et al., 2001). The most important advantage of porous carbons to zeolites is their hydrophobicity; however, they have a lower capacity for CO_2 uptake, which is due to the competitive adsorption of water (Adams et al., 1988; Hedin et al., 2010). Different carbon materials are evaluated for CO_2 separation, and the most common ones are AC (Walker et al., 1953; Heuchel et al., 1999), CMSs (Walker and Shelef, 1967; Okoye et al., 1997), CNTs (Yim et al., 2004; Matranga et al., 2003), and graphene heterostructures (Ghosh et al., 2008; Ohba and Kanoh, 2012; Taheri Najafabadi, 2015).

9.4.1.1 Activated Carbon

AC is a famous material for CO_2 separation as it is cheap, can be synthesized from available raw materials (used or natural carbon-based materials), and has a high surface area with an amorphous structure that is suitable for CO_2 separation. There is a weak interaction between adsorbed CO_2 and AC (due to the lack of active sites of AC); hence, low energy is required for its regeneration. On the other hand, as there is no electric field on the surface of AC, its CO_2 uptake is very low (Caglayan et al., 2013; Yin et al., 2013; David and Kopac, 2014; Deng et al., 2014; Heidari et al., 2014; Fiuza Jr et al., 2015; Montagnaro et al., 2015).

AC can be produced by pyrolysis of fly ash, biomass, or resins (as carbon-containing material) and any other carbonaceous materials. There is a uniform electric potential on the AC surface, which results in a lower enthalpy for CO_2 adsorption; consequently, AC in comparison with zeolite has lower CO_2 capacity. Due to the large surface area of ACs, at high pressures, they show greater adsorption capacities; hence, ACs can be used in different high-pressure gas-phase separation processes (Davini, 2002; Zhang et al., 2004).

In pre-combustion systems, AC is a good candidate for CO_2 separation from high-pressure flue gas. It was reported (Martin et al., 2010) that in pre-combustion configuration, the upper limit of AC capacity for CO_2 adsorption is 60–70 wt.%, whereas that in post-combustion configuration is 10–11 wt.%. For carbon-based adsorbents, in comparison with high surface area MOFs, volumetric carbon dioxide adsorption capacity is higher if material precursor and reaction conditions are selected carefully (Silvestre-Albero et al., 2011).

Different researchers investigated the performance of various synthesized and commercial ACs for CO_2 capture from low to high pressures. They also reported the application of PSA process with AC for CO_2 removal (Dong et al., 1990; Chen et al., 1997; Dreisbach et al., 1999; Mugge et al., 2000; Na et al., 2001; Radosz et al., 2008).

For CO_2–CH_4 and CO_2–N_2 separations, Kacem et al. (2015) made a comparison between AC performance and zeolite performance. At pressures above 4 bar, AC has higher CO_2 uptake than zeolites, and after regeneration, separated CO_2 by AC has higher purity. Moreover, in the presence of water vapour, the stability of AC is preserved, and no framework failure is observed (Xu et al., 2013).

Kikkinides and Yang (1993) reported the feasibility study of the PSA process for CO_2 separation and concentration by BPL AC and CMS. They used a feed stream containing 17% CO_2 with 4% O_2 and the rest as N_2, and at 25°C by AC, about 68.4% of CO_2 was recovered. Moreover, they reported that the equilibrium separation of CO_2 by AC is better than its kinetic separation by CMS. Due to the weaker interaction of CO_2 with AC, the adsorption heat of AC (–30 kJ mol^{-1}) is lower than that of zeolite (–36 kJ mol^{-1}).

Na et al. (2001) used a PSA process for CO_2 separation by a commercial AC. At 1 atm pressure, by increasing temperature from 15°C to 55°C, the CO_2 adsorption capacity was decreased from 3.2 to 1.6 mmol g^{-1}. At 55°C, with a feed stream containing 17% CO_2, 79% N_2, and 4% O_2, with PSA process, a maximum 34% recovery with a purity of 99.8% was obtained.

In 1998, Do et al. (1998) used Ajax AC for CO_2 separation at different temperatures and pressures up to 20 kPa. They reported that increasing adsorption temperature caused a considerable decrease in CO_2 adsorption capacity; namely, with increasing temperature from 25°C to 100°C, the CO_2 adsorption capacity was decreased from 0.75 to 0.11 mmol g^{-1}.

Siriwardane et al. (2001) studied the competitive adsorption of a mixture of 14.8% CO_2 and 85.2% N_2 on G-32H AC, and 13X and 4A molecular sieves. They reported that surface affinity of 13X is better for CO_2 and it can well be separated. Tang et al. (2004a and 2004b) and Maroto-Valer et al. (2005) activated anthracite coal at 890°C for 2 h and used it for pure CO_2 adsorption, by which 1.49 mmol g^{-1} CO_2 was adsorbed (based on thermogravimetric analysis). Wang et al. (2008) used coconut shells and bamboo chips for AC synthesis and evaluated their CO_2 adsorption in the presence of water. They reported that at low pressure, water has a detrimental influence on CO_2 adsorption. Radosz et al. (2008) evaluated a cheap process that uses carbon filters at low pressure for CO_2 separation. This process recovers about 90% of CO_2 with 90% purity. This process has a short sorption cycle and a fast sorption rate.

Researchers studied different routes to improve the adsorption capacity of ACs (Houshmand et al., 2013; Díez et al., 2015; Sethia and Sayari, 2015; Tseng et al., 2015). The incorporation of functional groups based on amines is one method. Mostazo-López et al. (2015) used a stepwise chemical treatment for the modification of AC surface. They successfully grafted amide and amine functional groups on the AC surface with only a 20% loss in the surface area. Gibson et al. (2015) impregnated polyamine on porous AC and found improved CO_2 capacity 12 times. Chitosan and triethylenetetramine (TETA) are other chemicals that can be impregnated on the surface of AC, and at 25°C and 40 bar, they increase CO_2 uptake by 60%–90%.

ACs are hydrophobic materials; hence, their effect in the presence of water is found to be reduced, and in under-hydrated conditions, a breakdown or a decrease in capacity does not occur. As the heat of adsorption of CO_2 on AC is lower, a lower regeneration temperature is required. At low pressure (<1 atm), for CO_2–N_2 adsorption, AC has moderate adsorption selectivity, and increasing pressure causes a decrease in selectivity. Physical adsorption of CO_2 by carbon-based materials is more energy efficient in comparison with metal oxides. As no new chemical bonds are formed between the sorbent and the sorbate, less energy is required for regeneration (Plaza et al., 2008).

9.4.1.2 Carbon Nanotubes

Recently, researchers focused on the use of CNTs for CO_2 adsorption as CNTs have promising electrical and thermal conductivities and high chemical and physical properties (Cinke et al., 2003; Su et al., 2009; Hsu et al., 2010; Lithoxoos et al., 2010; Sawant et al., 2012; Babu et al., 2013). Researchers investigated CO_2–N_2 adsorption on CNTs theoretically and experimentally (Zhao et al., 2002; Bienfait et al., 2004; Skoulidas et al., 2006; Huang et al., 2007; Dillon et al., 2008; Lu et al., 2008; Su et al., 2009; Hsu et al., 2010; Lithoxoos et al., 2010; Razavi et al., 2011). They reported that by choosing a suitable shape and pore size of CNT, CO_2 capture can be done successfully. By the application of a chemical functional group by the Fisher esterification method, CNT surface will be modified by which high adsorption storage capacity is obtained (Abuilaiwi et al., 2010). There are different surface functional groups and surface charge densities on CNTs, which cause large adsorption capacity by CNTs. These specific characteristics of CNT are obtained by the thermal treatment or the chemical modification of CNT. Recently, researchers focused on the chemical modification of CNT for its application in CO_2 separation from a flue gas stream. Functionalized CNT with amino groups is a case (Khelifa et al., 2004; Fatemi et al., 2011; Gui et al., 2013; Liu et al., 2014; Su et al., 2014). For instance, Su et al. (2011) reported the effect of the presence of 3-aminopropyltriethoxysilane (APTES) as a functional group in CNT at different adsorption temperatures. They reported that the higher the temperature, the lower the adsorption storage capacity, and the higher the water content, the higher the adsorption capacity. They concluded that CNT is a low-temperature adsorbent as at 20°C, the CO_2 adsorption capacity of APTES-CNT is 2.59 mmol g^{-1}. In order to reduce the regeneration time, Hsu et al. (2010) used a combination of thermal and vacuum adsorption systems. At 220°C, adsorption–regeneration cycles of APTES-CNT are twenty by keeping the adsorption capacity and physicochemical properties of CNT constant. In another research, polyethylene imine (PEI) functional group was used on the surface of single-walled carbon nanotube (SWCNT), and at 27°C, 2.1 mmol g^{-1} adsorption capacity was obtained (Dillon et al., 2008). It can be concluded that amine-functionalized CNTs are suitable CO_2 adsorbents, and for their regeneration, low energy is required. In some researches, the combined application of CNT and membrane was reported for CO_2 separation. In these cases, CNT acts as a mechanical reinforcing filler that provides high stability of adsorbent for CO_2 separation at 1.5 MPa with a selectivity of 43 and a permeability of 836 Barrers (Zhao et al., 2014).

Cinke et al. (2003) used SWCNT with a surface area of 1587 $m^2 g^{-1}$, a total pore volume of 1.55 $cm^3 g^{-1}$, and a micropore volume of 0.28 $cm^3 g^{-1}$ for CO_2 capture. They reported that in the temperature range 0°C–200°C, the adsorption capacity of SWCNTs for CO_2 is twice that of AC.

Lu et al. (2008) reported that at 25°C, raw CNT has lower CO_2 adsorption capacity in comparison with granular AC. Skoulidas et al. (2006) simulated diffusion and adsorption of CO_2–N_2 mixture at room temperature as a function of CNT diameter. They concluded that for diameters in the range of 1–5 nm, CO_2 diffusivities are not dependent on pressure and diffusion mechanism is not the Knudsen one. Huang et al. (2007) performed a Monte Carlo simulation by which they reported that the diameter of CNT has an influence on its CO_2 adsorption capacity. Razavi et al. (2011) reported that for the CO_2–N_2 mixture, CNTs among carbon-based materials have higher CO_2 selectivity. Moreover, CO_2 adsorption isotherm has Langmuir (type I) behaviour.

Researchers modified CNTs with amine functional groups and studied their application for CO_2 separation (Su et al., 2009; Kumar Mishra and Ramaprabhu, 2012; Lee et al., 2015a, 2015b). Lee et al. (2015a) used industrial-grade CNT, functionalized it with TEPA (tetraethylenepentamine), and investigated the effect of various parameters (such as regenerability, heat of adsorption, and CO_2 uptake) on their performance. At 70°C, this sorbent has 3.09 mmol g^{-1} CO_2 capacity. APTES, PEI, and other types of amines can be used for CNT impregnation (Gui et al., 2013; Ko et al., 2013).

9.4.1.3 Graphenes

In 2004, graphene was discovered. It has the largest surface area (2630 $m^2 g^{-1}$) among all carbon-based sorbents (Park et al., 1999; Patchkovskii, 2005). It is a planar sheet of C atoms that are extended in two dimensions.

As CO_2 is slightly acidic in nature, basic adsorbents have more affinity towards CO_2 capture (Boehm, 1994). For adsorbents that are functionalized with amine groups and specifically for those in which nitrogen is effectively incorporated in the support framework, high CO_2 uptake is observed (Arenillas et al., 2005; Hao et al., 2010; Wang et al., 2012; Zhao et al., 2012a, 2012b; Rezaei et al., 2013). One example is the incorporation of polyaniline (PANI), as a rich source of N_2 groups, on different supports (Yang et al., 2008; Blinova and Svec, 2012; Mishra and Ramaprabhu, 2012; Kemp et al., 2013).

Due to its high surface area and mechanical strength, graphene is a suitable support in this case. Moreover, it has a strong attraction for amine-containing molecule, and hydrothermally, it is stable. PANI–graphene nanocomposite shows excellent performance for CO_2 capture. Mishra and Ramaprabhu (2011) reported that at 11 bar pressure and temperatures of 25, 50, and 100°C, hydrogen-exfoliated graphene (HEG) has a CO_2 adsorption capacity of 21.6, 18, and 12 mmol g^{-1}, respectively. For PNI-HEG nanocomposites, at the same operating condition, CO_2 capacity of 75, 47, and 31 mmol g^{-1} is reported (Mishra and Ramaprabhu, 2012).

Polypyrrole (PPy) is a good CO_2 sorbent due to its basic nature and thermal stability, but it has a low surface area that can be compensated by its interface with graphene. Hence, the PPy–graphene composite exhibits high adsorption capacity. At 25°C and 1 bar, 4.3 mmolg^{-1} CO_2 adsorption is reported for PPy–graphene nanocomposite (Chandra and Kim, 2011; Chandra et al., 2012).

After 2011, researchers reported graphite-based separation of CO_2 (Asai et al., 2011; Meng et al., 2012; Hong et al., 2013; Gadipelli and Guo, 2015; Li et al., 2015a). Kemp et al. (2013) reported N-doped graphene composites with 1336 $m^2 g^{-1}$ surface area that have 2.7 mmol g^{-1} CO_2 capacity at 25°C and 1 atm with improved stability. Oh et al. (2014) investigated the application of borane-modified graphene that has 1.82 mmol g^{-1} CO_2 uptake at 1 atm and 25°C. Mesoporous graphene oxide (GO)–ZnO nanocomposite (Li et al., 2015b), GO (Wang et al., 2015), MIL-53, and its hybrid composite with graphene nanoplate (GNP) (Pourebrahimi et al., 2015), mesoporous TiO_2/GO nanocomposites (Chowdhury et al., 2015), MOF-5 aminated graphite oxide (AGO) (Zhao et al., 2013a), and UIO-66/GO composites (Cao et al., 2015) are the hybrid materials reported for CO_2 separation.

9.4.1.4 Carbon Molecular Sieves

CMSs are other sorbents that can be used for CO_2 capture. Due to their unique texture, kinetic separation of CO_2 is permitted. The best texture is the one with narrow pore size distribution (PSD), i.e. their pore mouths being of molecular size with a high micropore volume, by which good separation capacity and selectivity can be obtained (Foley, 1995).

Burchell et al. (1997) developed a molecular sieve with monolith carbon fibres, and at 30°C and 1 atm, they presented CO_2 uptake of 2.27 mmol g^{-1}. Alcaniz-Monge et al. (2011) synthesized carbon monoliths from nitrated coal tar pitch and evaluated and compared their performance using commercial Takeda 3A CMS. They announced that their synthesized monoliths have faster CO_2 kinetics.

9.4.2 Zeolites

Zeolites are aluminosilicates that have a crystalline porous structure. Their framework consists of tetrahedral Si or Al atoms that are connected by O atoms. Their rings have different pore sizes (5–12 Å) and structures (Chester and Derouane, 2009). Zeolites such as A, X, Y, and other natural ones such as chabazites (CHAs), ferrierites, mordenites, and clinopiles are good candidates for CO_2 capture. The molecular sieving effect (that is based on size difference) is the primary mechanism of CO_2 adsorption on zeolites. Another separation mechanism is based on polarization interaction between the CO_2 molecule and the electric field on the charged cations of the zeolite framework. Any alteration in polarity, pore size, and cations of zeolite can affect CO_2 separation by zeolites. Recently, amine moieties and other chemical functions are incorporated in zeolite framework (Suzuki et al., 1997; Hernandez-Huesca et al., 1999; Katoh et al., 2000; Harlick et al., 2003; Ko et al., 2003; Siriwardane et al., 2003; Merel et al., 2008; Remy et al., 2013; Shang et al., 2013).

Zeolites are extensively studied for CO_2 capture. The ions used in their ion-exchange process have an influence on their adsorptive equilibrium and energy. Kinetically, CO_2 adsorption on zeolite is very fast, which reaches its equilibrium state within a few minutes. Temperature and pressure influence CO_2 adsorption; any increase in temperature and any decrease in CO_2 partial pressure cause a decrease in CO_2 adsorption. As water vapour decreases zeolite capacity, its presence limits zeolite application. The CO_2 adsorption capacity of zeolites can be improved by optimizing basicity, the strength of the electrical field, and PSD, during zeolite synthesis. Generally, researchers used zeolite for adsorptive CO_2 separation in PSA or TSA process (Takamura et al., 2001; Gomes et al., 2002; Konduru et al., 2007; Zhang et al., 2008). Merel et al. (2008) reported a TSA process with zeolite 5A for CO_2 separation from a stream containing 10% CO_2 and 90% N_2. They reported 18% capture rate for CO_2. Konduru et al. (2007) used zeolite 13X in a TSA process for CO_2 separation from a mixture of 1.5% CO_2 and 98.5% N_2. After five cycles, 84% CO_2 recovery was reported.

Different researchers tried their best to improve zeolites for CO_2 capture. Cheung et al. (2013) controlled the pore size of NaKA zeolite by synthesizing nano-sized NaKA zeolites. They reported that adsorption kinetics is fast enough for CO_2 separation applications.

Loganathan et al. (2013) used pore expansion agents (MCM-41) for controlling pore size. They reported an average pore size of 30 nm and CO_2 uptake of 1.2 mmol g^{-1}. Remy et al. (2013) used low silica KFI zeolite (with Si/Al = 1.67) and performed ion-exchange process on it by K, Na, and Li. They reported that by ion exchange with Li, a large pore volume zeolite is obtained that has the highest CO_2 uptake, which is due to its strong electrostatic field. Ahmad et al. (2013) reported the application of melamine impregnation into β-zeolite by which at atmospheric pressure and 25°C, 3.7 mmol g^{-1} CO_2 uptake is obtained. Siriwardane et al. (2003) reported that the highest adsorption rate and the highest CO_2 adsorption capacity are obtained by natural zeolite that has the highest surface area with the highest Na content.

Khelifa et al. (2004) exchanged NaX zeolite (with Si/Al = 1.2) with Cr^{3+} and Ni^{2+}. Due to a weak CO_2 interaction, these exchanged zeolites have lower CO_2 adsorption capacities in comparison with parent NaX zeolite. Diaz et al. (2008a,b) evaluated that NaX, NaY, and zeolites ion-exchanged with Cs (as the most electropositive metal of periodic table) are used for CO_2 separation at high temperature (100°C), and Cs-treated zeolites are highly active for CO_2 separation.

Harlick and Tezel (2004) evaluated various types of synthetic zeolites (such as 5A, 13X, NaY, ZSM-5, and HiSiV-3000 (which is ZSM-5 with Si/Al > 1000)) for CO_2 capture. Based upon single-component isotherm data and in the pressure range of 0–2 atm, CO_2 adsorption capacity is in the order: 13X (with Si/Al = 2.2) > NaY (with Si/Al = 5.1) > H-ZSM-5–30 (with Si/Al = 30) > HiSiV-3000 > HY-5 (with Si/Al = 5). The strong interaction between CO_2 and zeolite is attributable to low Si/Al and the presence of Na (as cations) in zeolite structure.

Zukal et al. (2010) evaluated the performance of high-silica zeolites (with Si/Al > 60), namely TNU-9, IM-5, SSZ-74, ferrierite, ZSM-5, and ZSM-11, for CO_2 separation. At 100 kPa pressure, TNU-9 and IM-5 have the highest CO_2 adsorption capacity (2.61 and 2.42 mmol g^{-1}, respectively).

In post-combustion configuration, the feed contains lower CO_2, and the presented N_2 competes with CO_2 for adsorption site. In the case of CO_2–N_2 separation that needs high separation selectivity, zeolites are not suitable due to the weak interaction between the adsorbate and the adsorbent. In the case of cation presence in zeolite structure and for zeolites with low SiO_2/Al_2O_3 ratio, there adsorption would be enhanced. When cations are present in a zeolite structure, CO_2 will have a strong electrostatic interaction with zeolite. These adsorbents are good candidates for PSA, but due to the need for high energy for their regeneration, they are not advantageous (Harlick and Sayari, 2006).

Zhang et al. (2010) synthesized CHA zeolites with Si/Al < 2.5 and used Li, K, and Na (as alkali cations) and Ba, Mg, and Ca (as alkaline earth cations) for their exchange and evaluated their performance in VSA at temperatures lower than 120°C. At high temperature (> 100°C), CaCHA and NaCHA have good performance, whereas at low temperature, NaX zeolite has good performance.

Katoh et al. (2000) reported that the improved selectivity of these ion-exchanged zeolites is attributable to the presence of the strong bond between CO_2 and cation sites of adsorbent while N_2 interacts with zeolite wall.

For zeolites that are modified with amines, the nature and distribution of the cation and amine loading have an important role in preventing any probable blockage of porous structure with amine material (Bezerra et al., 2011; Lee et al., 2015c).

It was reported that zeolites are hydrophilic in nature, with low selectivity and a decrease in CO_2 adsorption capacity in the presence of gas moisture. This hydrophilic structure is a drawback for post-combustion configuration for CO_2 separation. The main reason is that CO_2 and water compete with each other for sorption sites, which influences the zeolite framework and structure (Li et al., 2008; Sayari and Belmabkhout, 2010; Marx et al., 2013). Esposito et al. (2015) reported that the more the water content, the less the cations present in the zeolite framework, and consequently, CO_2 uptake will decrease.

9.4.3 Metal–Organic Framework

Power supply industries produce and emit a large amount of carbon dioxide. For low carbon economy, CO_2 must be captured by the cost-effective methods. The application of MOFs is a suitable choice as they have large adsorption capacity and high CO_2 affinity. Moreover, they can store higher gas volumes (in comparison with traditional tanks). The most important characteristics of MOFs are their very high surface area and porosity. MOFs have organic ligands that can be modified by chemical routes; hence, their porous structure and surface functionality can flexibly be tuned. The specific selection of metal ion, functional groups, organic ligand, and activation method can tailor the functionality and pore size of MOF, which is a valuable advantage for them (Yaghi et al., 2003; Kitagawa et al., 2004; Li et al., 2011b).

In the field of CO_2 capture, MOFs are growing fast in research areas. As MOFs can have different structures and functional groups, their performance can be evaluated in a large scale with real flue gas conditions. On the other hand, their stability at high temperatures and humid condition, and various mechanical strength must be evaluated. Water stability is the most crucial target as MOF structure reacts with water vapour (which is present in the flue gas) by which MOF structure will be distorted, and surface area and porosity will be decreased, and consequently, CO_2 selectivity and its capacity will be decreased. On the other hand, if it is decided that flue gas be completely dehydrated, its separation cost will be increased. Hence, MOF must have stability in the presence of water. In MOFs, the coordination bond between the metal and the ligand may be hydrolysed with water, due to which ligand bond is displaced and the whole structure may be collapsed (Sumida et al., 2012; Jasuja et al., 2013a, 2013b; Bae et al., 2014; Burtch et al., 2014; Fracaroli et al., 2014; Zhang et al., 2014a). Consequently, the strength of the bond determines the stability of MOF. It was reported that thermodynamic factors, metal–ligand strength, and kinetic factors are specific ones that influence the structural stability of MOF by which water stability of MOF will be obtained.

Different researchers studied the application of MOFs for CO_2 separation (Sudima et al., 2012; Furukawa et al., 2013; Zhang et al., 2014b; Trickett et al., 2016; Yu et al., 2017). Different organic linkers and inorganic clusters can be used for their synthesis, and they have different pore size and surface area. The building blocks, pore volumes, and functionalities can be specifically selected. Due to these specifications, MOFs will have various applications. Among various available MOFs, ZIFs (Banerjee et al., 2008), MOF-74 (Britt et al., 2009), MIL-101 (Férey et al., 2005; Llewellyn et al., 2008), and HKUST-1 (Chui et al., 1999; Wang et al., 2002) are good ones for CO_2 adsorption.

It was reported that about 80,000 MOFs were synthesized (Bourrelly et al., 2005; Haque et al., 2010). MOF-200 and MOF-210 are the effective ones for CO_2 capture, as at 25°C and 500 kPa, they have 54.5 mmol g^{-1} CO_2 adsorption capacity with very high Langmuir and BET surface area; both of them have 10,400 $m^2 g^{-1}$ Langmuir surface area and have 4530 and 6240 $m^2 g^{-1}$ BET surface area, respectively. Moreover, the CO_2 uptake of MOF-200 and MOF-210 is higher than that for MOF-117 (which was 33.5 mmol g^{-1}) and MIL-101 (Cr) (which was 40 mmol g^{-1}) (Férey et al., 2005; Llewellyn et al., 2008). Another suitable MOF for CO_2 separation is NU-100 that presents 52.6 mmol g^{-1} CO_2 adsorption capacity at 25°C and 4000 kPa; this adsorbent has 6143 $m^2 g^{-1}$ BET surface area (Farha et al., 2010).

Generally, in the CO_2 adsorption process, a good adsorbent is the one that has high adsorption capacity at lower CO_2 partial pressures. For Mg-MOF-74, Bao et al. (2011) reported that at 25°C and 100 kPa, the CO_2 adsorption capacity is 8.61 mmol g^{-1}. On the other hand, at the same operating condition, the CO_2 adsorption capacity of 1.2 mmol g^{-1} is reported for MOF-5 and MOF-177 (Lu et al., 2010; Saha et al., 2010). It was recently reported that CPM-5 that is synthesized under microwave irradiation has 2.55 mmol g^{-1} CO_2 adsorption capacity at 25°C and 100 kPa. CPM-5 is a stable adsorbent for several weeks and has a high surface area of 2187 $m^2 g^{-1}$; the initial heat of adsorption is 36.1 kJ mol^{-1}. CO_2 diffusivities in CPM-5 at 0°C, 25°C, and 45°C are 1.86×10^{-12}, 7.04×10^{-12}, and $7.86 \times 10^{-12} m^2 s^{-1}$, respectively. Co-PL-1 is a MOF with a microporous three-dimensional structure. In its structure, the layers are made up of imidazole-4,5-dicarboxylic acid. The synthesis procedure for Co-PL-1 is a conventional hydrothermal one with three days of synthesis time. The microwave irradiation method can also be used for its synthesis, which needs 30 min time (MW-Co-Pl-1). At 25°C and 1 bar, MW-Co-PL-1 has 89 mg g^{-1} CO_2-uptake with 19.8% selectivity, and at 0.15 bar, it is 53 mg g^{-1} with 44% selectivity. During consecutive cycles of adsorption–regeneration, MW-Co-PL-1 is stable. Hence, it is concluded that the microwave synthesis method applies to large-scale zeolite production (Sabouni et al., 2012; Sabouni et al., 2013).

9.4.3.1 Hybrid Systems Based on MOFs

It was reported that hybrid systems, which are a combination of MOF and another solid sorbent, are more efficient than a single MOF sorbent. In hybrid systems, a synergy of two sorbents is used by which the overall performance of CO_2 separation is improved. Due to high surface area and easily functionalized sites, graphenes, CNTs, and ACs tune the final properties of the composite material.

Ge et al. (2013) reported the application of HKUST-1 in the structure of CNT. At 25°C, the hybrid system has a CO_2 saturation capacity of 7.83 mmol g^{-1}. In another research, multiwalled carbon nanotube (MWCNT) is dispersed in MIL-101 (Cr). The framework and the crystal structure of this hybrid composite are kept as MIL-101. For the MWCNT-MIL-101 composite, CO_2 capture is increased by 60%. The same findings are observed for MWCNT-MIL-53 (Cu) composite as well (Anbia and Hoseini, 2012; Anbia and Sheykhi, 2013).

HKUST-1 (Zhou et al., 2015), MOF-5 (Petit and Bandosz, 2009), and Cu-BTC (Bian et al., 2015) are the examples of hybrid systems of MOF and GO. In humid conditions, GO is a stabilizing agent for MOF and presents 3.3 mmol g^{-1} CO_2 capacity and good stability.

Zhao et al. (2013b) reported a hybrid system of graphite oxide with Cu-based MOF that has 50% increase in porosity with a unique structure which is suitable for CO_2 separation from the flue gas.

ILs are the poor sorbents for CO_2 separation due to their high viscosity. Nevertheless, a hybrid system of MOF-IL has good performance, and MOF is an ideal support for the incorporation of ILs into its porous structure. It was reported (Chen et al., 2011) that (BMIM) PF6 IL supported on IRMOF-1 is a suitable candidate for CO_2 adsorption and has high selectivity of CO_2–N_2.

Luo et al. (2013) published the first experimental result for IL/MOF composite. In their research, Bronsted acidic ILs were inserted into MIL-101 pores by a post-synthetic method with imidazole (IMIZ) or triethylenediamine (TEDA) as a solvent during the functionalization process. BET analysis confirmed the microporous structure of the resultant composite.

9.5 CHALLENGES AND OUTLOOK

Recently, researchers of all around the world have focused on the development of solid adsorbents for CO_2 separation from the flue gas streams with the aim of performing excellent performance and economical final cost. Comparison of AC, CNT, zeolite, and MOFs, as the common ones, is given in Table 9.2. It is worth

TABLE 9.2
A Comparison Between Different CO_2 Adsorbents

Specifications	Activated Carbon	Carbon Nanotube	Zeolites	Mixed-Organic Frameworks
CO_2/N_2 selectivity	Moderate	Moderate	Low	High
Regeneration energy	Lower temperature Better energy efficiency	Low energy	Significant	Low temperatures for regeneration
Capacity	Lower than zeolites at low pressures and gets high at high pressures	High	Moderate	High
Stability under moisture conditions	No breakthrough or decreased capacity in the presence of moisture	Increased capacity	Reduced capacity	Mainly unstable
Cost	Reasonable cost	Expensive	Low production cost	Expensive
Advantages	High conductivity High thermal and chemical stabilities Lightweight with high surface areas as well as large pore volumes Low energy consumption	High thermal and electrical conductivity High thermal and chemical stabilities High specific surface areas	Large micropores/ mesopores Medium CO_2 adsorption at ambient conditions	High specific surface areas High pore volume (55%–90%) Tunable pore size from microporous (<2 nm) to mesoporous (2–50 nm) scale Low densities High thermal and chemical stabilities Highly diverse structural chemistry
Disadvantages	Low adsorption and desorption temperatures Low CO_2 uptake compared to some types of zeolites and MOFs	Low adsorption capacity at high temperatures Synthesis is complicated	Due to moisture adsorption, CO_2 adsorption is poor in the presence of moisture High energy consumption Difficult readiness	Low performance Low economic efficiency Complicated and tedious synthesis procedure Moisture sensitivity Difficult application at high temperatures due to destroying the MOF construction

mentioning that the behaviour of each adsorbent must be evaluated in the presence of real flue gas. The presence of oxides of nitrogen and sulphur as impurities reduces the performance of adsorbents.

LIST OF ABBREVIATIONS

AC: activated carbon
AGO: aminated graphite oxide
APTES: 3-aminopropyltriethoxysilane
CMS: carbon molecular sieve
CNT: carbon nanotube
DEA: diethanolamine
EIA: Energy Information Administration
GNP: graphene nanoplate
GO: graphene oxide
HEG: hydrogen-exfoliated graphene
IL: ionic liquids
IMIZ: imidazole
MEA: monoethanolamine
MDEA: methyl diethanolamine
MOF: metal–organic framework
MWCNT: multiwalled carbon nanotubes
PANI: polyaniline
PEI: polyethylene imine
PPy: polypyrrole
PSA: pressure swing adsorption
PSD: pore size distribution
SWCNT: single-walled carbon nanotubes
TEDA: triethylenediamine
TEPA: tetraethylenepentamine
TETA: triethylenetetramine
TSA: temperature swing adsorption
WGSR: water–gas shift reaction
VSA: vacuum swing adsorption

REFERENCES

Abuilaiwi F. A., T. Laoui, M. Al-Harthi, A. A. Mautaz, "Modification and functionalization of multi-walled carbon nanotube (MWCNT) via Fischer esterification", *Arabian J. Sci. Eng.* 35(2010) 37–48.

Adams L. B., C. R. Hall, R. J. Holmes, R. A. Newton, "An examination of how exposure to humid air can result in changes in the adsorption properties of activated carbons", *Carbon* 26 (1988) 451–459.

Adams D., *Flue Gas Treatment for CO_2 Capture*. IEA Clean Coal Centre: London, 2010.

Ahmad K., O. Mowla, E. M. Kennedy, B. Z. Dlugogorski, J. C. Mackie, M. Stockenhuber, "A melamine-modified β-zeolite with enhanced CO_2 capture properties", *Energy Tech.* 1 (2013) 345–349.

Akhtar F., L. Andersson L, N. Keshavarzi, L. Bergström, "Colloidal processing and CO_2 capture performance of sacrificially templated zeolite monoliths", *Appl. Energy* 97 (2012) 289–296.

Alavinasab A., T. Kaghazchi, M. Takht Ravanchi, K. Shaabani, "Modeling of carbon dioxide absorption in a gas/liquid membrane contactor", *Desal. Water Treat.*, 29 (2011) 336–342.

Alcaniz-Monge J., J. P. Marco-Lozar, M. A. Lillo-Rodenas, "CO_2 separation by carbon molecular sieve monoliths prepared from nitrated coal tar pitch", *Fuel Process. Technol.* 92 (2011) 915–919.

Anbia M., V. Hoseini, "Development of MWCNT@MIL-101 hybrid composite with enhanced adsorption capacity for carbon dioxide", *Chem. Eng. J.* 191 (2012) 326–330.

Anbia M., S. Sheykhi, "Preparation of multi-walled carbon nanotube incorporated MIL-53-Cu composite metal–organic framework with enhanced methane sorption", *J. Ind. Eng. Chem.* 19 (2013) 1583–1586.

Araujo O. Q. F., J. L. Medeiros, R. M. B. Alves, *CO_2 Utilization: A Process Systems Engineering Vision*, Intech, 2014.

Arenillas A., K. M. Smith, T. C. Drage, C. E. Snape, "CO_2 capture using some fly ash- derived carbon materials", *Fuel* 84 (2005) 2204–2210.

Asai M., T. Ohba, T. Iwanaga, H. Kanoh, M. Endo, J. Campos-Delgado, M. Terrones, K. Nakai, K. Kaneko, "Marked adsorption irreversibility of graphitic nanoribbons for CO_2 and H_2O", *J. Am. Chem. Soc.* 133 (2011) 14880–14883.

Babu D. J., M. Lange, G. Cherkashinin, A. Issanin, R. Staudt, J. J. Schneider, "Gas adsorption studies of CO_2 and N_2 in spatially aligned double-walled carbon nanotube arrays", *Carbon* 61 (2013) 616–623.

Bae Y.S., J. Liu, C. E. Wilmer, H. Sun, A. N. Dickey, M. B. Kim, A. I. Benin, R. R. Willis, D. Barpaga, M. D. LeVan, R.Q. Snurr, "The effect of pyridine modification of Ni–DOBDC on CO_2 capture under humid conditions", *Chem. Comm.* 50 (2014) 3296–3298.

Banerjee R., A. Phan, B. Wang, C. Knobler, H. Furukawa, M. Okeeffe, O.M. Yaghi, "High-throughput synthesis of zeolitic imidazolate frameworks and application to CO_2 capture", *Sci.* 319 (2008) 939–943.

Bao Z., L. Yu, Q. Ren, X. Lu, S. Deng, "Adsorption of CO_2 and CH_4 on a magnesium-based metal organic framework", *J. Col. Inter. Sci.* 353 (2011) 549–556.

Bezerra D. P., R. S. Oliveira, R. S. Vieira, C. L. Cavalcante Jr., D. C. Azevedo, "Adsorption of CO_2 on nitrogen-enriched activated carbon and zeolite 13X", *Adsorption* 17 (2011) 235–246.

Bian Z., J. Xu, S. Zhang, X. Zhu, H. Liu, J. Hu, "Interfacial Growth of metal organic framework/graphite oxide composites through pickering emulsion and their CO_2 capture performance in the presence of humidity", *Langmuir* 31 (2015) 7410–7417.

Bienfait M., P. Zeppenfeld, N. Dupont-Pavlovsky, M. Muris, M. R. Johnson, T. Wilson, M. De Pies, O. E. Vilches, "Thermodynamics and structure of hydrogen, methane, argon, oxygen, and carbon dioxide adsorbed on single-wall carbon nanotube bundles", *Phys. Rev. B* 350 (2004) 423–426.

Blinova N. V., Svec F., "Functionalized polyaniline-based composite membranes with vastly improved performance or separation of carbon dioxide from methane", *J. Membr. Sci.* 423 (2012) 514–521.

Boehm H. P., "Some aspects of the surface-chemistry of carbon-blacks and other carbons", *Carbon*, 32 (1994) 759–769.

Bourrelly S., P.L. Llewellyn, C. Serre, F. Millange, T. Loiseau, G. Férey, "Different adsorption behaviors of methane and carbon dioxide in the isotypic nanoporous metal terephthalates MIL-53 and MIL-47", *J. Am. Chem. Soc.* 127 (2005) 13519–13521.

Britt D., H. Furukawa, B. Wang, T. G. Glover, O.M. Yaghi, "Highly efficient separation of carbon dioxide by a metal-organic framework replete with open metal sites", *Proc. Natl. Acad. Sci. U. S. A.* 106 (2009), 20637–20640.

Burchell T. D., R. R. Judkins, M. M. Rogers, A. M. Williams, "A novel process and material for the separation of carbon dioxide and hydrogen sulfide gas mixtures" *Carbon* 35 (1997) 1279–1294.

Burtch N. C., H. Jasuja, K. S. Walton, "Water stability and adsorption in metal-organic frameworks", *Chem. Rev.* 114 (2014) 10575–10612.

Caglayan B. S., A. Erhan Aksoylu, "CO_2 adsorption on chemically modified activated carbon", *J. Hazard. Mater.* 252–253 (2013) 19–28.

Cao Y., Y. Zhao, Z. Lv, F. Song, Q. Zhong, "Preparation and enhanced CO_2 adsorption capacity of UiO-66/graphene oxide composites", *J. Ind. Eng. Chem.* 27 (2015), 102–107.

Chandra V., K. S. Kim, "Highly selective adsorption of Hg^{2+} by a polypyrrole-reduced graphene oxide composite", *Chem. Commun.* 47 (2011) 3942–3944.

Chandra V., S. U. Yu, S. H. Kim, Y. S. Yoon, D. Y. Kim, A. H. Kwon, "Highly selective CO_2 capture on N-doped carbon produced by chemical activation of poly- pyrrole functionalized graphene sheets", *Chem. Commun.* 48 (2012) 735–737.

Chen J. H., D. S. H. Wong, C. S. Tan, R. Subramanian, C. T. Lira, M. Orth, "Adsorption and desorption of carbon dioxide onto and from activated carbon at high pressures", *Ind. Eng. Chem. Res.* 36 (1997) 2805–2815.

Chen Y., Z. Hu, K.M. Gupta, J. Jiang, "Ionic liquid/metal–organic framework composite for CO_2 capture: a computational investigation", *J. Phys. Chem. C* 115 (2011) 21736–21742.

Chester A. W., E. G. Derouane, *Zeolite Characterization and Catalysis*, Springer, 2009.

Cheung O., Z. Bacsik, Q. Liu, A. Mace, N. Hedin, "Adsorption kinetics for CO_2 on highly selective zeolites NaKA and nano-NaKA", *Appl. Energy* 112 (2013) 1326–1336.

Choi S, J. H. Drese, C. W. Jones, "Adsorbent materials for carbon dioxide capture from large anthropogenic point sources", *Chem. Sus. Chem.* 2 (2009) 796–854.

Chowdhury S., G. K. Parshetti, R. Balasubramanian, "Post-combustion CO_2 capture using mesoporous TiO_2/graphene oxide nanocomposites", *Chem. Eng. J.* 263 (2015) 374–384.

Chui S.S.Y., S.M.F. Lo, J.P.H. Charmant, A.G. Orpen, I.D. Williams, "A chemically functionalizable nanoporous material $[Cu_3(TMA)_2(H_2O)_3]_n$", *Science* 283 (1999), 1148–1150.

Cinke M., J. Li, C. W. Bauschlicher Jr, A. Ricca, M. Meyyappan, "CO_2 adsorption in single-walled carbon nanotubes", *Chem. Phys. Lett.* 376 (2003)761–766.

Davini P., "Flue gas treatment by activated carbon obtained from oil-fired fly ash", *Carbon* 40 (2002) 1973–1979.

David E., J. Kopac, "Activated carbons derived from residual biomass pyrolysis and their CO_2 adsorption capacity", *J. Anal. Appl. Pyrolysis* 110 (2014) 322–332.

D'Alessandro D. M., B. Smit, J. R. Long, "Carbon dioxide capture: Prospects for new materials", *Angew. Chem. Int. Ed.* 49 (2010) 6058–6082.

Deng S., H. Wei, T. Chen, B. Wang, J. Huang, G. Yu, "Superior CO_2 adsorption on pine nut shell-derived activated carbons and the effective micropores at different temperatures", *Chem. Eng. J.* 253 (2014) 46–54.

Diaz E., E. Munoz, A. Vega, S. Ordonez, "Enhancement of the CO_2 retention capacity of Y zeolites by Na and Cs treatment: Effect of adsorption temperature and water treatment", *Ind. Eng. Chem. Res.* 47 (2008a) 412–418.

Diaz E., E. Munoz, A. Vega, S. Ordonez, "Enhancement of the CO_2 retention capacity of X zeolites by Na- and Cs- treatment", *Chemosphere* 70 (2008b) 1375–1382.

Díez N., P. Álvarez, M. Granda, C. Blanco, R. Santamaría, R. Menéndez, "CO_2 adsorption capacity and kinetics in nitrogen-enriched activated carbon fibers prepared by different methods", *Chem. Eng. J.* 281 (2015) 704–712.

Dillon E.P., C. A. Crouse, A. R. Barron, "Synthesis, characterization, and carbon dioxide adsorption of covalently attached polyethyleneimine-functionalized single-wall carbon nanotubes", *ACS Nano* 2 (2008) 156–164.

Do D. D., K. Wang, "A new model for the description of adsorption kinetics in heterogeneous activated carbon", *Carbon* 36 (1998) 1539–1554.

Dong F., H. Lou, M. Goto, T. Hirose, "The Petlyuk PSA process for separation of ternary gas mixtures: Exemplification by separating a mixture of CO_2-CH_4-N_2", *Sep. Purif. Technol.* 15 (1990) 31–40.

Dreisbach F., R.Staudt, J. U. Keller, "High pressure adsorption data of methane, nitrogen, carbon dioxide and their binary and ternary mixtures on activated carbon", *Adsorption* 5 (1999) 215–227.

Esposito S., A. Marocco, G. Dellagli, B. De Gennaro, M. Pansini, "Relationships between the water content of zeolites and their cation population", *Micro. Meso. Mat.* 202 (2015) 36–43.

Farha O.K., A. Yazaydin, I. Eryazici, C.D. Malliakas, B.G. Hauser, M. G. Kanatzidis, S.T. Nguyen, R.Q. Snurr, J.T. Hupp, "De novo synthesis of a metal-organic framework material featuring ultrahigh surface area and gas storage capacities" *Nat. Chem.* 2 (2010) 944–948.

Fatemi S., M. Vesali-Naseh, M. Cyrus, J. Hashemi, "Improving CO_2/CH_4 adsorptive selectivity of carbon nanotubes by functionalization with nitrogen-containing groups", *Chem. Eng. Res. Des.* 89(2011) 1669–1675.

Férey C., C. Mellot-Draznieks, C. Serre, F. Millange, J. Dutour, S. Surblé, I. Margiolaki," A chromium terephthalate-based solid with unusually large pore volumes and surface area", *Science* 309 (2005) 2040–2042.

Figueroa J. D., T. Fout, S. Plasynski, H. Mcllvried, R. D. Srivastava, "Advances in CO_2 capture technology, the U.S. Department of Energy carbon sequestration program", *Int. J. Greenhouse GasCon.* 2 (2008) 9–20.

Fiuza Jr R. A., R. Medeiros de Jesus Neto, L. B. Correia, H. M. Carvalho Andrade, "Preparation of granular activated carbons from yellow mombin fruit stones for CO_2 adsorption", *J. Environ. Manage.* 161(2015) 198–205.

Foley H. C., "Carbogenic molecular sieves: synthesis, properties and applications", *Microporous Mater.* 4 (1995) 407–433.

Fracaroli A. M., H. Furukawa, M. Suzuki, M. Dodd, S. Okajima, F. Gándara, J. A. Reimer, O. M. Yaghi, "Metal–organic frameworks with precisely designed interior for carbon dioxide capture in the presence of water", *J. Am. Chem. Soc.* 136 (2014) 8863–8866.

Furukawa H., K. E. Cordova, M. O'Keeffe, O. M. Yaghi, "The chemistry and applications of metal–organic frameworks", *Science* 341 (2013) 1230444.

Gadipelli S., Z. X. Guo, "Graphene-based materials: Synthesis and gas sorption, storage and separation", *Prog. Mater. Sci.* 69 (2015) 1–60.

Ge L., L.Wang, V. Rudolph, Z. Zhu, "Hierarchically structured metal–organic framework/vertically-aligned carbon nanotubes hybrids for CO_2 capture", *RSC Adv.* 3 (2013) 25360–25366.

Gholami G., M. Soleimani, M. Takht Ravanchi, "Application of carbon membranes for gas separation: Review", *Ind. Res. Tech.*, 3 (2013) 53–58.

Gibson J.A.A., A.V. Gromov, S. Brandani, E.E.B. Campbell, "The effect of pore structure on the CO_2 adsorption efficiency of polyamine impregnated porous carbons", *Micro. Meso. Mat.* 208 (2015) 129–139.

Gomes V.G., K. W. K. Yee, "Pressure swing adsorption for carbon dioxide sequestration from exhaust gases", *Sep. Purif. Technol.* 28 (2002) 161–171.

Ghosh A., K. S. Subrahmanyam, K. S. Krishna, S. Datta, A. Govindaraj, S. K. Pati, "Uptake of H_2 and CO_2 by graphene", *J. Phys. Chem. C* 112 (2008) 15704–15707.

Gui M. M., Y. X. Yap, S. P. Chai, A. Mohamed, "Multi-walled carbon nanotubes modified with (3-aminopropyl)triethoxysilane for effective carbon dioxide adsorption", *Int. J. Greenhouse Gas Con.* 14 (2013) 65–73.

Hao G. P., W. C. Li, D. Qian, A. H. Lu, "Rapid synthesis of nitrogen-doped porous carbon monolith for CO_2 capture", *Adv. Mater.* 22 (2010) 853–857.

Haque E., N. Khan, H.J. Park, S.H. Jhung, "Synthesis of a metal-organic framework material, iron terephthalate, by ultrasound, microwave, and conventional electric heating: a kinetic study", *Chem. Eur. J.* 16 (2010) 1046–1052.

Harlick P. J. E., F. H. Tezel, "Adsorption of carbon dioxide, methane and nitrogen: Pure and binary mixture adsorption for ZSM-5with SiO_2/Al_2O_3 ratio of 280", *Sep. Purif. Technol.* 33 (2003) 199–210.

Harlick P. J. E., F. H. Tezel, "An experimental adsorbent screening study for CO_2 removal from N_2", *Micropor. Mesopor. Mat.* 76 (2004) 71–79.

Harlick P. J. E., A. Sayari, "Applications of pore-expanded mesoporous silicas, triamine silane grafting for enhanced CO_2 adsorption", *Ind. Eng. Chem. Res.* 45 (2006) 3248–3255.

Hedin N., Chen L. J., Laaksonen A., "Sorbents for CO_2 capture from flue gas-aspects from materials and theoretical chemistry", *Nanoscale* 2 (2010) 1819–1841.

Heidari A., H. Younesi, A. Rashidi, A. A. Ghoreyshi, "Evaluation of CO_2 adsorption with eucalyptus wood based activated carbon modified by ammonia solution through heat treatment", *Chem. Eng. J.* 254 (2014) 503–513.

Hernandez-Huesca R., L. Diaz, G. Aguilar-Armenta, "Adsorption equilibria and kinetics of CO_2, CH_4, and N_2 in natural zeolites", *Sep. Purif. Technol.* 15 (1999) 163–173.

Heuchel M., Davies G. M., Buss E., Seaton N. A., 'Adsorption of carbon dioxide and methane and their mixtures on an activated carbon: Simulation and experiment, *Langmuir* 15 (1999) 8695–8705.

Hong S.M., S. H. Kim, K. B. Lee, "Adsorption of carbon dioxide on 3-aminopropyl-triethoxysilane modified graphite oxide", *Energy Fuels* 27 (2013) 3358–3363.

Houshmand A., M.S. Shafeeyan, A. Arami-Niya, W.M.A.W. Daud, "Anchoring a halogenated amine on the surface of a microporous activated carbon for carbon dioxide capture", *J. Taiwan Inst. Chem. Eng.* 44 (2013) 774–779.

Huang L., L. Zhang, Q. Shao, L. Lu, X. Lu, S. Jiang, W. Shen, "Simulations of binary mixture adsorption of carbon dioxide and methane in carbon nanotubes: Temperature, pressure, and pore size effects", *J. Phys. Chem. C* 111 (2007) 11912–11920.

Hsu S-C., C. Lu, F. Su, W. Zeng, W. Chen, "Thermodynamics and regeneration studies of CO_2 adsorption on multi-walled carbon nanotubes", *Chem. Eng. Sci.* 65 (2010) 1354–1361.

Ho M.T., G.W. Allinson, D.E. Wiley, "Reducing the cost of CO_2 capture from flue gases using pressure swing adsorption", *Ind. Eng. Chem. Res.* 47 (2008) 4883–4890.

Jasuja H., N.C. Burtch, Y.G. Huang, Y. Cai, K. S. Walton, "Kinetic water stability of an isostructural family of zinc-based pillared metal–organic frameworks", *Langmuir* 29 (2013a) 633–642.

Jasuja H., K. S. Walton, "Effect of catenation and basicity of pillared ligands on the water stability of MOFs", *Dalton Trans.* 42 (2013b) 15421–15426.

Jin H., X. Zhang, "Chemical looping combustion for power generation and carbon dioxide (CO_2) capture", In Zheng L. (Ed.), *Oxy-Fuel Combustion for Power Generation and Carbon Dioxide (CO_2) Capture*, 1st Ed., Elsevier, UK (2011).

Kacem M., M. Pellerano, A. Delebarre, "Pressure swing adsorption for CO_2/N_2 and CO_2/CH_4 separation: Comparison between activated carbons and zeolites performances", *Fuel Proces. Technol.* 138 (2015) 271–283.

Katoh M., T. Yoshikawa, T. Tomonari, K. Katayama, T. Tomida, "Adsorption characteristics of ion-exchanged ZSM-5 zeolites for CO_2/N_2 mixtures", *J. Colloid Inter. Sci.* 226 (2000) 145–150.

Kemp K. C., V. Chandra, M. Saleh, K. S. Kim, "Reversible CO_2 adsorption by an activated nitrogen doped graphene/polyaniline material", *Nanotechnol.* 24 (2013) 235703.

Khelifa A., L. Benchechida, Z. Derriche, "Adsorption of carbon dioxide by X zeolites exchanged with Ni^{2+} and Cr^{3+}: isotherms and isosteric heat", *J. Colloid Interface Sci.* 278 (2004) 9–17.

Kikkinides E. S., R. T. Yang, "Concentration and recovery of CO_2 from flue gas by pressure swing adsorption", *Ind. Eng. Chem. Res.* 32 (1993) 2714–2720.

Kitagawa S., R. Kitaura, S. I. Noro, "Functional porous coordination polymers", *Ang. Chem. Int. Ed.*, 43 (2004) 2334–2375.

Ko D., R. Siriwardane, L.T. Biegler, "Optimization of a pressure swing adsorption process using zeolite 13X for CO_2 sequestration" *Ind. Eng. Chem. Res.* 42 (2003) 339–348.

Ko Y. G., H.J. Lee, H. C. Oh, U.S. Choi, "Amines immobilized double-walled silica nanotubes for CO_2 capture", *J. Haz. Mat.* 250–251 (2013) 53–60.

Konduru N., P. Lindner, N. M. Assaf-Anid, "Curbing the greenhouse effect by carbon dioxide adsorption with zeolite 13X", *AIChEJ.* 53 (2007) 3137–3143.

Kumar Mishra A., S. Ramaprabhu, "Polyaniline/multi-walled carbon nanotubes nanocomposite-an excellent reversible CO_2 capture candidate", *RSC Adv.* 2 (2012) 1746–1750.

Lee M.S., S.J. Park, "Silica-coated multi-walled carbon nanotubes impregnated with polyethyleneimine for carbon dioxide capture under the flue gas condition", *J. Sol. State Chem.* 226 (2015a) 17–23.

Lee M.S., S.Y. Lee, S. J. Park, "Preparation and characterization of multi-walled carbon nanotubes impregnated with polyethyleneimine for carbon dioxide capture", *Int. J. Hydrogen Energy* 40 (2015b) 3415–3421.

Lee C. H., D. H. Hyeon, H. Jung, W. Chung, D. H. Jo, D. K. Shin, S. H. Kim, "Effects of pore structure and PEI impregnation on carbon dioxide adsorption by ZSM-5 zeolites", *Ind. Eng. Chem. Res.* 23 (2015c) 251–256.

Li G., P. Xiao, P. Webley, J. Zhang, R. Singh, M. Marshall, "Capture of CO_2 from high humidity flue gas by vacuum swing adsorption with zeolite13X", *Adsorption* 14 (2008) 415–422.

Li J. R., R. J. Kuppler, H. C. Zhou, "Selective gas adsorption and separation in metal-organic frameworks", *Chem. Soc. Rev.* 38 (2009) 1477–1504.

Li G., P. Xiao, D. Xu, P. A. Webley, "Dual mode roll-up effect in multi component non-isothermal adsorption processes with multilayered bed packing", *Chem. Eng. Sci.* 66 (2011a) 1825–1834.

Li J. R., Y. Ma, M. C. McCarthy, J. Sculley, J. Yu, H. K. Jeong, P. B.Balbuena, H-C. Zhou, "Carbon dioxide capture-related gas adsorption and separation in metal-organic frameworks", *Coord. Chem. Rev.* 255 (2011b) 1791–1823.

Li F., X. Jiang, J. Zhao, S. Zhang, "Graphene oxide: A promising nanomaterial for energy and environmental applications", *Nano Energy*, 16 (2015a) 488–515.

Li W., X. Jiang, H. Yang, Q. Liu, "Solvothermal synthesis and enhanced CO_2 adsorption ability of mesoporous graphene oxide-ZnO nanocomposite", *Appl. Surf. Sci.* 356 (2015b) 812–816.

Lithoxoos G. P., A. Labropoulos, L. D. Peristeras, N. Kanellopoulos, J. Samios, I. G. Economou, "Adsorption of N_2, CH_4, CO and CO_2 gases in single walled carbon nanotubes: a combined experimental and Monte Carlo molecular simulation study", *J. Supercrit. Fluids* 55 (2010) 510–523.

Liu Q., Y. Shi, S. Zheng, L. Ning, Q. Ye, M. Tao, Y. He, "Amine-functionalized low-cost industrial grade multi-walled carbon nanotubes for the capture of carbon dioxide", *J. Energy Chem.* 23 (2014) 111–118.

Llewellyn P.L., S. Bourrelly, C. Serre, A. Vimont, M. Daturi, L. Hamon, G. De Weireld, J.S. Chang, D.Y. Hong, Y. Kyu Hwang, S. Hwa Jhung, G. Férey, "High uptakes of CO2 and CH4 in mesoporous metal-organic frameworks MIL-100 and MIL-101", *Langmuir* 24 (2008) 7245–7250.

Loganathan S., M. Tikmani, A. K. Ghoshal, "Novel pore-expanded MCM-41 for CO_2 capture: synthesis and characterization", *Langmuir* 29 (2013) 3491–3499.

Lu C., H. Bai, B. Wu, F. Su, J. F. Hwang, "Comparative study of CO_2 capture by carbon nanotubes, activated carbons, and zeolites", *Energy Fuels* 22 (2008) 3050–3056.

Lu Ch-M, J. Liu, K. Xiao, A. T. Harris, "Microwave enhanced synthesis of MOF-5 and its CO_2 capture ability at moderate temperatures across multiple capture and release cycles", *Chem. Eng. J.* 156 (2010) 465–470.

Luo Q.X., M. Ji, M.H. Lu, C. Hao, J.S. Qiu, Y.Q. Li, "Organic electron-rich N-heterocyclic compound as a chemical bridge: building a Brönsted acidic ionic liquid confined in MIL-101 nanocages", *J. Mat. Chem. A* 1 (2013) 6530–6534.

Matranga C., L. Chen, M. Smith, E. Bittner, J. K. Johnson, B. Bockrath, "Trapped CO_2 in carbon nanotube bundles", *J. Phys. Chem. B* 107 (2003) 12930–12941.

Martin C.F., M. G. Plaza, J. J. Pis, F. Rubiera, C. Pevida, T. A. Centeno, "On the limits of CO_2 capture capacity of carbons", *Sep. Purif. Technol.* 74 (2010) 225–229.

Maroto-Valer M. M., Z. Tang, Y. Zhang, "CO_2 capture by activated and impregnated anthracites", *Fuel Process. Technol.* 86 (2005)1487–1502.

Marx D., L. Joss, M. Hefti, R. Pini, M. Mazzotti, "The role of water in adsorption-based CO_2 capture systems", *Energy Procedia* 37 (2013) 107–114.

Medeiros J. L., L. C. Barbosa, Araujo O. Q. F. "An equilibrium approach for CO_2 and H_2S absorption with aqueous solutions of alkanolamines: Theory and parameter estimation", *Ind. Eng. Chem. Res.* 52 (2013) 9203–9226.

Meng L.Y., S.J. Park, "Effect of exfoliation temperature on carbon dioxide capture of graphene nanoplates", *J. Col. Inter. Sci.* 38 (2012) 285–290.

Merel J., M. Clausse, F. Meunier, "Experimental investigation on CO_2 post-combustion capture by indirect thermal swing adsorption using 13X and 5A zeolites", *Ind. Eng. Chem. Res.* 47 (2008) 209–215.

Mishra A. K., Ramaprabhu S., "Carbon dioxide adsorption in graphene sheets", *AIP Adv.* 1 (2011) 032152–7.

Mishra A. K., Ramaprabhu S., "Nano structured polyaniline decorated graphene sheets for reversible CO_2 capture", *J. Mater. Chem*, 22 (2012) 3708–3712.

Montagnaro F., A. Silvestre-Albero, J. Silvestre-Albero, F. Rodríguez-Reinoso, A. Erto, A. Lancia, M. Balsamo, "Post-combustion CO_2 adsorption on activated carbons with different textural properties", *Micro. Meso. Mat.* 209(2015) 157–164.

Mostazo-López M. J., R. Ruiz-Rosas, E. Morallón, D. Cazorla-Amorós, "Generation of nitrogen functionalities on activated carbons by amidation reactions and Hofmann rearrangement: Chemical and electrochemical characterization", *Carbon* 91 (2015) 252–265.

Mugge J., H. Bosch, T. Reith, "Gas adsorption kinetics inactivated carbon." In Adsorption Science and Technology" *2nd Pacific Basin on Adsorption Science and Technology*, Brisbane, Australia, (2000).

Na B. K., K. K. Koo, H. M. Eum, H. Lee, H. K. Song, "CO_2 recovery from flue gas by PSA process using activated carbon", *Korean J. Chem. Eng.* 18 (2001) 220–227.

Nakamura T., "Recovery and sequestration of CO2 from stationary combustion systems by photosynthesis of microalgae quarterly technical progress report", *National Energy report* (2003).

Noh G., S. R. Docherty, E. Lam, X. Huang, D. Mance, J. L. Alfke, C. Copéret, "CO_2 hydrogenation to CH_3OH on supported Cu nanoparticles: Nature and role of Ti in bulk oxides vs. as isolated surface sites", *J. Phys. Chem. C*, 123 (2019) 31082–31093.

Oh J., Y.H. Mo, V.D. Le, S. Lee, J. Han, G. Park, Y.H. Kim, S.E. Park, S. Park, "Borane-modified graphene-based materials as CO_2 adsorbents", *Carbon* 79 (2014) 450–456.

Ohba T., Kanoh H., "Intensive edge effects of nanographenes in molecular adsorptions", *J. Phys. Chem. Lett.* 3 (2012) 511–516.

Okoye I. P., Benham M., Thomas K. M. "Adsorption of gases and vapors on carbon molecular sieves", *Langmuir* 13 (1997) 4054–4059.

Olajire A. A., "CO_2 capture and separation technologies for end-of-pipe applications - A review", *Energy* 35 (2010) 2610–2628.

Park C., Anderson P. E., Chambers A., Tan C. D., Hidalgo R., Rodriguez N. M. "Further studies of the interaction of hydrogen with graphite nanofibers", *J. Phys. Chem. B*, 103 (1999) 10572–10581.

Patchkovskii S., Tse J. S., Yurchenko S. N., Zhechkov L., Heine T., Seifert G., "Graphene nanostructures as tunable storage media for molecular hydrogen", *PNAS102* (2005) 10439–10444.

Petit C., T. J. Bandosz, "MOF–graphite oxide composites: Combining the uniqueness of graphene layers and metal–organic frameworks", *Adv. Mat.*, 21 (2009) 4753–4757.

Plaza M.G., C. Pevida, B. Arias, J. Fermoso, A. Arenillas, F. Rubiera, J. J. Pis, "Application of thermogravimetric analysis to the evaluation of aminated solid sorbents for CO_2 capture", *J. Therm. Anal. Calorimeters* 92 (2008) 601–606.

Pourebrahimi S., M. Kazemeini, E. Ganji Babakhani, A. Taheri, "Removal of the CO_2 from flue gas utilizing hybrid composite adsorbent MIL-53 (Al)/GNP metal-organic framework", *Micro. Meso. Mat.* 218 (2015) 144–152.

Radosz M., X. Hu, K. Krutkrarnelis, Y. Shen, "Flue-gas carbon capture on carbonaceous sorbents: Toward a low-cost multifunctional carbon filter for "green" energy producers", *Ind. Eng. Chem. Res.* 47 (2008) 3783–3794.

Ramdin M., A. Amplianitis, S. Bazhenov, A. Volkov, V. Volkov, T. J. Vlugt, T. W. de Loos, "Solubility of CO_2 and CH_4 in Ionic Liquids: Ideal CO_2/CH_4 Selectivity", *Ind. Eng. Chem. Res.* 53 (2014) 15427–15435.

Razavi S.S., S.M. Hashemianzadeh, H. Karimi, "Modeling the adsorptive selectivity of carbon nanotubes for effective separation of CO_2/N_2 mixtures", *J. Mol. Model.* 17 (2011) 1163–1172.

Remy T., S. A. Peter, L. Van Tendeloo, S. Van der Perre, Y. Lorgouilloux, C. E. Kirschhock, G. V. Baron, J. F. Denayer, "Adsorption and separation of CO_2 on KFI zeolites: Effect of cation type and Si/Al ratio on equilibrium and kinetic properties", *Langmuir* 29 (2013) 4998–5012.

Rezaei F., R. P. Lively, Y. Labreche, G. Chen, Y. F. Fan, W. J. Koros, "Aminosilane- grafted polymer/silica hollow fiber adsorbents for CO_2 capture from flue gas", *ACS Appl. Mater. Interfaces* 5 (2013) 3921–3931.

Sabouni R., H. Kazemian, S. Rohani, "Microwave synthesis of the CPM-5 metal organic framework", *Chem. Eng. Technol.* 35 (2012) 1085–1092.

Sabouni R., H. Kazemian, S. Rohani, "Carbon dioxide adsorption in microwave-synthesized metal organic framework CPM-5: Equilibrium and kinetics study", *Micro. Meso. Mat.* 175 (2013) 85–91.

Saha D., Z. Bao, F. Jia, S. Deng, "Adsorption of CO_2, CH_4, N_2O, and N_2 on MOF-5, MOF-177, and Zeolite 5A", *Environ. Sci. Technol.* 44 (2010) 1820–1826.

Sawant S. Y., R. S. Somani, H. C. Bajaj, S. S. Sharma, "A dechlorination pathway for synthesis of horn shaped carbon nanotubes and its adsorption properties for CO_2, CH_4, CO and N_2", *J. Hazard. Mater.* 227–228 (2012) 317–326.

Sayari A., Y. Belmabkhout, "Stabilization of amine-containing CO_2 adsorbents: Dramatic effect of water vapor", *J. Am. Chem. Soc.* 132 (2010) 6312–6314.

Sethia G., A. Sayari, "Comprehensive study of ultra-microporous nitrogen-doped activated carbon for CO_2 capture", *Carbon* 93 (2015) 68–80.

Shang J., G. Li, R. Singh, P. Xiao, J. Z. Liu, P. A. Webley, "Determination of composition range for "molecular trapdoor" effect in chabazite zeolite", *J. Phys. Chem. C* 117 (2013) 12841–12847.

Sharma T., S. Sharma, H. Kamyab, A. Kumar, "Energizing the CO_2 utilization by chemo-enzymatic approaches and potentiality of carbonic anhydrases: A review", *J. Clean. Prod.* 247 (2019) 119138.

Silvestre-Albero J., A. Wahby, E. Sepŭlveda, M. Martínez-Escandell, K. Kaneko, F. Rodríguez-Reinoso, "Ultra high CO_2 adsorption capacity on carbon molecular sieves at room temperature", *Chem. Commun.* 47 (2011) 6840–6842.

Siriwardane R. V., M. S. Shen, E. P. Fisher, J. A. Poston, "Adsorption of CO_2 on molecular sieves and activated carbon", *Energ. Fuels* 15 (2001) 279–284.

Siriwardane R. V., M. S. Shen, E. P. Fisher, "Adsorption of CO_2, N_2, and O_2 on natural zeolites", *Energ Fuels* 17 (2003) 571–576.

Skoulidas A. I., D. S. Sholl, J. K. Johnson, "Adsorption and diffusion of carbon dioxide and nitrogen through single-walled carbon nanotube membranes" *J. Chem. Phys.* 124 (2006) 054708.

Smit B., "Carbon capture and storage: introductory lecture", *Faraday Discuss* 192 (2016) 9–25.

Song C. F., Y. Kitamura, S. H. Li, "Evaluation of stirling cooler system for cryogenic CO_2 capture", *Appl. Energy* 98 (2012) 491–501.

Songolzadeh M., M. Takht Ravanchi, M. Soleimani, "Carbon dioxide capture and storage: a general review on adsorbents", *World Acad. Sci. Eng. Technol.* 70 (2012) 225–232.

Songolzadeh M., M. Soleimani, M. Takht Ravanchi, R. Songolzadeh, "Carbon Dioxide separation from flue gases; a technological review emphasizing on reduction in greenhouse gas emissions", *Sci. World J.*, 2014 (2014) 1–34.

Songolzadeh M., M. Soleimani, M. Takht Ravanchi, "Using modified Avrami kinetic and two component isotherm equation for modeling of CO_2/N_2 adsorption over a 13X Zeolite bed", *J. Nat. Gas Sci. Eng.*, 27 (2015) 831–841.

Songolzadeh M., M. Soleimani, M. Takht Ravanchi, "Evaluation of metal type in MIL-100 structure to synthesize a selective adsorbent for the basic N-compounds removal from liquid fuels", *Micro. Meso. Mat.*, 274 (2019) 54–60.

Su F., C. Lu, W. Cnen, H. Bai, J. F. Hwang, "Capture of CO_2 from flue gas via multi-walled carbon nanotubes", *Sci. Total Env.* 407 (2009) 3017–3023.

Su F., C. Lu, H-S. Chen, "Adsorption, desorption, and thermodynamic studies of CO_2 with high-amine-loaded multi-walled carbon nanotubes", *Langmuir* 27 (2011)8090–8098.

Su F., C. Lu, A. J. Chung, C. H. Liao, "CO_2 capture with amine-loaded carbon nanotubes via a dual-column temperature/vacuum swing adsorption", *Appl. Energy* 113(2014)706–712.

Sumida K., D. L. Rogow, J. A. Mason, T. M. McDonald, E. D. Block, Z. R. Herm, T. H. Bae, J. R. Long, "Carbon dioxide capture in metal–organic frameworks", *Chem. Rev.* 112 (2012) 724–781.

Suzuki T., A. Skoda, M. Suzuki, J. Izumi, "Adsorption of carbon dioxide onto hydrophobic zeolite under high moisture", *J. Chem. Eng. Jpn.* 30 (1997) 954–958.

Taheri Najafabadi A. "Emerging applications of grapheme and its derivatives in carbon capture and conversion: Current status and future prospects", *Renew. Sus. Energy Rev.* 41 (2015) 1515–1545.

Takamura Y., S. Narita, J. Aoki, S. Hironaka, S. Uchida, "Evaluation of dual-bed pressure swing adsorption for CO_2 recovery from boiler exhaust gas", *Sep. Purif. Technol.* 24 (2001) 519–528.

Takht Ravanchi M., S. Sahebdelfar, F. Tahriri Zangeneh, "Carbon dioxide sequestration in petrochemical industries with the aim of reduction in greenhouse gas emissions", *Front. Chem. Sci. Eng.*, 5 (2011) 173–178.

Takht Ravanchi M., S. Sahebdelfar, "Carbon dioxide capture and utilization in petrochemical industry: Potentials and challenges" *Appl. Pet. Res.* 4 (2014) 63–77.

Tang Z., M. M. Maroto-Valer, Y. Zhang, "CO_2 capture using anthracite based sorbents", *Div. Fuel Chem., Am. Chem. Soc.Prepr. Symp.* 49 (2004a) 298–299.

Tang Z., Y. Zhang, M. M. Maroto-Valer, "Study of the CO_2 adsorption capacities of modified activated anthracites", *Div. Fuel Chem.,Am. Chem. Soc.Prepr. Symp.* 49 (2004b) 308–309.

Tarka J. T., J.P. Ciferno, M.L. Gray, D. Fauth, "CO_2 capture systems using amine enhanced solid sorbents", *5th Annual Conference on Carbon Capture and Sequestration*, Pittsburg (2006).

Trickett C. A., A. Helal, B. A. AlMaythalony, Z. H. Yamani, K. E. Cordova, O. M. Yaghi, "The chemistry of metal–organic frameworks for CO_2 capture, regeneration and conversion", *Nat. Mater. Rev.* 2 (2016) 17045.

Tseng R.L., F.C. Wu, R.S. Juang, "Adsorption of CO_2 at atmospheric pressure on activated carbons prepared from melamine-modified phenol–formaldehyde resins", *Sep. Pur. Technol.* 140 (2015) 53–60.

Walker P. L., R. J. Foresti, C. C. Wright, "Surface area studies of carbon-carbon dioxide reaction", *Ind. Eng. Chem.* 45 (1953) 1703–1710.

Walker P. L., Shelef M., "Carbon dioxide sorption on carbon molecular sieves", *Carbon* 5 (1967) 7–11.

Wang Q.M., D. Shen, M. Bulow, M.L. Lau, S. Deng, F.R. Fitch, N.O. Lemcoff, J. Semanscin, "Metallo-organic molecular sieve for gas separation and purification", *Micro. Meso. Mat.* 55 (2002) 217–230.

Wang Y., Y. Zhou, C. Liu, L. Zhou, "Comparative studies of CO_2 and CH_4 sorption on activated carbon in presence of water", *Col. Surf.* A322 (2008) 14–18.

Wang X. R., H. Q. Li, X. J. Hou, "Amine-functionalized metal organic frame work as a highly selective adsorbent for CO_2 over CO", *J. Phys. Chem. C*, 116 (2012) 19814–19821.

Wang J., X. Mei, L. Huang, Q. Zheng, Y. Qiao, K. Zang, S. Mao, R. Yang, Z. Zhang, Y. Gao, Z. Guo, Z. Huang, Q. Wang, "Synthesis of layered double hydroxides/graphene oxide nanocomposite as a novel high-temperature CO_2 adsorbent", *J. Energy Chem.* 24 (2015) 127–137.

Xu D., P. Xiao, J. Zhang, G. Li, G. Xiao, P. A. Webley, Y. Zhai, "Effects of water vapor on CO_2 capture with vacuum swing adsorption using activated carbon", *Chem. Eng. J.* 230 (2013) 64–72.

Yaghi O. M., M. O'Keeffe, N. W. Ockwig, H. K. Chae, M. Eddaoudi and J. Kim, "Reticular synthesis and the design of new materials" *Nature*, 423 (2003) 705–714.

Yan S., M. Fang, Z. Wang, Z. Luo, "Regeneration performance of CO_2-rich solvents by using membrane vacuum regeneration technology: Relationships between absorbent structure and regeneration efficiency", *Appl. Energy* 98 (2012) 357–367.

Yang, R. T., *Adsorbents: Fundamentals and Applications*. WileyInter Science: Hoboken, NJ, 2003.

Yang H., Xu Z., Fan M., Gupta R., Slimane R. B., Bland A. E., "Progress in carbon dioxide separation and capture: A Review", *J. Environ. Sci.* 20 (2008) 14–27.

Yim W. L., Byl O., Yates J. T., Johnson J. K., "Vibrational behavior of adsorbed CO_2 on single-walled carbon nanotubes", *J. Chem. Phys.* 120 (2004) 5377–5386.

Yin G., Z. Liu, Q. Liu, W. Wu, "The role of different properties of activated carbon in CO_2 adsorption", *Chem. Eng. J.* 230 (2013) 133–140.

Yong Z, V. G. Mata, A. E. Rodrigues, "Adsorption of carbon dioxide on chemically modified high surface area carbon-based adsorbents at high temperature", *Adsorption* 7 (2001) 41–50.

Yu C. H., C. H. Huang, C. S. Tan, "A review of CO_2 capture by absorption and adsorption", *Aero. Air Qual. Res.* 12 (2012) 745–769.

Yu J., L. H. Xie, J. R. Li, Y. Ma, J. M. Seminario, P. B. Balbuena, "CO_2 capture and separations using MOFs: Computational and experimental studies", *Chem. Rev.* 117 (2017) 9674–9754.

Zhang Y., M. M. Maroto-Valer, Y. Zhang, "Microporous activated carbons produced from unburned carbon fly ash and their application for CO_2 capture", *Div. Fuel Chem., Am. Chem. Soc. Prepr. Symp.* 49 (2004) 304–305.

Zhang J., P. A. Webley, P. Xiao, "Effect of process parameters on power requirements of vacuum swing adsorption technology for CO_2 capture from flue gas", *Energy Con. Manage.* 49 (2008) 346–356.

Zhang Z., W. Zhang, X. Chen, Q. Xia, Z. Li, "Adsorption of CO_2 on zeolite 13X and activated carbon with high surface area", *Sep. Sci. Technol.* 45 (2010) 710–719.

Zhang M, Y. Guo, "Rate based modeling of absorption and regeneration for CO_2 capture by aqueous ammonia solution", *Appl. Energ.* 111 (2013) 142–152.

Zhang W., Y. Hu, J. Ge, H-L. Jiang, S.H. Yu, "A facile and general coating approach to moisture/water-resistant metal–organic frameworks with intact porosity", *J. Am. Chem. Soc.* 136 (2014a) 16978–16981.

Zhang Z., Z. Z. Yao, S. Xiang, B. L. Chen, "Perspective of microporous metal-organic frameworks for CO_2 capture and separation", *Energy Env. Sci.* 7 (2014b) 2868–2899.

Zhao J., A. Buldum, J. Han, J. P. Lu, "Gas molecule adsorption in carbon nanotubes and nanotube bundles", *Nanotechnology* 13 (2002) 195–200.

Zhao D., D. Yuan, A. Yakovenko, H-C. Zhou, "A NbO-type metal-organic framework derived from a polyyne-coupled di-isophthalate linker formed in situ", *Chem. Commun.* 46 (2010) 4196–4198.

Zhao Y. F., Zhao L., Yao K. X., Yang Y., Zhang Q., Han Y., "Novel porous carbon materials with ultrahigh nitrogen contents for selective CO_2 capture", *J. Mater. Chem.* 22 (2012a) 19726–19731.

Zhao Y. F., Liu X., Yao K. X., Zhao L., Han Y., "Superior capture of CO_2 achieved by introducing extra-framework cations into N-doped microporous carbon", *Chem. Mater.*, 24 (2012b) 4725–4734.

Zhao Y., M. Seredych, Q. Zhong, T.J. Bandosz, "Aminated graphite oxides and their composites with copper-based metal-organic framework: In search for efficient media for CO_2 sequestration", *RSC Adv.* 3 (2013a) 9932–9941.

Zhao Y., H. Ding, Q. Zhong, "Synthesis and characterization of MOF-aminated graphite oxide composites for CO_2 capture", *Appl. Surf. Sci.* 284 (2013b) 138–144.

Zhao Y. Y., B. T. Jung, L. Ansaloni, W. S. W. Ho, "Multi-walled carbon nanotube mixed matrix membranes containing amines for high pressure CO_2/H_2 separation", *J. Membr. Sci.* 459 (2014) 233–243.

Zhou L., J. Fan, X. Shang, "CO_2 Capture and separation properties in the ionic liquid 1-n-butyl-3-methylimidazolium nonafluorobutylsulfonate", *Materials* 7 (2014) 3867–3880.

Zhou X., W. Huang, J. Miao, Q. Xia, Z. Zhang, H. Wang, Z. Li, "Enhanced separation performance of a novel composite material GrO@MIL-101 for CO_2/CH_4 binary mixture", *Chem. Eng. J.* 266 (2015) 339–344.

Zhuang Q., R. Pomalis, L. Zheng, B. Clements, "Ammonia-based carbon dioxide capture technology: Issues and solutions", *Energy Procedia*, 4 (2011) 1459–1470.

Zukal A., J. Mayerova, M. Kubu, "Adsorption of carbon dioxide on high-silica zeolites with different framework topology", *Top. Cat.* 53 (2010) 1361–1366.

10 Novel Composite Materials for CO_2 Fixation

Priya Banerjee, Uttariya Roy, Avirup Datta, and Aniruddha Mukhopadhyay

CONTENTS

10.1 INTRODUCTION

In recent times, it is being widely anticipated that an expanding global population will result in an increased demand and requirement of planetary resources for their basic needs (food, clothes, and shelter). Human race is now largely dependent on fossil fuels as their major energy source for meeting their requirements. Over the previous decades, combustion of fossil fuels has resulted in an alarming increase in atmospheric CO_2 levels, which in turn has catalysed the rate of increase in global temperatures (Gao et al., 2014). Power plants, motor vehicles, and industrial technologies are primary sources of CO_2 emissions across the world. Among all other industries, cement and petrochemical industries are considered as the most significant emitters of CO_2 on a global scale (Songolzadeh et al., 2014). Although the density of CO_2 in the atmosphere is only 0.0034%, it is reported that CO_2 is responsible for 50% of the total greenhouse effect experienced by this planet (Olah et al., 2002).

According to the Intergovernmental Panel on Climate Change (IPCC), high levels of atmospheric CO_2 lead to an increase in average Earth's temperature and an increase in sea level (Crombie et al., 2011; Mohammadi et al., 2013). IPCC has also suggested that it is essential to achieve a 50% to 80% reduction of existing global atmospheric greenhouse gas concentrations by 2050 in order to prevent the critical conditions of global warming (Songolzadeh et al., 2014). This target may be achieved by using alternative sources of energy. However, in scenarios of the present energy demand, it is difficult to achieve a complete substitution of fossil fuels by available sources of alternative energy (Sharma et al., 2019).

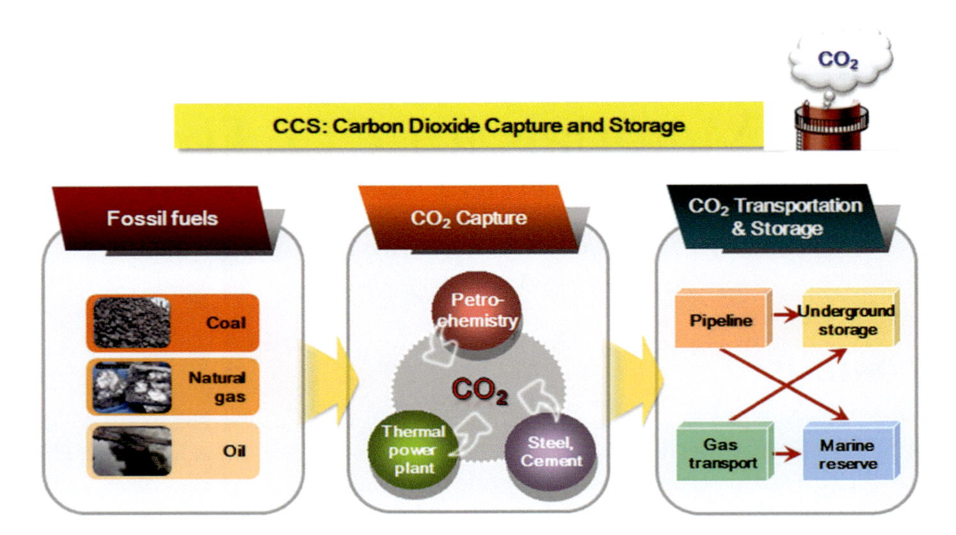

FIGURE 10.1 Summary of the basic concepts of CCS. (Reproduced with permission from Lee and Park, 2014.)

Another promising approach for reducing existing levels of atmospheric CO_2 is the carbon-capture and sequestration (CCS) technology (Beyzavi et al., 2015). The basic concept of CCS is illustrated in Figure 10.1. The limitations such as cost of energy, susceptibility to corrosion, and ineffectiveness of wet-scrubbing units have facilitated the exploration of substitutive strategies for CCS. According to contemporary studies, several approaches for CO_2 fixation (using nanomaterials, porphyrin, metal–organic frameworks (MOFs), activated carbon (AC), etc.) have been investigated worldwide.

Implementation of CCS strategies aids in the stabilization of atmospheric levels of greenhouse gases and provides provisions for controlled fossil fuel usage (Lee and Park, 2014). CCS technologies also help reduce CO_2 emissions as well as the total cost incurred for achieving the same (Lee and Park, 2014). This study discusses different aspects of CCS with special emphasis on novel composite materials being investigated for their CCS potential.

10.2 MECHANISMS OF CO_2 FIXATION

Increasing level of atmospheric CO_2 is considered as a major environmental issue of this century. Hence, it has become imperative to develop efficient processes for avoiding the accumulation of CO_2 in the atmosphere. There are various natural processes through which this carbonaceous gas is fixed in the atmosphere. Photosynthesis is one such process where carbon is stored in the leaf as starch. Approximately 385×10^9 tons of CO_2 is fixed per annum, and the gross amount is greater than a factor of two (Beer et al., 2010). The biological fixation of CO_2 includes six established metabolic pathways, as well as major pathways, that generate and utilize H_2, reduce N_2 to ammonia, oxidize water, and reduce oxygen. Natural conversion and fixation of CO_2 occurs by the formation of C–H and C–C bonds as well as by the breakage of C–O bonds (Furukawa et al., 2013).

The biological systems involved in CCS may range from a single microbe to entire ecosystems by playing pivotal roles in different biogeochemical cycles. Involvement of all the metabolic pathways necessitates assimilation and usage of energy, and organisms should execute these energetic conversions in an efficient and regulated manner. However, in areas undergoing rapid urbanization and industrialization, the levels of CO_2 emissions are extremely high and beyond the solo capacity of biological system-mediated control (Sharma et al., 2019).

Recent studies have reported different processes for the reduction of atmospheric CO_2 levels. However, most strategies reported in recent times require an improvement in terms of efficiency and cost reduction (Ben-Mansour et al., 2016). The significant limitations of conventional processes of CCS are as follows:

- Massive requirement of energy (for providing pure O_2 during the combustion of oxy-fuel, compression of sequestered CO_2, and regeneration of sorbents used for post-combustion capture),
- Large costs of pre-combustion capture,
- Inadequate technical expertise,
- Lack of a single holistic strategy for CCS,
- Non-investigation of the applicability and feasibility of existing processes on commercial scales.

Classification and application of different strategies for CCS are given in Table 10.1.

TABLE 10.1
Different Technologies for CCS and Their Application

Technologies	Classification	Applied Areas
CO_2-capture technologies	Post-combustion Pre-combustion Oxy-fuel combustion	Direct capture from (thermoelectric) power plants and industries (steel mills, refineries, etc.)
CO_2 transportation technologies	CO_2 compression and transportation	Transportation of captured CO_2 to a storage site via a pipeline or other means of transport
CO_2 storage technologies	Characterization and evaluation of bedrock layers Drilling and injection Observation and prediction of movement of bedrock layers Evaluation of the effect on the environment and post-management	Deposit to a bedrock layer (basins, oilfields, etc.) in the ground and below the sea and its management

Source: Reproduced with permission from Lee and Park (2014).

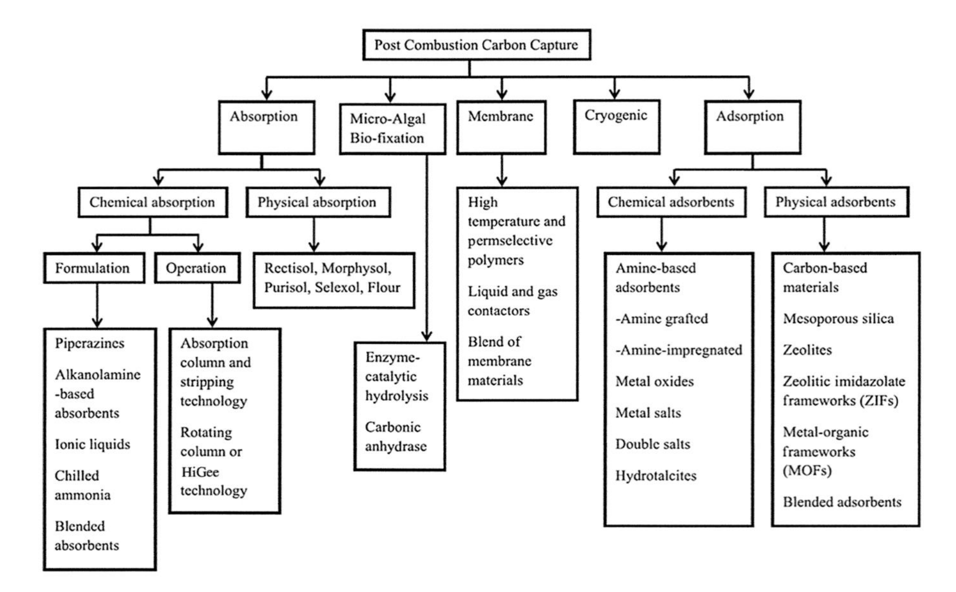

FIGURE 10.2 Different types of post-combustion carbon-capture strategies. (Reproduced with permission from Ben-Mansour et al., 2016.)

In comparison with other processes of CCS reported so far, post-combustion CCS is found to be more advantageous due to its convenient incorporation within the present industrial infrastructures and/or any technology of combustion as well as its convenience for maintenance and monitoring (Ben-Mansour et al., 2016). A comparative analysis of different post-combustion CCS processes is shown in Figure 10.2. Amidst all the other strategies depicted in Figure 10.2, adsorption has a major advantage in terms of the regeneration of adsorbents by convenient thermal or pressure adjustments or simple chemical treatment, thereby rendering the same suitable for subsequent reuse (Ben-Mansour et al., 2016). According to recent reports, nanomaterials of both microporous and mesoporous nature are considered as appropriate adsorbents as well as catalysts for CCS from gaseous mixtures owing to their high pore volume, specific surface area, and thermal stability (Sneddon et al., 2014). Such recently reported nanoparticles and nanocomposite materials have been reviewed in this study for the CCS potential.

10.3 CO_2 FIXATION USING NOVEL COMPOSITE MATERIALS

10.3.1 Nanoparticle-Based Composite Materials

Recent research has shown that nanomaterials (both microporous and mesoporous in nature) having greater surface area, porosity, and suitable functional groups can be the ideal adsorbent for the selective detachment of CO_2 gas from polluted air (Sneddon et al., 2014). The schematic representation depicting the comparative analysis of pore sizes of various nanoporous material investigated for CO_2 capture has been presented in Figure 10.3. Such adsorbents are user-friendly and cause less corrosion-associated problems. The major advantages of these materials are sustainability and probability of an equal balance from the perspective of CO_2 fingerprint in a product's life cycle (Primo et al., 2012). Niu et al. (2016) suggested silica nanoparticles integrated with polyethylenimine as a potential nanocomposite having an average surface area and density higher than those of other nanotubes by six and two times, respectively. Xiong et al. (2012) studied that 1-vinyl-3-(2-methoxy-2-oxyl ethyl) imidazolium chloride nanoparticle synthesized from a mixture of 4-vinylbenzyl-triphenylphosphorous chloride and ethylene glycol dimethacrylate was found to efficiently catalyse the cycloaddition reaction of CO_2 for the formation of epoxides owing to its high stability, efficient performance, and reusability.

Ceria nanoparticles have been reported to be effective catalysts for the incorporation of CO_2 into aliphatic α, ω-diamines (Juarez et al., 2010a; Primo et al., 2014). These nanoparticles (8 nm size) have been synthesized by the calcination of alginate aerogel beads using a biopolymer-templating process (Primo et al., 2014). These nanomaterials can also be used for the production of aromatic dicarbamates using dimethyl carbonate through the trans-alkylation of cyclic carbonates (Juarez et al., 2010b). Ceria is also used as a potential catalytic agent for the formation of dimethyl carbonate from the atmospheric carbon dioxide and alcohols (Tamura et al., 2013).

Harada and Hatton (2015) have reported colloidal clustering of MgO nanomaterials with alkaline metallic nitrates and a mixture of nitrogen oxides to be a competent substance for the quick and promising intake of CO_2. Salts of nitrite applied for coating the MgO nanoparticles were found to increase the critical surface thickness of the composite, thereby inhibiting the detrimental morphological changes in the adsorbent caused by repeated cycles of adsorption and desorption.

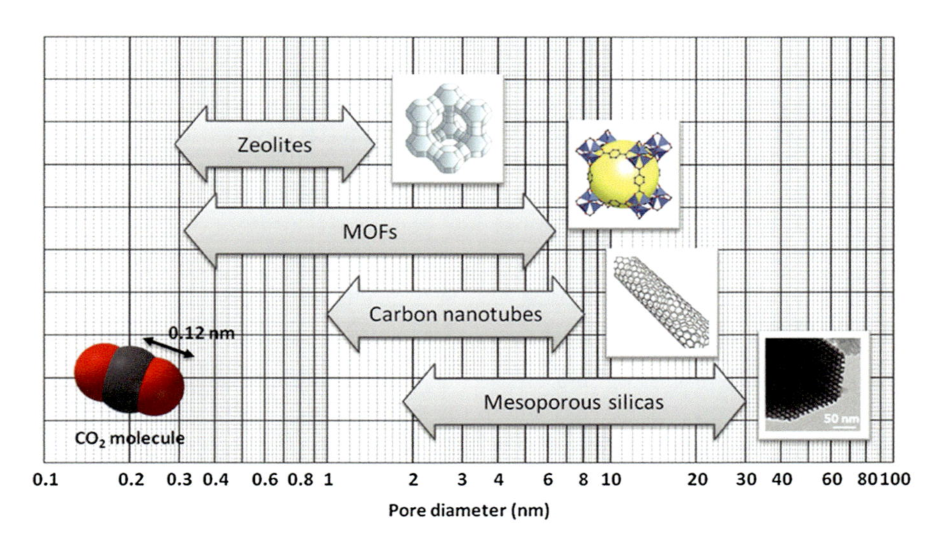

FIGURE 10.3 Schematic representation of the comparative analysis of pore sizes of different widely applied nanoporous materials. (Reproduced with permission from Sneddon et al., 2014.)

Salts of various transition metals such as Pd, Cu, Ag, and Co were reported to act as homogenous catalysts in the carboxylative cyclic reactions at temperatures ranging from 40°C to 60°C and CO_2 pressure of 5–14 MPa (Cui et al., 2015). Functional polymers have been found to be suitable substrates for such nanoporous catalysts, due to their controllable structure, porous nature, and the presence of various functional groups. Biswas et al. (2014) reported polymer-based Cu-Ni nanomaterials as pseudo-homogeneous reactants for alkyne–azide click coupling reactions having magnetic reusability. Noh et al. (2019) reported that Cu nanoparticles dispersed on Ti-containing supports were used for the hydrogenation of CO_2 to CH_3OH. Best results were obtained using TiO_2 supports as Lewis acid sites located on TiO_2 surfaces promoted the methanol synthesis by stabilizing surface intermediates and demonstrated higher Cu nanoparticle-loading capacities. Herbois et al. (2015) reported Ru nanoparticle – β-cyclodextrin-based nanocomposite materials – for the hydrogenation of biofuranic compounds. Yu et al. (2012) studied that Ag-nanoparticles supported by N-heterocyclic carbene reportedly catalysed CO_2 and demonstrated high activity, strength, and recyclability under the conditions of optimum temperature and pressure.

Palladium nanoparticle-impregnated films reportedly acted as dip catalysts in the Suzuki–Miyaura reaction (Hariprasad and Radhakrishnan, 2012). In a separate study, water-compatible Pd nanoparticles immobilized on a microporous polymer matrix were found to exhibit better activity and low Pd-leaching characteristic in carbon cross-coupling reactions (Yang et al., 2013). Pd nanoparticles immobilized on a cross-linked polymer reportedly used CO_2 as a carbon source for the formulation of amines (Li et al., 2016). The cross-linked polymer having nitrogenous porous sites and excess surface area was found to increase the CO_2 fixation efficiency of palladium nanoparticles (Li et al., 2016).

Many researchers also suggested that metal oxide composites with carbonaceous substances exhibited better efficacy for size-specific adsorption of gases (Chowdhury et al., 2015, Hong et al., 2012, Li et al., 2015, Mishra and Ramaprabhu, 2014). Li et al. (2016) reported a novel porous nanomaterial (N-rGO-ZnO) composed of zinc oxide and N-doped reduced graphene oxide (rGO). Incorporation of ZnO was found to increase the microporosity as well as the CO_2 and N_2 adsorption capacities of rGO. Carbon dioxide-uptake efficiency and nitrogen-uptake efficiency of N-rGO-ZnO were also found to increase in a temperature-dependent manner (Li et al., 2016).

10.3.2 Porphyrin-Based Composite Materials

One of the most promising and environmentally favourable approaches for carbon dioxide fixation is the conversion of the same to cyclic carbonates through the chemical interaction with epoxides. These cyclic carbonates may serve as the value-added intermediates in organic synthesis, promising polymer precursors, aprotic polar solvents, solvent of electrolytes for secondary batteries, etc. (Kim et al., 2013). Porphyrins have been reported as significant catalysts for this reaction over the previous decade.

Porphyrins have benzene ring structure, have the light-absorbing capacity, and are available in abundance. According to Ema et al. (2014), well-designed bifunctional porphyrin catalysts exhibit high catalytic performance as a result of the integrated action of the nucleophilic halide and the Lewis acid metal centre. Chen et al. (2010) have reported that under optimized conditions, porous metalloporphyrins synthesized via free-radical polymerization have demonstrated high capacity and recyclability in the CO_2/epoxide cycloaddition reaction. Komatsu et al. (1991) had reported the synthesis of Cr-porphyrin polymers via the Sonogashira coupling of porphyrins and 1, 4-diiodobenzene in the presence of silica materials. Under normal conditions of temperature, these microporous Cr (III)-F porphyrin composites have reportedly exhibited potential catalytic activities for CO_2 capture with the subsequent conversion of epoxides to cyclic carbonates (Komatsu et al., 1991). In a recent study, Liu et al. (2018) have also reported 83% concentration of CO_2 using catalytic space-engineered porous porphyrin MOFs (as shown in Figure 10.4).

Recent investigation reports have suggested that the transition of CO_2 into highly beneficial chemicals using porphyrin-based organic polymer (POP) has attracted great attention from both the academic and industrial sectors owing to its unique properties such as high surface areas, adjustable porous configuration, and enhanced mechanical strength. Therefore, different types of modified and novel POPs have been investigated for CO_2 capture and subsequent transition (Chen et al., 2016). Moreover, the integration of different metal ions with porphyrin matrices was found to render the composite higher stability than the parent materials and also permitted a spatial distribution of the attached functional groups (Ema et al., 2014). Hence, it may be concluded that proper functionalization is capable of enhancing the catalytic performance of metal–porphyrin composites for CO_2 capture.

10.3.3 Metal–Organic Framework (MOF)-Based Composite Materials

MOFs are microporous in nature and composed of metallic clusters or ions attached to organic ligands (Furukawa et al., 2013; Qiu and Zhu, 2009). According to recent studies, MOFs have also been considered as potential catalysts for CO_2 fixation

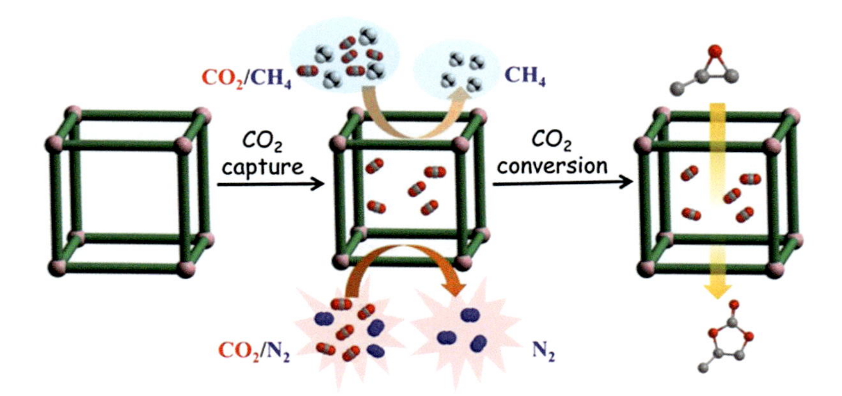

FIGURE 10.4 Schematic representation of integrated capture and conversion of CO_2 by M-porous porphyrin MOF-1 (M=Rh, Ir) using coordination space engineering (Reproduced with permission from Liu et al., 2018.)

owing to their crystalline nature, high surface area, mechanical integrity, exposed channels, and long-lasting porosity. Moreover, their structure tunability and affinity for CO_2 have also rendered MOFs as suitable catalytic agents for the transformation of CO_2 (Beyzavi et al., 2015). Porous MOFs act as potential agents for the catalytic interactions owing to the high surface area, distinct pores, presence of modified functional groups on their surface, and promising recyclable catalytic efficiency for the separation of CO_2 (Gao et al., 2014; Chen et al., 2010).

The activity of MOFs for catalysing the conversion of CO_2 has therefore gained enormous attention. Many different MOFs have already been reported as promising materials for CO_2 fixation and its subsequent catalysis through several chemical reactions such as terminal alkyne activation, hydroboration, and addition of epoxides (Kumar et al., 2016). Contemporary studies have focused on the development of different novel MOF-based composites that exhibit greater surface areas and higher gas-loading capacities (Sumida et al., 2011, Xiang et al., 2011). Such substances are found to demonstrate the high sorption enthalpy, which in turn facilitates the adsorption of the gaseous CO_2. Adsorptive techniques that facilitate the physisorption of CO_2 onto porous MOFs have acquired momentum as the output of facile regeneration approach (Kumar et al., 2016). Moreover, the different metal centres and functional ligands present in MOFs facilitate the synthesis of innovative complex materials capable of enhanced gas storage (Kang et al., 2015). Enhancement of MOF activity has been accomplished via chemical catenation, modification of chemical bonding, and participation of electrostatic force (Liu et al., 2012). A schematic representation of selective CO_2 capture by different MOF-based adsorbents is shown in Figure 10.5.

MOFs have been investigated as potential supports for achieving uniform dispersion of metal (oxide) particles, thereby yielding novel composite materials having greater surface areas (Kumar et al., 2016). Kang et al. (2015) reported the novel composite materials synthesized by dispersing carboxyl-based carbon nanotubes in MOF nanomaterials. This composite reportedly demonstrated greater CO_2 fixation with enhanced adsorption enthalpies in comparison with each parent constituent (Kang et al., 2015).

MOFs with structural versatility and modularity have also been explored as novel functional heterogeneous catalytic agents (Gao et al., 2014; Kim et al., 2013). Other contemporary investigations have also reported different strategies, including multipoint reactions, segmentation of pore space, associative crystallization, exchange of ion and ligand, charge polarization, improvement of the local electric field, regulation of morphology, and permeation of partial framework, for increasing the efficiency of CO_2 fixation using different MOF-based composites (Yu et al., 2017).

10.3.4 Activated Carbon-Based Composite Materials

Many researchers have reported that ACs having high basicity possess the potential of sequestering CO_2 from space-limited chambers. As CO_2 is acidic in nature, it interacts with basic ACs via dipole interactions, hydrogen bonding, and covalent bonding (Creamer and Gao, 2016). AC-based filters have been frequently utilized for the separation of different gases such as SO_2, H_2S, NO_x, HCl, and HF and vapours of various volatile organic compounds (e.g. formic acid, acetic acid) released from

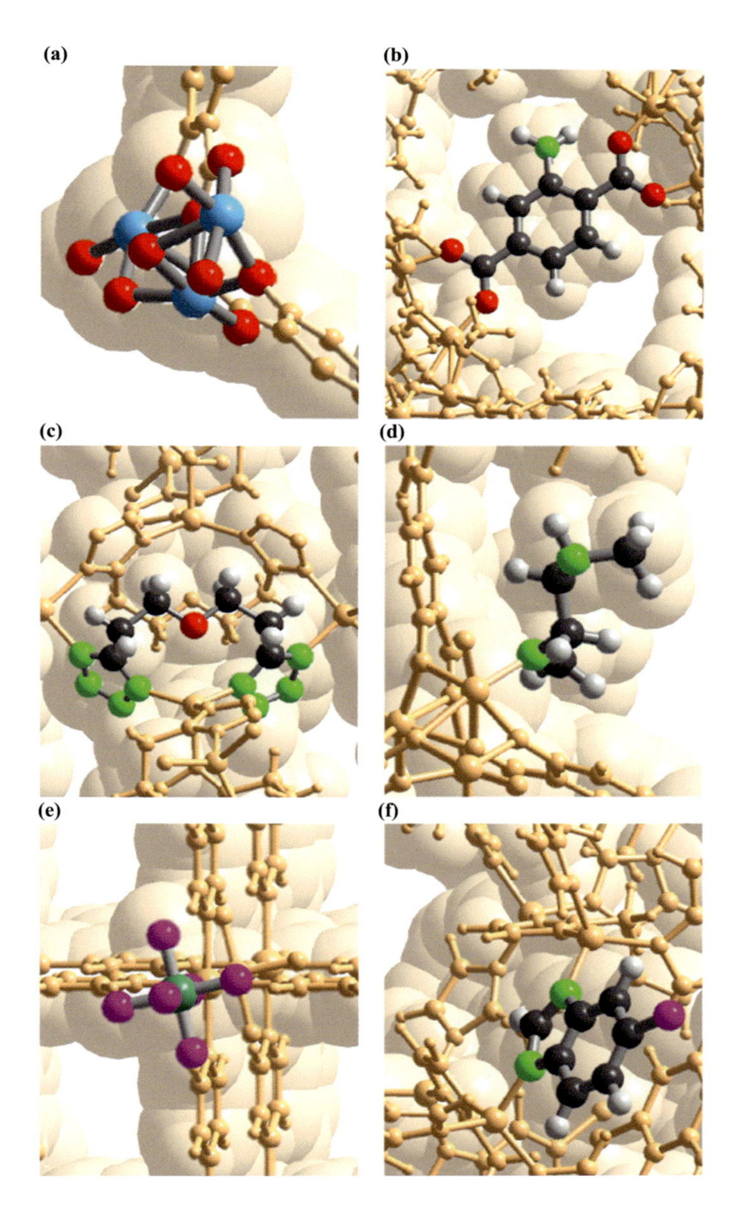

FIGURE 10.5 Illustrative representation of significant structural design structures of effective MOF adsorbents for selective capture of CO_2. (Reproduced with permission from Trickett et al., 2017.) (a) Coordinatively linked unsaturated metal sites; (b) covalently associated polar functionalities; (c) multiatomic amines having non-coordinating atoms which besides serving as linker components, freely interact with CO_2 as well; (d) alkyl amines bonded to coordinatively linked unsaturated metal sites or covalently linked with the organic linker; (e) specific interactions between non-metallic moieties present within the pores of a MOF, which induce a robust quadrupolar linkage with CO_2; (f) hydrophobic nature and/or pore dynamics that facilitate selective CO_2 capture in an aqueous environment.

different electronics, pharmaceuticals, and biochemical industries (Meeyoo et al., 1997; Przepiorski et al., 2004). A number of studies have also associated enhanced rates of CO_2 adsorption with surface modification of ACs (Przepiorski et al., 2004). Moreover, alkaline ACs also reportedly act as catalysts in chemical reactions (Raymundo-Pinero et al., 2002). However, treatment with additives can also block the pores of AC, thereby resulting in reduced efficiency of the same (Nakamura et al., 1996; Molina-Sabio et al., 1994). Various studies have reported an efficient modification of ACs using gaseous or liquid substances (Shim et al., 2001; Lozano-Castello et al., 2001; Illan-Gomez et al., 1996; Ahmadpour et al., 1996). Incorporation of nitrogenous groups on AC surface also reportedly results in increased basicity of ACs (Plaza et al., 2009; Shafeeyan et al., 2010). In a study by Mangun et al. (2001), fibres of AC were modified with dry ammonia (NH_3) at high temperature (500°C–800°C). Such modification introduced new nitrogen-containing C–N and C=N groups in the AC fibres (as revealed by Fourier transform infrared spectroscopy). Much modification was found to enhance CO_2 uptake with a parallel rise in temperature till 400°C. A further rise in temperature, however, caused a decline in rates of CO_2 uptake by closing the pores of the modified AC fibres. Similar studies by Boehm et al. (1984) suggested that the decomposition of NH_3 at high temperatures liberated different radicals (NH_2, NH, and H), which interacted with the treated carbon to produce functional groups such as amino ($-NH_2$), cyano (–CN), pyrrolic, pyridinic, and quaternary nitrogen. The results obtained by Figueiredo et al. (1999) also demonstrated that the pore size distribution of ACs was altered by ammonical heat treatment.

ACs synthesized by Sevilla and Fuertes (2011) demonstrated high CO_2 sequestration capacity (4.8 mmol g^{-1}). Weak interactions established between CO_2 and AC surface in this process facilitated a convenient regeneration of the ACs reported herein. ACs modified with oxides and hydroxides of metals such as Fe, Al, Ca, and Mg were found to demonstrate better CO_2 uptake in comparison with unmodified ACs (Creamer and Gao, 2016). However, these metallic compounds should be added in optimum quantities, as an excess of the same may reduce the effective surface area, which in turn decreases the active sites of adsorption. Moreover, AC-based composites have been found to enhance both physical and chemical routes of adsorption (Creamer and Gao, 2016).

10.4 CONCLUSION

Many promising adsorbents play a vital role in most of CCS processes designed for CO_2 detachment from different sources. Recent research work has significantly concentrated on designing potential adsorbents having increased adsorption quality, constancy, and recyclability. This study reviews different composite materials discussed in contemporary pieces of literature on CO_2 fixation and its subsequent transformation to the value-added products. Greater surface area, presence of various functional groups, porosity, reusability, stability, and greater catalytic activities of various nanocomposites have facilitated enhanced CCS efficiencies. The composite materials described in this chapter can be applied as catalysts, scavengers, and sorbents for CCS. However, all the promising approaches for CO_2 conversion and sequestration are still considered as temporary remedies for reducing the assimilation of carbon in the surroundings. Therefore, the human demand for energy has to be met by other unconventional resources (solar energy, tidal energy, wind energy, hydro energy, etc) as alternatives to fossil fuels in order to prevent any harmful and significantly negative effects on the environment. Nonetheless, all the approaches discussed in this review for CO_2 fixation and conversion into the value-added products can be reckoned as the latest and most promising technologies in the concerned field.

Colour code: black, red, green, purple, white, dark green, and blue represent carbon, oxygen, nitrogen, fluorine or chlorine, hydrogen, silicon, and magnesium, respectively. The adjacent MOF structures are depicted in pale orange.

REFERENCES

Ahmadpour, A., and D. D. Do. The preparation of active carbons from coal by chemical and physical activation. *Carbon* 34, no. 4 (1996): 471–479.

Beer, C., M. Reichstein, E. Tomelleri, P. Ciais, M. Jung, N. Carvalhais, C. Rödenbeck, M.A. Arain, D. Baldocchi, G.B. Bonan and A. Bondeau. Terrestrial gross carbon dioxide uptake: Global distribution and covariation with climate. *Science* 329, no. 5993 (2010): 834–838.

Ben-Mansour, R., M. A. Habib, O. E. Bamidele, M. Basha, N. A. A. Qasem, A. Peedikakkal, T. Laoui, and M. Ali. Carbon capture by physical adsorption: Materials, experimental investigations and numerical modeling and simulations–a review. *Applied Energy* 161 (2016): 225–255.

Beyzavi, M. H., C. J. Stephenson, Y. Liu, O. Karagiaridi, J. T. Hupp and O. K. Farha. Metal–organic framework-based catalysts: Chemical fixation of CO_2 with epoxides leading to cyclic organic carbonates. *Frontiers in Energy Research* 2 (2015): 63.

Biswas, M., A. Saha, M. Dule, and T. K. Mandal. Polymer-assisted chain-like organization of CuNi alloy nanoparticles: Solvent-adoptable pseudohomogeneous catalysts for alkyne–azide click reactions with magnetic recyclability. *The Journal of Physical Chemistry C* 118, no. 38 (2014): 22156–22165.

Boehm, H. P., G. Mair, T. Stoehr, A. R. De Rincón and B. Tereczki. Carbon as a catalyst in oxidation reactions and hydrogen halide elimination reactions. *Fuel* 63, no. 8 (1984): 1061–1063.

Chen, Y., R. Luo, Q. Xu, W. Zhang, X. Zhou and H. Ji. State-of-the-art aluminum porphyrin-based heterogeneous catalysts for the chemical fixation of CO_2 into cyclic carbonates at ambient conditions. *ChemCatChem* 9, no. 5 (2017): 767–773.

Chen, B., S. Xiang and G. Qian. Metal– organic frameworks with functional pores for recognition of small molecules. *Accounts of Chemical Research* 43, no. 8 (2010): 1115–1124.

Chowdhury, S., G. K. Parshetti and R. Balasubramanian. Post-combustion CO_2 capture using mesoporous TiO_2/graphene oxide nanocomposites. *Chemical Engineering Journal* 263(2015): 374–384.

Creamer, A. E. and B. Gao. Carbon-based adsorbents for postcombustion CO_2 capture: A critical review. *Environmental Science & Technology* 50, no. 14 (2016): 7276–7289.

Crombie, M., S. Imbus and I. Miracca. CO_2 capture project phase 3—Demonstration phase. *Energy Procedia* 4(2011): 6104–6108.

Cui, M., Q. Qian, Z. He, J. Ma, X. Kang, J. Hu, Z. Liu and B. Han. Synthesizing Ag nanoparticles of small size on a hierarchical porosity support for the carboxylative cyclization of propargyl alcohols with CO_2 under ambient conditions. *Chemistry–A European Journal* 21, no. 45 (2015): 15924–15928.

Ema, T., Y. Miyazaki, J. Shimonishi, C. Maeda and J. Hasegawa. Bifunctional porphyrin catalysts for the synthesis of cyclic carbonates from epoxides and CO_2: Structural optimization and mechanistic study. *Journal of the American Chemical Society* 136, no. 43 (2014): 15270–15279.

Figueiredo, J. L., M. F. R. Pereira, M. M. A. Freitas and J. J. M. Orfao. Modification of the surface chemistry of activated carbons. *Carbon* 37, no. 9 (1999): 1379–1389.

Furukawa, H., K. E. Cordova, M. O'Keeffe, and O. M. Yaghi. The chemistry and applications of metal-organic frameworks. *Science* 341, no. 6149 (2013): 1230444.

Gao, W. Y., Y. Chen, Y. Niu, K. Williams, L. Cash, P. J. Perez, L. Wojtas, J. Cai, Y. S. Chen and S. Ma. Crystal engineering of an nbo topology metal–organic framework for chemical fixation of CO_2 under ambient conditions. *Angewandte Chemie International Edition* 53, no. 10 (2014): 2615–2619.

Harada, T. and T. A. Hatton. Colloidal nanoclusters of MgO coated with alkali metal nitrates/nitrites for rapid, high capacity CO_2 capture at moderate temperature. *Chemistry of Materials* 27, no. 23 (2015): 8153–8161.

Hariprasad, E. and T. P. Radhakrishnan. Palladium nanoparticle-embedded polymer thin film "dip catalyst" for Suzuki–Miyaura reaction. *ACS Catalysis* 2, no. 6 (2012): 1179–1186.

Herbois, R., S. Noël, B. Léger, S. Tilloy, S. Menuel, A. Addad, B. Martel, A. Ponchel and E. Monflier. Ruthenium-containing β-cyclodextrin polymer globules for the catalytic hydrogenation of biomass-derived furanic compounds. *Green Chemistry* 17, no. 4 (2015): 2444–2454.

Hong, W. G., B. H. Kim, S. M. Lee, H. Y. Yu, Y. J. Yun, Y. Jun, J. B. Lee and H. J. Kim. Agent-free synthesis of graphene oxide/transition metal oxide composites and its application for hydrogen storage. *International Journal of Hydrogen Energy* 37, no. 9 (2012): 7594–7599.

Illán-Gómez, M. J., A. Garcia-Garcia, C. Salinas-Martinez de Lecea, and A. Linares-Solano. Activated carbons from Spanish coals. 2. Chemical activation. *Energy & Fuels* 10, no. 5 (1996): 1108–1114.

Juárez, R., P. Concepción, A. Corma, and H. García. Ceria nanoparticles as heterogeneous catalyst for CO_2 fixation by ω-amino alcohols. *Chemical Communications* 46, no. 23 (2010a): 4181–4183.

Juarez, R., S. F. Parker, P. Concepcion, A. Corma, and H. Garcia. Heterolytic and heterotopic dissociation of hydrogen on ceria-supported gold nanoparticles. Combined inelastic neutron scattering and FT-IR spectroscopic study on the nature and reactivity of surface hydrogen species. *Chemical Science* 1, no. 6 (2010b): 731–738.

Kang, Z., M. Xue, D. Zhang, L. Fan, Y. Pan and S. Qiu. Hybrid metal-organic framework nanomaterials with enhanced carbon dioxide and methane adsorption enthalpy by incorporation of carbon nanotubes. *Inorganic Chemistry Communications* 58(2015): 79–83.

Kim, M. H., T. Song, U. R. Seo, J. E. Park, K. Cho, S. M. Lee, H. J. Kim, Y. J. Ko, Y. K. Chung and S. U. Son. Hollow and microporous catalysts bearing Cr (III)–F porphyrins for room temperature CO_2 fixation to cyclic carbonates. *Journal of Materials Chemistry A* 5, no. 45 (2017): 23612–23619.

Komatsu, M., T. Aida and S. Inoue. Novel visible-light-driven catalytic carbon dioxide fixation. Synthesis of malonic acid derivatives from CO_2, an. alpha.,. beta.-unsaturated ester or nitrile, and diethylzinc catalyzed by aluminum porphyrins. *Journal of the American Chemical Society* 113, no. 22 (1991): 8492–8498.

Kumar, S., G. Verma, W. Y. Gao, Z. Niu, L. Wojtas and S. Ma. Anionic metal–organic framework for selective dye removal and CO_2 fixation. *European Journal of Inorganic Chemistry* 2016, no. 27 (2016): 4373–4377.

Lee, S.Y. and S.J. Park. A review on solid adsorbents for carbon dioxide capture. *Journal of Industrial and Engineering Chemistry* 23 (2015): 1–11.

Li, W., H. Yang, X. Jiang and Q. Liu. Highly selective CO_2 adsorption of ZnO based N-doped reduced graphene oxide porous nanomaterial. *Applied Surface Science* 360 (2016): 143–147.

Li, W., X. Jiang, H. Yang and Q. Liu. Solvothermal synthesis and enhanced CO_2 adsorption ability of mesoporous graphene oxide-ZnO nanocomposite. *Applied Surface Science* 356 (2015): 812–816.

Liu, J., Y. Z. Fan, X. Li, Y. W. Xu, L. Zhang, and C. Y. Su. Catalytic space engineering of porphyrin metal–organic frameworks for combined CO_2 capture and conversion at a low concentration. *ChemSusChem* 11, no. 14 (2018): 2340–2347.

Liu, J., P. K. Thallapally, B. P. McGrail, D. R. Brown and J. Liu. Progress in adsorption-based CO_2 capture by metal–organic frameworks. *Chemical Society Reviews* 41, no. 6 (2012): 2308–2322.

Lozano-Castello, D., M. A. Lillo-Rodenas, D. Cazorla-Amorós and A. Linares-Solano. Preparation of activated carbons from Spanish anthracite: I. Activation by KOH. *Carbon* 39, no. 5 (2001): 741–749.

Mangun, C. L., K. R. Benak, J. Economy and K. L. Foster. Surface chemistry, pore sizes and adsorption properties of activated carbon fibers and precursors treated with ammonia. *Carbon* 39, no. 12 (2001): 1809–1820.

Meeyoo, V., D. L. Trimm and N. W. Cant. Adsorption-reaction processes for the removal of hydrogen sulphide from gas streams. *Journal of Chemical Technology & Biotechnology: International Research in Process, Environmental and Clean Technology* 68, no. 4 (1997): 411–416.

Mishra, A. K. and S. Ramaprabhu. Enhanced CO_2 capture in Fe_3O_4-graphene nanocomposite by physicochemical adsorption. *Journal of Applied Physics* 116, no. 6 (2014): 064306.

Mohammadi, A., M. Soltanieh, M. Abbaspour and F. Atabi. What is energy efficiency and emission reduction potential in the Iranian petrochemical industry? *International Journal of Greenhouse Gas Control* 12(2013): 460–471.

Molina-Sabio, M., V. Perez and F. Rodriguez-Reinoso. Impregnation of activated carbon with chromium and copper salts: Effect of porosity and metal content. *Carbon* 32, no. 7 (1994): 1259–1265.

Nakamura, T., S. Tanada, N. Kawasaki, T. Hara, J. Fujisawa, and K. Shibata. Hydrogen sulfide removal by iron containing activated carbon. *Toxicological & Environmental Chemistry* 55, no. 1–4 (1996): 279–283.

Niu, M., H. Yang, X. Zhang, Y. Wang and A. Tang. Amine-impregnated mesoporous silica nanotube as an emerging nanocomposite for CO_2 capture. *ACS Applied Materials & Interfaces* 8, no. 27 (2016): 17312–17320.

Noh, G., S. R. Docherty, E. L. X. Huang, D. Mance, J. L. Alfke, and C. Copéret. CO_2 Hydrogenation to CH_3OH on Supported Cu Nanoparticles: Nature and Role of Ti in Bulk Oxides vs. as Isolated Surface Sites. *The Journal of Physical Chemistry C* (2019) (DOI: 10.1021/acs.jpcc.9b09631).

Olah, G. A., B. Török, J. P. Joschek, I. Bucsi, P. M. Esteves, G. Rasul and G. K. S. Prakash. Efficient chemoselective carboxylation of aromatics to arylcarboxylic acids with a superelectrophilically activated carbon dioxide– Al_2Cl_6/Al system. *Journal of the American Chemical Society* 124, no. 38 (2002): 11379–11391.

Plaza, M. G., C. Pevida, B. Arias, M. D. Casal, C. F. Martín, J. Fermoso, F. Rubiera, and J. J. Pis. Different approaches for the development of low-cost CO_2 adsorbents. *Journal of Environmental Engineering* 135, no. 6 (2009): 426–432.

Przepiórski, J., M. Skrodzewicz and A. W. Morawski. High temperature ammonia treatment of activated carbon for enhancement of CO_2 adsorption. *Applied Surface Science* 225, no. 1–4 (2004): 235–242.

Primo, A., E. Aguado and H. Garcia. CO_2-fixation on aliphatic α, ω-diamines to form cyclic ureas, catalyzed by ceria nanoparticles that were obtained by templating with alginate. *ChemCatChem* 5, no. 4 (2013): 1020–1023.

Qiu, S. and G. Zhu. Molecular engineering for synthesizing novel structures of metal–organic frameworks with multifunctional properties. *Coordination Chemistry Reviews* 253, no. 23–24 (2009): 2891–2911.

Raymundo-Pinero, E., D. Cazorla-Amorós, A. Linares-Solano, J. Find, U. Wild and R. Schlögl. Structural characterization of N-containing activated carbon fibers prepared from a low softening point petroleum pitch and a melamine resin. *Carbon* 40, no. 4 (2002): 597–608.

Shafeeyan, M. S., W. M. A. W. Daud, A. Houshmand, and A. Shamiri. A review on surface modification of activated carbon for carbon dioxide adsorption. *Journal of Analytical and Applied Pyrolysis* 89, no. 2 (2010): 143–151.

Sevilla, M., and A. B. Fuertes. Sustainable porous carbons with a superior performance for CO_2 capture. *Energy & Environmental Science* 4, no. 5 (2011): 1765–1771.

Sharma, T., S. Sharma, H. Kamyab, and A. Kumar. Energizing the CO_2 utilization by chemo-enzymatic approaches and potentiality of carbonic anhydrases: A review. *Journal of Cleaner Production* (2019): 119138.

Shim, J., S. Park and S. Ryu. Effect of modification with HNO_3 and NaOH on metal adsorption by pitch-based activated carbon fibers. *Carbon* 39, no. 11 (2001): 1635–1642.

Sneddon, G., A. Greenaway and H. H. P. Yiu. The potential applications of nanoporous materials for the adsorption, separation, and catalytic conversion of carbon dioxide. *Advanced Energy Materials* 4, no. 10 (2014): 1301873.

Songolzadeh, M., M. Soleimani, M. T. Ravanchi, and R. Songolzadeh. Carbon dioxide separation from flue gases: A technological review emphasizing reduction in greenhouse gas emissions. *The Scientific World Journal* 2014 (2014).

Sumida, K., D. L. Rogow, J. A. Mason, T. M. McDonald, E. D. Bloch, Z. R. Herm, T. Bae, and J. R. Long. Carbon dioxide capture in metal–organic frameworks. *Chemical Reviews* 112, no. 2 (2011): 724–781.

Tamura, M., K. Noro, M. Honda, Y. Nakagawa and K. Tomishige. Highly efficient synthesis of cyclic ureas from CO_2 and diamines by a pure CeO_2 catalyst using a 2-propanol solvent. *Green Chemistry* 15, no. 6 (2013): 1567–1577.

Trickett, C. A., A. Helal, B. A. Al-Maythalony, Z. H. Yamani, K. E. Cordova and O. M. Yaghi. The chemistry of metal–organic frameworks for CO_2 capture, regeneration and conversion. *Nature Reviews Materials* 2, no. 8 (2017): 17045.

Xiang, Z., Z. Hu, D. Cao, W. Yang, J. Lu, B. Han and W. Wang. Metal–organic frameworks with incorporated carbon nanotubes: Improving carbon dioxide and methane storage capacities by lithium doping. *Angewandte Chemie International Edition* 50, no. 2 (2011): 491–494.

Xiong, Y., Y. Wang, H. Wang, R. Wang, and Z. Cui. Novel one-step synthesis to cross-linked polymeric nanoparticles as highly active and selective catalysts for cycloaddition of CO_2 to epoxides. *Journal of Applied Polymer Science* 123, no. 3 (2012): 1486–1493.

Yang, Y., S. Ogasawara, G. Li and S. Kato. Water compatible Pd nanoparticle catalysts supported on microporous polymers: Their controllable microstructure and extremely low Pd-leaching behaviour. *Journal of Materials Chemistry A* 1, no. 11 (2013): 3700–3705.

Yu, D., M. X. Tan, and Y. Zhang. Carboxylation of terminal alkynes with carbon dioxide catalyzed by poly (N-heterocyclic carbene) - supported silver nanoparticles. *Advanced Synthesis & Catalysis* 354, no. 6 (2012): 969–974.

Yu, J., L. Xie, J. Li, Y. Ma, J. M. Seminario and P. B. Balbuena. CO_2 capture and separations using MOFs: Computational and experimental studies. *Chemical Reviews* 117, no. 14 (2017): 9674–9754.

11 Microalgae-Based Biorefinery for Utilization of Carbon Dioxide for Production of Valuable Bioproducts

Rahul Kumar Goswami, Komal Agrawal, Sanjeet Mehariya, Antonio Molino, Dino Musmarra, and Pradeep Verma

CONTENTS

11.1 INTRODUCTION

In the past two decades, major environmental risk has been due to the uncontrolled release of CO_2 from different industrial sectors in the atmosphere, which causes a greenhouse effect. CO_2 is not harmful when it is present in a required concentration, because it maintains the flora and fauna of the Earth, but a higher concentration of CO_2 has shown an adverse effect on the Earth. The increased concentration of CO_2 is due to the industrialization and deforestation. Recently, it has become a global issue, which has affected the environment, governmental policy, and governance (Zhao and Su 2014; Mehariya et al. 2020; 2018; Siciliano et al. 2018). Many approaches have been used to maintain CO_2 concentration, such as physical and chemical methods, but these approaches are not economically feasible and eco-friendly. However, biological-based CO_2 mitigation is safe, eco-friendly, and cost-effective; additionally, it helps in renewal energy generation. Thus, researchers are focused on the biological mitigation of CO_2 from the environment for the generation and utilization of renewable energy. Many plants, algae, fungi, and bacteria are used for the biological CO_2 mitigation. Among them, algae can be a suitable candidate for CO_2 mitigation because they are phototrophs, easy to cultivate, and can fix atmospheric CO_2 efficiently and cost-effectively (Del Campo et al. 2000; 2001; Blanco et al. 2007; Sánchez et al. 2008; Fernández-Sevilla, Acién Fernández, and Molina Grima 2010; Mehariya et al. 2019a; Molino, Mehariya, et al. 2019). Moreover, microalgal biomass contains several intracellular compounds, which have several human health benefits, and leftover biomass can be transformed into bioenergy. This chapter discusses CO_2 emission and its adverse effect on the environment. Moreover, this chapter also discusses algal-based CO_2 mitigation in detail with their carbon fixation strategy, photo-bioreactor design, and utilization of CO_2 in the conversion of microalgal biomass to bioenergy and further valuable bioproducts (Zeng et al. 2011).

11.1.1 CO_2 EMISSION

The human activities or industrial revolution causes the emission of CO_2 in the environment. The oil-producing countries cause higher CO_2 emission per person (Figure 11.1). Moreover, highly populated countries result in high per capita emissions, such as Australia (17 tonnes) followed by the United States (16.2 tonnes) and Canada (15.6 tonnes). Asia is the largest CO_2 emitter, which accounts for 53% of the global emission. Moreover, in Asia, China is the highest carbon emitter (10 billion tonnes/year), which accounts for one-third of the global emission (Ritchie and Roser 2018). Moreira and Pires (2016) reported that CO_2 concentration in the ecosystem is increasing by the rate of 2 ppm year^{-1}; currently, CO_2 emission is around 30,000 Mt year^{-1} worldwide. In 2010, Environmental Protection Agency (EPA) of

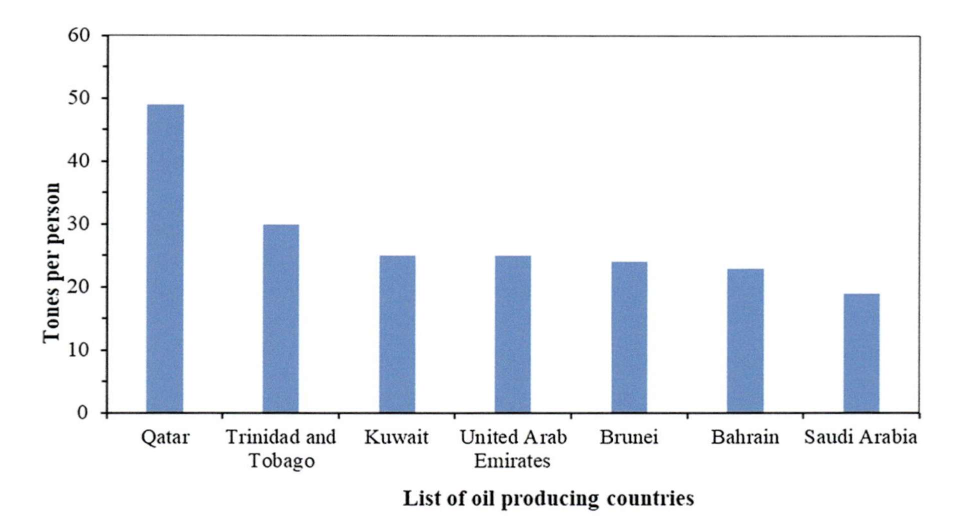

FIGURE 11.1 Graphical representation of CO_2 emission in oil-producing countries (Ritchie and Roser 2018).

the United States reported that 71% of greenhouse gas (GHG) emission is due to energy consumption and production in transportation sector globally (Wilbanks and Fernandez 2014).

11.1.2 Harmful Effect of CO_2

According to Usui and Ikenouchi (1997), CO_2 emission contributes to more than 68% of GHG, which is the most common reason for global warming. Currently, global warming is the most serious problem because Antarctic ice sheets are losing their mass, leading to an increase in sea level along with frequent heatwaves for a longer period (Pires 2017).

11.2 CO_2-CAPTURE SYSTEM FROM ATMOSPHERE

CO_2 can be available in free form, eluted from different systems; therefore, various approaches are used to capture CO_2. These approaches can be based on different capture mechanisms and its key benefit is to reduce CO_2 from the ecosystem, and CO_2 can be used for energy generation or as a substrate for chemical production (Stepan et al. 2002).

11.2.1 Mechanism of Sequestration

In the atmosphere, CO_2 is present in mixture along with other gases, which demands different capture mechanisms. CO_2 can be sequestrated in pre-combustion and post-combustion mechanisms. In the pre-combustion process, CO_2 capture occurs prior to complete combustion of fossil fuels, whereas in post-combustion process,

CO_2 capture is carried out after the complete combustion of fossil fuels (Le Moullec and Kanniche 2011). However, in both cases, the physical and chemical methods can be used. In the physical method, membrane filter systems are used, whereas in the chemical method, CO_2 is immobilized by the adsorption materials like lithium oxide, then neutralized by alkaline solution, and converted into bicarbonate ions or other useful chemicals (Lackner 2003). However, in the biological system, many photoautotrophic organisms trap CO_2 and convert it into organic compounds via photosynthesis catalysed by nicotinamide adenine dinucleotide phosphate (NADPH) and adenosine triphosphate (ATP) (Iverson 2006; Zhao and Su 2014).

11.2.2 Different Available Technologies for CO_2 Capturing, Their Efficacy, and Drawbacks

In the physical method, membrane separation techniques are mostly used because they are low-carbon emission processes, which can be operated in a continuous manner and in which a gas mixture continuously passes through the membrane and CO_2 binds to the membrane. However, in the geological injection system, CO_2 is injected into geological reservoirs like oil-depleted wells (Lackner 2003). In the chemical process, CO_2 is immobilized using absorption material like lithium hydroxide and neutralized by alkaline-mediated treatment, or it is chemically reacted with heavy metal oxides to form stable carbonate ions (Lackner 2003). The physical and chemical methods have many disadvantages: large space, geological structure, expensive treatment or equipment, chemicals, and chances of leakage (Lackner 2003). In the biological system, e.g. forestation and oceanic fertilization, macroalgae, microalgae, and cyanobacteria are used to mitigate CO_2. These phototrophic organisms contain chloroplast and have capabilities to fix CO_2 by the photosynthetic pathway. Forestation requires a large area, but CO_2 sequestration rate is lower. However, microalgal biomitigation techniques are far better than other biological mitigation techniques. The microalgae are the photosynthetic organisms, which have 50% cell dry weight of carbon (Kumar et al. 2010). The key advantage and disadvantage of different available CO_2 capture methods are discussed in Table 11.1.

11.3 MICROALGAL CULTIVATION MODES

Microalgae use CO_2 as C and light as an energy source for their metabolism in the presence of water and nutrients. During the metabolic activity, microalgae accumulate several bioactive compounds in their cells, which have several industrial applications (Belay et al. 1993). Microalgae are generally considered as auto-phototrophs, which are the main oxygen-evolving photosynthetic microbes. Moreover, globally 50% CO_2 is captured by microalgae during the photosynthesis (Zhou et al. 2017). Microalgal cells can grow in a broad range of temperatures (20°C–30°C) and fresh or marine water (Molino, Mehariya, et al. 2019). Microalgal cells can grow faster as compared to plants; it lacks root, stem, and leaves but it contains chlorophyll pigments, which are essential for photosynthesis. This pigment helps in the conversion of sunlight into chemical energy, and it helps to utilize carbon source and convert into glucose (Cheah et al. 2015). It is a physiochemical process that converts CO_2 into

TABLE 11.1
Advantages and Disadvantages of Several CO_2 Mitigation Techniques (Zhou et al. 2017; White et al. 2003; Lal 2008; Beedlow et al. 2004; Olajire 2013; Kita and Ohsumi 2004; De Silva, Ranjith, and Perera 2015; Tang, Yang, and Bian 2014; Salek et al. 2013; Zhao and Su 2014; Lackner 2003)

Class	Techniques	Mechanisms	Advantages	Disadvantages
Physical system	1. Membrane separation filter	CO_2 flue gas passes through the membrane filter	1. Increase mass transfer	1. Membrane blocking 2. Costly process
	2. Oceanic injection	Injection of CO_2 into the bottom of sea	1. Maximum CO_2-holding capacity	1. Energy incompetent 2. Dangerous for sea organisms 3. Required high-cost injection equipment 4. Gas leakage problem
	3. Geological injection	Injection of CO_2 in geologic reservoirs such as depleted well and coal seams	1. Reserve CO_2 2. Possible recovery of oil	1. Require geological injection area 2. Gas leakage problem 3. High cost
	4. Adsorption	Using molecular sieve or other adsorption materials	1. Low waste generation 2. Adaptable to other CO_2 mitigation process	1. Co-adsorption of additional toxic compounds 2. Energy incompetent
Chemical system	1. Chemical absorption	Bind CO_2 and neutralize into carbonic acid and then convert into less toxic carbonate ions.	1. Harmless and permanent techniques of sequestration 2. Supply of rich base ions such as K^+ and Na^+	1. High energy required 2. High cost
	2. Minerals' carbonation	Reacting heavy metals or metal oxides to CO_2 and converting into stable carbonate and bicarbonate ions	1. Harmless techniques 2. Abundantly available metal oxides (MgO, CaO)	1. Large equipment required for processing 2. Require large amount of minerals and reagent in processing, which increases the cost

(*Continued*)

TABLE 11.1 (*Continued*)
Advantages and Disadvantages of Several CO_2 Mitigation Techniques (Zhou et al. 2017; White et al. 2003; Lal 2008; Beedlow et al. 2004; Olajire 2013; Kita and Ohsumi 2004; De Silva, Ranjith, and Perera 2015; Tang, Yang, and Bian 2014; Salek et al. 2013; Zhao and Su 2014; Lackner 2003)

Class	Techniques	Mechanisms	Advantages	Disadvantages
Biological system	1. Forestation or plantation	Utilizing atmospheric CO_2 and converting into OC molecule (photosynthesis)	1. Natural CO_2 sequestration 2. Chemical free 3. Provide organic compounds and oxygen	1. Large area required 2. Limited CO_2 sequestration 3. High cost
	2. Oceanic organism Fertilization	Triggered growth of photosynthetic microorganisms such as macroalgae and cyanobacteria by providing nutrients	1. Increasing rate of CO_2 mitigation 2. Required less amount of nutrients for their growth 3. Utilization of carbonate ions	1. Effect on oceanic phytoplankton and animals 2. Increase the production of methane
	3. Microalgae-based mitigation	Utilization of CO_2 with the help of sunlight by the photosynthetic pathway Utilization of CO_2 with the help of sunlight by the photosynthetic pathway	1. High photosynthetic activity than tree 2. Faster mitigation than other method 3. Co-production of the value-added product	1. Dependent on light source 2. Sensitive to other toxic compounds

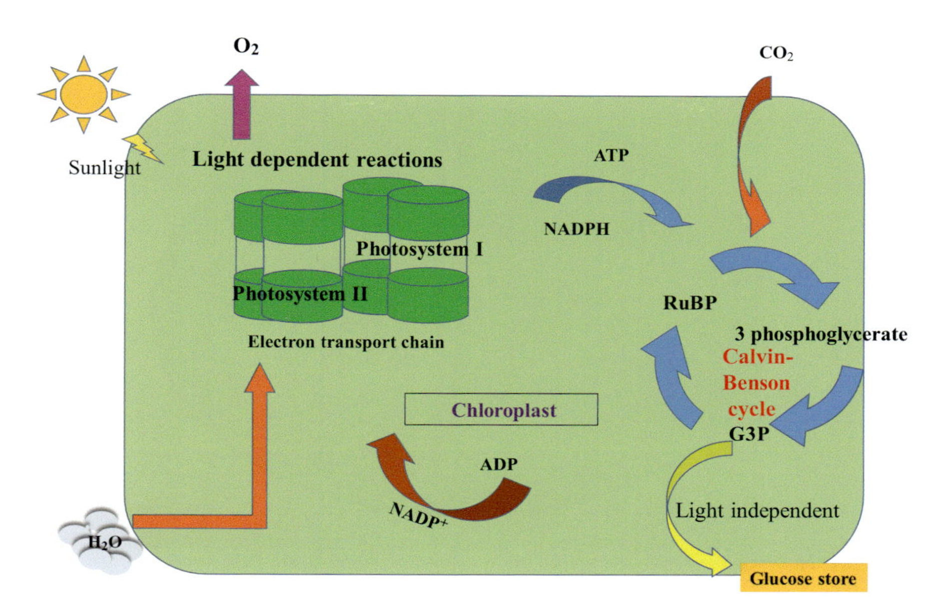

FIGURE 11.2 Schematic representation of photo-biochemical principles of microalgal-based CO_2 fixation (Zhao and Su 2014).

organic compounds, especially glucose or carbohydrates. The microalgal photosynthesis follows both mechanisms: light dependent and light independent (Figure 11.2). In the light-dependent mechanism, microalgal cells capture the sunlight as an energy source for the generation of ATP and NADPH (energy carrier) from ADP to $NADP^+$ through the electron transport chain. In the light-independent mechanism, the cell fixes CO_2 and produces the intermediate molecule of carbohydrates using the earlier-generated ATP and NADPH (Zhao and Su 2014).

However, microalgal cells may grow under different modes of cultivation based on the availability of space, environmental conditions, and types of species as discussed in the subsections.

11.3.1 Photoautotrophic

The main benefit of the photoautotrophic cultivation is that it requires only sunlight and CO_2 for their growth (Ummalyma and Sukumaran 2014). It utilizes atmospheric CO_2 through Calvin cycle, which involves different stages such as (i) carboxylation, (ii) reduction, and (iii) regeneration, as shown in Figure 11.3 (Zhao and Su 2014).

11.3.2 Heterotrophic

Heterotrophic growth occurs in the absence of light, in which microalgal cells consume organic carbon (OC) from media to obtain energy and carbons. This OC should be tiny molecules, so it passes through the cell wall, which is metabolized in the form of sugars and fats followed by phosphogluconate pathway (PP) (Zhou et al. 2017).

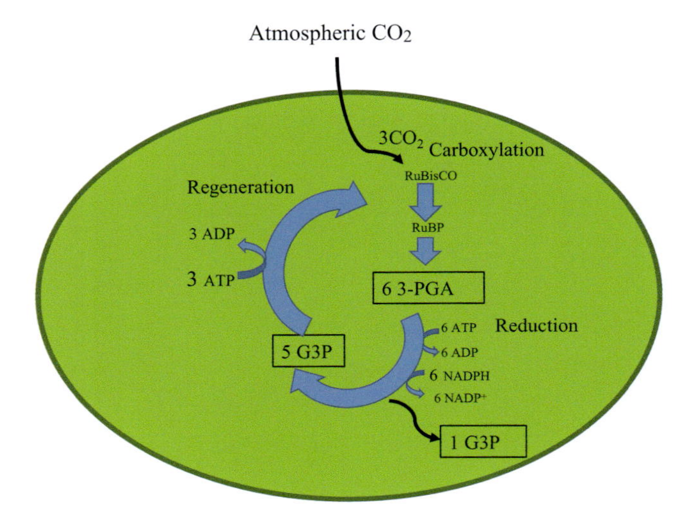

FIGURE 11.3 Systemic reaction of Calvin cycle for atmospheric CO_2 fixation.

Thraustochytrids, marine microalgae (obligate heterotrophs), are mainly grown in the absence of light and consume OC to produce lipids. Various *Chlorella* sp., such as *C. minutissima* (Bhatnagar et al. 2010), *C. protothecoides* (Shi et al. 1999), *C. zofingiensis* (Ip and Chen 2005), and *C. vulgaris* (Liang, Sarkany, and Cui 2009), can grow heterotrophically as well as photo-autotrophically. Therefore, the heterotrophic growth of microalgae offers several benefits; it can be cultivated in the presence of OC (glucose, acetic acid) without light (Chen et al. 2015; Rahaman et al. 2011). This cultivation has high cell density, which allows for simplified biomass recovery (Morales-Sánchez et al. 2015). Moreover, in this cultivation mode, microalgae can be cultivated in traditional bioreactors at the industrial scale. In bioreactors, several operational factors (carbon and oxygen levels, pH, sterile conditions, and temperature) can be regulated (Perez-Garcia et al. 2011). The optimized operational can help to reduces volume-to-surface (V/S) ratio of the bioreactor, which helps in the easier fabrication of bioreactor (Zhan, Rong, and Wang 2017). Therefore, the heterotrophic mode of cultivation is more economical than the photoautotrophic mode of cultivation due to higher biomass productivity. However, the main disadvantage of the heterotrophic cultivation is that atmospheric CO_2 cannot be fixed during growth, so CO_2 mitigation does not occur. The economic analysis of heterotrophic cultivation suggests that the major cost belongs to the equipment and its installation, which reduces the feasibility of the heterotrophic cultivation at the commercial scale (Lowrey, Brooks, and McGinn 2015).

11.3.3 Mixotrophic

In the mixotrophic cultivation, the microalgal cell utilizes OC and inorganic carbon (IOC) for their growth in the presence or absence of light during their growth period (Zhan, Rong, and Wang 2017). In this cultivation, the microalgal cell requires sunlight and CO_2 for their growth by energy driven by ATP and NADPH, which are formed in the photosynthetic process. However, in the dark conditions, it consumes OC present in the media. Pulz and Scheibenbogen (2007) found that in the mixotrophic cultivation, biomass productivity of *C. vulgaris* is double that of the autotrophic and heterotrophic growth. It is a two-stage cultivation method: in the first stage, microalgae grow in heterotrophic mode and consume the OC, whereas in the second stage, phototrophic cultivation starts after the depletion of OC (Zhan, Rong, and Wang 2017). The main advantages of the mixotrophic cultivation are increased photosynthetic rate and the higher concentration of pigments as compared to the autotrophic and heterotrophic cultivations (Zhan, Rong, and Wang 2017). In the mixotrophic cultivation, there is an increase in the cell density and the biomass concentration; therefore, it reduces the cultivation cost. Heredia-Arroyo et al. (2011) found that the mixotrophic cultivation of *C. vulgaris* significantly improved biomass production. The key advantages and disadvantages of each cultivation mode of microalgae are discussed in Table 11.2.

11.4 INFLUENCING FACTORS FOR MICROALGAL GROWTH

Microalgal growth can be influenced by different physiochemical factors such as nutrient source, light (colour and strength), and temperature. To attain the higher growth rate and CO_2-capture efficiency, these factors need to be optimized, which may vary for each microalgal species. However, each factor is discussed in detail in the subsections.

11.4.1 CO_2 Concentration

Literature showed that CO_2 concentration and its dissolution rate in the medium greatly influence the growth rate of microalgae. However, carbon source may be a limiting factor in microalgal growth: if CO_2 concentration is low, the microalgal growth may become slow. However, a higher concentration of CO_2 (5% or above) may inhibit the microalgal growth. Moreover, ideal CO_2 concentration for cell evolution and its tolerance limit can be different for each microalgal strain. In some microalgal strains, higher than 1% v/v CO_2 concentration shows a negative effect on their growth and biomass production (Zhan, Rong, and Wang 2017).

11.4.2 Toxic Gases

Microalgae can uptake toxic gas when it is cultivated using flue gas from the industrial exhausts, which contains several toxic gases (So_x, NO_x) and heavy metals (Hg^o). Moreover, if the concentration of SO_2 is more than 100 ppm in the flue gas mixture, it causes growth inhibition; e.g., *Chlorella* sp. KR1, a high-performance microalgal species, does not survive during cultivation with CO_2 (15%) in the presence of SO_2 (150 ppm). Some microalgae can survive in these concentrations, but their lag phase becomes longer. However, a higher concentration of SO_2 flue gas can affect the carbon fixation efficiency and microalgal growth rate, which can reduce the biomass production (Kumar et al. 2014). Although SO_2 directly not

TABLE 11.2
The Key Advantages and Disadvantages of Different Modes of Cultivation (Zhan, Rong, and Wang 2017; Heredia-Arroyo et al. 2011)

Mode of cultivation	Carbon Source	Energy Source	Advantages	Disadvantages
Autotrophic	CO_2 from the environment	Light/sunlight	Low cost Cultivation without using external sources Highest production rate of phytopigments Biomitigation of CO_2 from the environment	Low growth rate or biomass Optimization problem Dependent on weather conditions
Heterotrophic	OC source	OC source	Highest biomass production as compared to autotrophs Highest lipid accumulation Easy-to-design bioreactor No requirement of external light source	Cannot fix the atmospheric CO_2 No role in biomitigation of CO_2 Need for sterile condition or media Higher cost than the other cultivation methods Dark condition may demise the phytochemical production
Mixotrophic	OC and utilize CO_2 present in the atmosphere	OC and light	Higher growth rate and biomass production compared to other cultivation method. Sustaining phytochemical production or lipid accumulation Growth and increase in biomass in both dark and light conditions	Less CO_2 production Role in biomitigation of CO_2 Reduce energy conversion efficiency Maintenance of sterile condition Not all microalgae possess this type of cultivation

affects the microalgal growth, it decreases the pH of culture media (Matsumoto et al. 1997). However, buffer or salt is externally added into the medium in order to maintain its pH. The NO_x is present in the flue gas in a different form, and it varies from area to area. Both NO (90%–95%) and NO_2 (10%–15%) do not influence the pH of the medium, but the presence of NO in growth media can hamper the microalgal growth. Microalgae can consume an optimum/lower amount of NO from the growth medium for their cell growth (Kumar et al. 2010). However, the maximum concentration (300 ppm) of NO can decrease the growth of microalgae (Lee and Lee 2002). Mercury, a heavy metal, is released during the combustion in coal-fired power plants, and it may be absorbed by fabric filter and electrostatic precipitators. Moreover, Hg^o can be transformed into Hg^{2+} in the presence of oxygen, and it may be separated by washing. Many researchers suggest that the extremely trace amount of mercury can inhibit the growth of microalgae due to the disruption of chlorophyll (Li et al. 2012).

11.4.3 Light

Light is necessary for the microalgal cell development and fixation of CO_2, which can be provided by natural sunlight to both open pond and closed cultivation systems. There is a correlation between microalgal growth and light intensity. If the intensity of light increases, then there is an improvement in the photosynthetic efficiency. However, if the intensity of light reaches the saturation point, then there is inhibition or a reduction in the photosynthetic efficiency, which is known as the photoinhibition (Sung et al. 1999). Similarly, when the light intensity is low, the activity of Rubisco may not be accelerated and the photosynthetic efficiency is reduced (Xu et al. 2001). However, the optimum light intensity can vary from 100 to 200 $\mu E\ m^{-2} s^{-1}$ for microalgal growth (Sung et al. 1999). Besides the light intensity, the day–night cycles also influence the growth of microalgae. Generally, continuous illumination is provided to microalgae for better growth (24 h). Zhao and Su (2014) suggested that some microalgal species cannot grow in continuous illumination, so it requires different day–night cycles for their growth, such as 12:12 h and 16:8 h.

11.4.4 Temperature

Temperature is necessary for the growth of microalgal cell for biomass production and CO_2 fixation. Generally, the optimum temperature for the growth of microalgae is 15°C–30°C. Microalgal growth can be inhibited at a lower temperature as it does not act in favour of RuBisCO activity, and this prevents the acceleration of photosynthesis. High temperature can have an effect on the rate of respiration and metabolic process of microalgae (Xu et al. 2001), as it can inhibit and increase the lag phase of microalgae (Sung et al. 1999). Few microalgal species can survive above the optimum temperature; e.g., the optimum temperature for the growth of *Chlorella* sp. T-1 is 35°C (Maeda et al. 1995), whereas that for *Chlorella* ZY-1 (Yue and Chen 2005) and *Chlorella* KR-1 (Sung et al. 1998) is 40°C.

11.4.5 Nutrients

Phosphorus (P) and nitrogen (N) are the important elements for the synthesis of nucleic acids and proteins, which are associated with the metabolic activity of microalgal cells (Kumar et al. 2010). In microalgal cultivation, growth medium usually contains N in the form of nitrate (NO_3^-), nitrite (NO_2), and ammonium (NH_4^+), and P in the form of hydrogen phosphate (HPO_4^{2-}) and dihydric phosphate ($H_2PO_4^{2-}$). When N and P are present in appropriate concentration, there is an increase in the growth of microalgae; however, when their concentrations are high in the medium, there will be a toxic effect on the growth of microalgae, and when their concentrations are extremely low, there is an inhibition in the growth of microalgae (Zhao and Su 2014).

11.4.6 pH

The optimum pH may vary for each microalgal species, such as marine microalgae (pH 7.9–8.3) and freshwater microalgae (pH 6.0–8.0) (de Morais and Costa 2007). Besides, pH changes occur in the medium due to CO_2 concentration and water-soluble pollutants. Atmospheric CO_2 does not have a significant role in pH changes due to the lower CO_2 concentration in air. CO_2 uptake by microalgal cells increases the pH (9.5–10), whereas the existence of SO_2 in flue gas (100–250 ppm) decreases the pH (3.5–2.5) (Zhao and Su 2014). Most microalgae have different optimum pH range for their growth: pH 6.8 is favourable for the growth of *Synechococcus* PCC7942 (Kajiwara et al. 1997), whereas pH 4.0 is suitable for the growth of *Chlorella* sp.KR-1 (Sung et al. 1998).

11.5 RELATIONSHIP BETWEEN THE CARBON CONCENTRATION AND MICROALGAL GROWTH

The carbon concentration and microalgal growth are directly related to each other, but the biomass of microalgae directly depends on the carbon concentration.

11.5.1 Role of CO_2 in Microalgal Growth

During the microalgal cultivation in photo-bioreactor using pure carbon source through the medium, CO_2 can be directly utilized by microalgae for biomass production. But when microalgae are cultivated using atmospheric CO_2 in natural conditions, there is a reduction in the growth rate of microalgae, thus mitigating bad odours from the atmosphere (Razzak et al. 2017). However, the toxic compounds, high aeration ratio, and changing weather conditions may inhibit the CO_2 mitigation efficiency (Hanagata et al. 1992; Kodama 1993; Mandalam and Palsson 1998). Therefore, to overcome these factors, pre-adaption of microalgal species is required. Some microalgal strains can resist and survive at higher CO_2 concentration up to 70% (Zhao and Su 2014) or even 100% (Maeda et al. 1995). However, 1% (v/v) or above concentration of CO_2 shows a negative effect on their growth and biomass production (Zhan, Rong, and Wang 2017). Therefore, the optimum concentration of CO_2 is needed for each microalgal species to fix CO_2. However, acclimatization with a higher concentration of CO_2 improves the higher CO_2 fixation rate (Borowitzka 1994).

11.5.2 CO_2 Fixation from Atmosphere by Microalgae and Their Efficacy

Biomitigation of CO_2 from the environment by the microalgal system is a good approach. Microalgae can mitigate CO_2 from post-combustion of the flue gases. Microalgae can entrap CO_2 from the polluted environment and remove trace contaminants from wastewater. Microalgae can utilize nutrients (C, N, and P) from wastewater and produce a significant amount of oxygen, which can be utilized by bacteria during the degradation of organic materials from wastewater. Kodama (1993) reported that *Chlorococcum littorale* (marine green alga) can tolerate higher CO_2 concentration (40%). The mechanism of photosynthesis can be described by the following empirical equation (Zhao and Su 2014):

$$6CO_2 + 6H_2O \xrightarrow{\text{Light energy } (hv)} C_6H_{12}O_6 + 6O_2 \tag{11.1}$$

The microalgae can follow CCM (CO_2-concentrating mechanism) pathway, like other photosynthetic organisms, and CCM pathway directly enhances the rate of photosynthesis and reduces the rate of photorespiration (Aizawa and Miyachi 1986; Badger and Price 1994; Zhao and Su 2014). Generally, IOC dissolves in water in the form of CO_3^{2-}, CO_2, HCO_3^-, and H_2CO_3 when the dynamic ionization equilibrium is reached, but only CO_2 and HCO_3^- are the main IOC forms, which can be utilized by microalgal cells in different ways. But it prefers CO_2 and HCO_3^- (Moroney, Bartlett, and Samuelsson 2001; Colman et al. 2002). HCO_3^- has been used by direct way, e.g. cation exchange and active transport (Miller and Colman 1980; Amoroso et al. 1998), as well as via an indirect way, which catalyses HCO_3^- as CO_2 and OH^- by periplasmic carbonic anhydrase (PCA) (Zhao and Su 2014). In this process, the carbonic anhydrase (CA) and pyrenoid play a vital role in (Dissolved inorganic carbons) DIC conversion in microalgal CCM. CA includes PCA, cytosolic carbonic anhydrase (CCA), and chloroplast carbonic anhydrase (chCA). The main function of PCA regulates the balance between HCO_3^- and CO_2. With a continuous source of CO_2 to the cells of microalgae, chCA may enhance the transport of CO_2 and HCO_3^- from the plasma membrane to the chloroplasts (Zhao and Su 2014). chCA is an important CA in the CCM process, which is shown in Figure 11.4.

11.5.3 Measurement of CO_2 Fixation

Microalgal-based CO_2 fixation mostly depends on CO_2 consumption during the microalgal growth, which can be observed by determining the CO_2 concentration at the inlet or outlet in photo-bioreactor. Moreover, various microalgal strains have different capacity of CO_2 fixation as reported in Table 11.3. Besides, carbon content gives a more correct value of the total CO_2 utilized during the microalgal cell growth

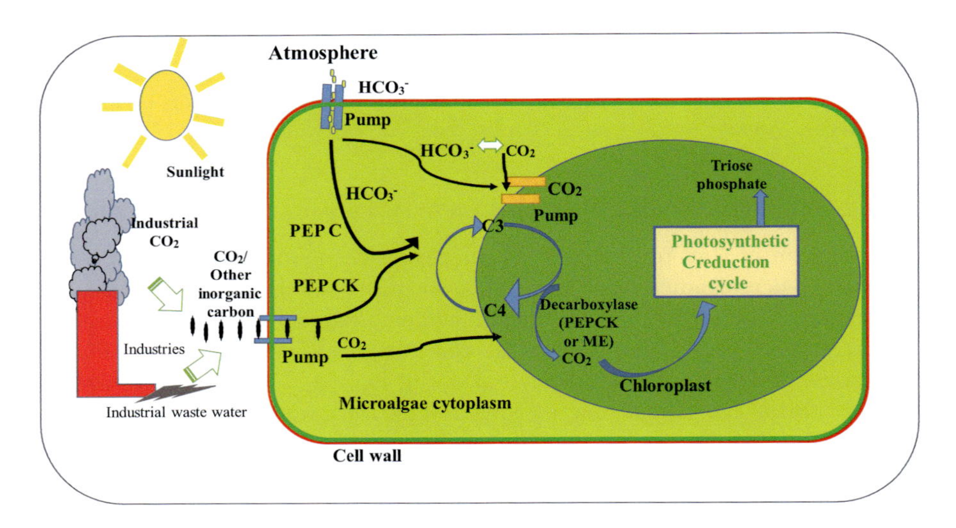

FIGURE 11.4 Systemic representation of photochemical process of CO_2 fixation from the atmosphere by CCM process (Zhao and Su 2014).

TABLE 11.3
CO_2 Fixation (%) Efficacy of Various Microalgal Species (Salih 2011; Judd et al. 2015)

Microalgal Species	CO_2 Fixation (%)	References
Anabaena sp.	90	Judd et al. (2015)
Botryococcus braunii	88	Judd et al. (2015)
Chlamydomonas sp.	15	Hirata et al. (1996)
Chlorella sp.	40	Nagase et al. (1998)
Chlorella vulgaris	Up to 95	Li et al. (2013)
Chlorococcum littorale	60	Kodama (1993)
Cyanidium caldarium	100	Seckbach, Gross, and Nathan (1971)
Dunaliella tertiolecta	15	Miura et al. (1993)
Eudorina sp.	20	Yoshihara et al. (1996)
Euglena gracilis	45	Nakano et al. (1996)
Nannochloris sp.	15	Matsumoto et al. (1995)
Scenedesmus sp.	80	Hanagata et al. (1992)
Synechococcus elongatus	60	Miyairi (1995)
Tetraselmis sp.	15	Hirata et al. (1996)

if growth medium does not contain any other OC source (Tang et al. 2011). On this basis, CO_2-capture efficiency can be calculated using the following equation:

$$RCO_2 = C_C P\left(\frac{MCO_2}{M_C}\right) \tag{11.2}$$

where RCO_2 represents the CO_2 fixation rate (g $L^{-1}day^{-1}$), C_C carbon content of the microalgal cell % (w/w), *P* biomass productivity (g $L^{-1}day^{-1}$), MCO_2 molecular weight of carbon dioxide, and M_C molecular weight of carbon.

11.6 UTILIZATION OF MICROALGAL BIOMASS FOR DIFFERENT APPLICATIONS

Microalgae capture sunlight and convert it into chemical energy through photosynthesis; therefore, CO_2 fixation ability is ten times greater as compared to terrestrial plants (Raja et al. 2008). Many different microalgal species are present, but few species of microalgae are commercially available and its biomass can be utilized for bioactive compounds, feed, and biofuel production. As shown in Figure 11.5, microalgae can consume CO_2 as C source for their growth and different types of bioactive compounds are produced using their biomass (Sathasivam et al. 2019). Some microalgae are commercially used for the production of bioactive compounds: *Dunaliella*, *Spirulina*, *Chlorella*, *Haematococcus*, *Botryococcus*, *Phaeodactylum*, *Nannochloris*, *Tetraselmis*, *Arthrospira*, etc (Sathasivam et al. 2019; Molino et al. 2020). Microalgal products contain 40% lipids, 50% proteins, and 10% carbohydrates. After the isolation of the value-added proteins, carbohydrates, pigments, and the remaining lipids are used for the production of biofuels.

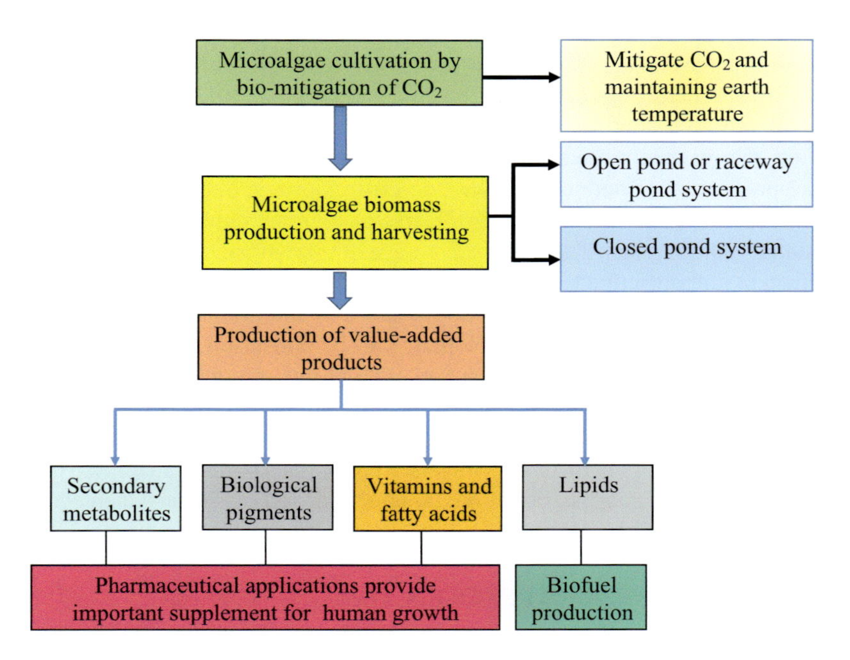

FIGURE 11.5 Systemic representation of overall process of microalgal biomitigation, microalgal harvesting, and production value (Sathasivam et al. 2019).

11.6.1 Production of Bioactive Compounds

Microalgal biomass contains different high-value-added compounds such as PUFAs (poly-unsaturated fatty acids), phycobiliproteins, carotenoids, polysaccharides, various types of important pigments, phycotoxins, and nutraceutical compounds, which are used as protein supplements and in aquaculture and human nutrition (Del Campo, García-González, and Guerrero 2007). Becker (2004) reported that microalgal-derived biomass contains several minerals and vitamins such as astaxanthin, biotin, iodine, iron, lutein, niacin, nicotinate, pantothenic acid, potassium, vitamin (A, B1, B2, C, and E), and anthocyanins, which are summarized in Table 11.4.

11.6.2 Production of Biofuels

Many countries have made efforts to decrease the utilization of fossil fuels to meet energy demands and improve energy conversion efficiency. Biofuels are an alternative source to fulfil the energy requirements. There is an extensive research on the field of biofuel generations. The first-generation biofuels are synthesized from food, cereals, or food products and are already commercialized in developed countries such as Brazil, Europe, and the United States (Milano et al. 2016). The drawbacks of the first-generation biofuels are that they utilize agricultural food materials, crops, and foods and they directly or indirectly compete with food and agricultural lands (Adenle, Haslam, and Lee 2013; Milano et al. 2016). However, the second-generation biofuels are produced from agricultural residual biomass like wood processing wastes (lignocellulose) and non-edible crops such as *Jatropha curcas*, *Madhuca indica* seeds, and tobacco seed oil (TSO). Therefore, it does require food crops or farmland; however, their conversion process is too high and their conversion rate is very slow (Adenle, Haslam, and Lee 2013; Alam et al. 2012; Milano et al. 2016). Consequently, researchers are focused towards an alternative possibility to fulfil growing energy requirement; therefore, microalgae can be an alternative source for biofuel production (Brennan and Owende 2010; Ndimba et al. 2013; Milano et al. 2016). Generally, they are present in diverse or extreme environments with simple nutritional requirements. The major advantage of microalgal cultivation is that it can be grown in non-agricultural land (Milano et al. 2016; Ndimba et al. 2013; Slade and Bauen 2013). Microalgae convert sunlight and CO_2 into biomass in the presence of water and nutrients. Furthermore, biomass can be converted into biofuel, which is considered as the third-generation biofuel. The biofuel production from microalgal biomass mainly depends on microalgal species, and their carbon and lipid contents. Many different biochemical and thermochemical processes can convert microalgal biomass into biofuels via different conversion processes such as anaerobic digestion (AD), fermentation, transesterification, liquefaction, and pyrolysis (Chen et al. 2015; Mehariya et al., 2019b).

11.7 FUTURE RESEARCH PROSPECTIVE

Biomitigation of CO_2 by microalgae is a virtuous approach compared to other conventional methods, because it can mitigate CO_2 and other pollutants from the environment. The mixotrophic cultivation is more beneficial than any other

TABLE 11.4
Microalgal-Derived Different Value-Added Compounds and Their Commercial Application

Species Name	Metabolite Produced	Commercial Application	References
Chlorella sorokiniana	α-Carotene	Lower risk of premature death, food colourant, antioxidant	Matsukawa et al. (2000)
Dunaliella salina	β-Carotene	Anti-cancerous property, prevent night blindness and prevent liver fibrosis	Raja, Hemaiswarya, and Rengasamy (2007); Chu (2012)
Chlorella zofingiensis *Haematococcus pluvialis*	Astaxanthin	Anti-inflammatory, antioxidant, and anti-cancerous activities, cardiovascular health	Ip and Chen (2005); Molino, Rimauro, et al. (2018); Sanzo et al. (2018); Molino, Mehariya, et al. (2018); Molino et al. (2020)
Muriellopsis sp. *Chlorella protothecoides* *Scenedesmus almeriensis* *Chlorococcum citriforme* *Neospongiococcus gelatinosum*	Lutein	Prevent macular degeneration	Del Campo et al. (2000, 2001); Blanco et al. (2007); Sánchez et al. (2008); Fernández-Sevilla, Acién Fernández, and Molina Grima (2010); Mehariya et al. (2019a); Molino, Mehariya, et al. (2019)
Ankistrodesmus spiralis *Aphanizomenon flosaquae* *Chloromonas nivalis* *Chlorella luteoviridis* *Chlorella minutissima* *Chlorella sorokiniana* *Chlorella sphaerica* *Scenedesmus* sp. *Stichococcus* sp.	Mycosporine-like amino acid	Sun cream, UV protectant properties	Xiong, Kopecky, and Nedbal (1999); Chu (2012); Duval, Shetty, and Thomas (1999); Karsten, Lembcke, and Schumann (2007)

(Continued)

TABLE 11.4
Microalgal-Derived Different Value-Added Compounds and Their Commercial Application

Species Name	Metabolite Produced	Commercial Application	References
Dunaliella viridis *Dunaliella salina* *Chlamydomonas* sp. *Chlamydomonas pulsatilla* *Chlamydomonas reinhardtii* *Chlamydomonas submarinum*	Glycerol	Moisture to the skin and preservation of microbes	Hadi, Shariati, and Afsharzadeh (2008); Ahmad and Hellebust (1986); Kaçka and Dönmez (2008); Miyasaka et al. (1998); León and Galván (1999); Blackwell and Gilmour (1991)
Nannochloropsis gaditana *Nannochloropsis oceanica* *Nannochloropsis* sp. *Nannochloropsis salina* *Pinguiococcus pyrenoidosus* *Thraustochytrium* sp. *Chlorella minutissima* *Dunaliella salina* *Pavlova* sp.	Eicosapentaenoic acid	Reducing cardiovascular diseases, high blood pressure, beneficial effects on depression, rheumatoid arthritis, and asthma	Adarme-Vega et al. (2012); Patil et al. (2007); Yongmanitchai and Ward (1991); Bhosale, Rajabhoj, and Chaugule (2010); Hu et al. (2008); Carvalho and Malcata (2005); Molino, Martino, et al. (2019); Leone et al. (2019)
Pinguiococcus pyrenoidosus *Pavlova lutheri* *Nannochloropsis* sp.	Docosahexaenoic acid (DHA)	In correct eye and brain development in infants and maintaining cardiovascular health in adults	Hu et al. (2008); Carvalho and Malcata (2005); Guihéneuf et al. (2009); Sathasivam et al. (2019)

cultivation methods due to efficient CO_2 mitigation as well as the removal of nutrients from wastewater. Moreover, in the near future, microalgal-based CO_2 sequestration unit needs to be installed for direct exploitation of flue gas from industries to the cultivation of microalgae. This approach can help in the reduction of GHG emissions, and the produced microalgal biomass may be used for the recovery of bioactive compounds. Moreover, after the extraction of biomolecules, the residual biomass may be exploited for biofuel production via biochemical or thermochemical process. Microalgal cultivation system using wastewater can also be utilized for direct liquefaction without biomass harvesting. Therefore, microalgal-based biorefinery needs to be developed in the near future for the reduction of GHG and the production of biofuels.

REFERENCES

Adarme-Vega, T Catalina, David K Y Lim, Matthew Timmins, Felicitas Vernen, Yan Li, and Peer M Schenk. 2012. "Microalgal Biofactories: A Promising Approach towards Sustainable Omega-3 Fatty Acid Production." *Microbial Cell Factories* 11 (1): 96. https://doi.org/10.1186/1475–2859–11–96.

Adenle, Ademola A, Gareth E Haslam, and Lisa Lee. 2013. "Global Assessment of Research and Development for Algae Biofuel Production and Its Potential Role for Sustainable Development in Developing Countries." *Energy Policy* 61: 182–95. https://doi.org/10.1016/j.enpol.2013.05.088.

Ahmad, Iftikhar, and Johan A Hellebust. 1986. "The Role of Glycerol and Inorganic Ions in Osmoregulatory Responses of the Euryhaline Flagellate *Chlamydomonas pulsatilla* Wollenweber." *Plant Physiology* 82 (2): 406–10. https://doi.org/10.1104/pp.82.2.406.

Aizawa, Katsunori, and Shigetoh Miyachi. 1986. "Carbonic Anhydrase and CO_2 Concentrating Mechanisms in Microalgae and Cyanobacteria." *FEMS Microbiology Reviews* 2 (3): 215–33. https://doi.org/10.1111/j.1574-6968.1986.tb01860.x.

Alam, Firoz, Abhijit Date, Roesfiansjah Rasjidin, Saleh Mobin, Hazim Moria, and Abdul Baqui. 2012. "Biofuel from Algae- Is It a Viable Alternative?" *Procedia Engineering* 49: 221–27. https://doi.org/10.1016/j.proeng.2012.10.131.

Amoroso, Gabi, Dieter Sültemeyer, Christoph Thyssen, and Heinrich P Fock. 1998. "Uptake of HCO_3^- and CO_2 in Cells and Chloroplasts from the Microalgae *Chlamydomonas Reinhardtii* and *Dunaliella Tertiolecta*." *Plant Physiology* 116 (1): 193–201.

Badger, Murray R, and G Dean Price. 1994. "The Role of Carbonic Anhydrase in Photosynthesis." *Annual Review of Plant Biology* 45 (1): 369–92.

Becker, Wolfgang. 2004. "18 Microalgae in Human and Animal Nutrition." *Handbook of Microalgal Culture: Biotechnology and Applied Phycology*, edited by A. Richmond, 312–351. Oxford, UK: Wiley-Blackwell. https://doi.org/10.1002/9780470995280.ch18

Beedlow, Peter A., David T. Tingey, Donald L. Phillips, William E. Hogsett, and David M. Olszyk. 2004. "Rising Atmospheric CO_2 and Carbon Sequestration in Forests." *Frontiers in Ecology and the Environment* 2 (6): 315. https://doi.org/10.2307/3868407.

Belay, Amha, Yoshimichi Ota, Kazuyuki Miyakawa, and Hidenori Shimamatsu. 1993. "Current Knowledge on Potential Health Benefits of *Spirulina*." *Journal of Applied Phycology* 5 (2): 235–41. https://doi.org/10.1007/BF00004024.

Bhatnagar, Ashish, Monica Bhatnagar, Senthil Chinnasamy, and K C Das. 2010. "*Chlorella minutissima* - A Promising Fuel Alga for Cultivation in Municipal Wastewaters." *Applied Biochemistry and Biotechnology* 161 (1): 523–36. https://doi.org/10.1007/s12010-009-8771–0.

Bhosale, Rahul A, M P Rajabhoj, and B B Chaugule. 2010. "*Dunaliella salina* Teod. as a Prominent Source of Eicosapentaenoic Acid." *International Journal on Algae* 12 (2): 185–189.

Blackwell, John R, and D James Gilmour. 1991. "Determination of Intracellular Volume and Internal Solute Concentrations of the Green Alga *Chlorococcum submarinum*." *Archives of Microbiology* 157 (1): 80–85. https://doi.org/10.1007/BF00245340.

Blanco, Antonio M, José Moreno, José A Del Campo, Joaquín Rivas, and Miguel G Guerrero. 2007. "Outdoor Cultivation of Lutein-Rich Cells of *Muriellopsis* sp. in Open Ponds." *Applied Microbiology and Biotechnology* 73 (6): 1259–66.

Borowitzka, M A. 1994. "Large-Scale Algal Culture Systems: The next Generation." *Australasian Biotechnology* 4 (4): 212–15.

Brennan, Liam, and Philip Owende. 2010. "Biofuels from Microalgae—A Review of Technologies for Production, Processing, and Extractions of Biofuels and Co-Products." *Renewable and Sustainable Energy Reviews* 14 (2): 557–77. https://doi.org/10.1016/j.rser.2009.10.009.

Carvalho, Ana P, and F Xavier Malcata. 2005. "Optimization of $ω$-3 Fatty Acid Production by Microalgae: Crossover Effects of CO_2 and Light Intensity under Batch and Continuous Cultivation Modes." *Marine Biotechnology* 7 (4): 381–88. https://doi.org/10.1007/s10126-004-4047–4.

Cheah, Wai Yan, Pau Loke Show, Jo Shu Chang, Tau Chuan Ling, and Joon Ching Juan. 2015. "Biosequestration of Atmospheric CO_2 and Flue Gas-Containing CO_2 by Microalgae." *Bioresource Technology* 184: 190–201. https://doi.org/10.1016/j.biortech.2014.11.026.

Chen, Huihui, Dong Zhou, Gang Luo, Shicheng Zhang, and Jianmin Chen. 2015. "Macroalgae for Biofuels Production: Progress and Perspectives." *Renewable and Sustainable Energy Reviews* 47: 427–37. https://doi.org/10.1016/j.rser.2015.03.086.

Chu, Wan-Loy. 2012. "Biotechnological Applications of Microalgae." *IeJSME* 6 (1): S24–37.

Colman, Brian, I Emma Huertas, Shabana Bhatti, and Jeffrey S Dason. 2002. "The Diversity of Inorganic Carbon Acquisition Mechanisms in Eukaryotic Microalgae." *Functional Plant Biology* 29 (3): 261–70.

Del Campo, José A, Mercedes García-González, and Miguel G Guerrero. 2007. "Outdoor Cultivation of Microalgae for Carotenoid Production: Current State and Perspectives." *Applied Microbiology and Biotechnology* 74 (6): 1163–74.

Del Campo, José A, José Moreno, Herminia Rodríguez, M Angeles Vargas, Joaquín Rivas, and Miguel G Guerrero. 2000. "Carotenoid Content of Chlorophycean Microalgae: Factors Determining Lutein Accumulation in *Muriellopsis* sp.(Chlorophyta)." *Journal of Biotechnology* 76(1): 51–59.

Del Campo, José A, Herminia Rodrıguez, José Moreno, M Angeles Vargas, Joaquın Rivas, and Miguel G Guerrero. 2001. "Lutein Production by *Muriellopsis* sp. in an Outdoor Tubular Photobioreactor." *Journal of Biotechnology* 85(3): 289–95.

Duval, Brian, Kalidas Shetty, and William H Thomas. 1999. "Phenolic Compounds and Antioxidant Properties in the Snow Alga *Chlamydomonas nivalis* after Exposure to UV Light." *Journal of Applied Phycology* 11 (6): 559. https://doi.org/10.1023/A:1008178208949.

Fernández-Sevilla, José M, F G Acién Fernández, and E Molina Grima. 2010. "Biotechnological Production of Lutein and Its Applications." *Applied Microbiology and Biotechnology* 86 (1): 27–40. https://doi.org/10.1007/s00253-009-2420–y.

Guihéneuf, Freddy, Virginie Mimouni, Lionel Ulmann, and Gérard Tremblin. 2009. "Combined Effects of Irradiance Level and Carbon Source on Fatty Acid and Lipid Class Composition in the Microalga Pavlova Lutheri Commonly Used in Mariculture." *Journal of Experimental Marine Biology and Ecology* 369 (2): 136–43. https://doi.org/10.1016/j.jembe.2008.11.009.

Hadi, M R, M Shariati, and S Afsharzadeh. 2008. "Microalgal Biotechnology: Carotenoid and Glycerol Production by the Green Algae *Dunaliella* Isolated from the Gave-Khooni Salt Marsh, Iran." *Biotechnology and Bioprocess Engineering* 13 (5): 540. https://doi.org/10.1007/s12257-007-0185–7.

Hanagata, Nobutaka, Toshifumi Takeuchi, Yoshiharu Fukuju, David J Barnes, and Isao Karube. 1992. "Tolerance of Microalgae to High CO_2 and High Temperature." *Phytochemistry* 31 (10): 3345–48. https://doi.org/https://doi.org/10.1016/0031-9422(92)83682–O.

Heredia-Arroyo, Tamarys, Wei Wei, Roger Ruan, and Bo Hu. 2011. "Mixotrophic Cultivation of *Chlorella vulgaris* and Its Potential Application for the Oil Accumulation from Non-Sugar Materials." *Biomass and Bioenergy* 35 (5): 2245–53. https://doi.org/10.1016/j.biombioe.2011.02.036.

Hirata, KMJSK, M Phunchindawan, J Tukamoto, S Goda, and K Miyamoto. 1996. Hirata "Cryopreservation of Microalgae Using Encapsulation-Dehydration." *Cryo-Letters* 17 (5): 321–328.

Hu, Qiang, Milton Sommerfeld, Eric Jarvis, Maria Ghirardi, Matthew Posewitz, Michael Seibert, and Al Darzins. 2008. "Microalgal Triacylglycerols as Feedstocks for Biofuel Production: Perspectives and Advances." *The Plant Journal* 54 (4): 621–39.

Ip, Po-Fung, and Feng Chen. 2005. "Production of Astaxanthin by the Green Microalga *Chlorella zofingiensis* in the Dark." *Process Biochemistry* 40 (2): 733–38. https://doi.org/10.1016/j.procbio.2004.01.039.

Iverson, T. M. (2006). "Evolution and unique bioenergetic mechanisms in oxygenic photosynthesis." *Current Opinion in Chemical Biology*, 10(2), 91–100. https://doi.org/https://doi.org/10.1016/j.cbpa.2006.02.013

Judd, Simon, Leo J P van den Broeke, Mohamed Shurair, Yussuf Kuti, and Hussein Znad. 2015. "Algal Remediation of CO_2 and Nutrient Discharges: A Review." *Water Research* 87: 356–66.

Kaçka, Aşkın, and Gönül Dönmez. 2008. "Isolation of *Dunaliella* Spp. from a Hypersaline Lake and Their Ability to Accumulate Glycerol." *Bioresource Technology* 99 (17): 8348–52. https://doi.org/10.1016/j.biortech.2008.02.042.

Kajiwara, Susumu, Hidenao Yamada, Narumasa Ohkuni, and Kazuhisa Ohtaguchi. 1997. "Design of the Bioreactor for Carbon Dioxide Fixation by *Synechococcus* PCC7942." *Energy Conversion and Management* 38: S529–32. https://doi.org/10.1016/S0196–8904(96)00322–6.

Karsten, U, S Lembcke, and R Schumann. 2007. "The Effects of Ultraviolet Radiation on Photosynthetic Performance, Growth and Sunscreen Compounds in Aeroterrestrial Biofilm Algae Isolated from Building Facades." *Planta* 225 (4): 991–1000. https://doi.org/10.1007/s00425-006-0406–x.

Kita, Jun, and Takashi Ohsumi. 2004. "Perspectives on Biological Research for CO_2 Ocean Sequestration." *Journal of Oceanography*. https://doi.org/10.1007/s10872-004-5762–1.

Kodama, M. 1993. "A New Species of Highly CO_2 Tolerant Fast Growing Marine Microalga Suitable for High-Density Culture." *J Mar Biotechnol* 1: 21–25.

Kumar, Amit, Sarina Ergas, Xin Yuan, Ashish Sahu, Qiong Zhang, Jo Dewulf, F. Xavier Malcata, and Herman van Langenhove. 2010. "Enhanced CO_2 Fixation and Biofuel Production via Microalgae: Recent Developments and Future Directions." *Trends in Biotechnology*. https://doi.org/10.1016/j.tibtech.2010.04.004.

Kumar, Prasun, Mamtesh Singh, Sanjeet Mehariya, Sanjay K S Patel, Jung-Kul Lee, and Vipin C Kalia. 2014. "Ecobiotechnological Approach for Exploiting the Abilities of Bacillus to Produce Co-Polymer of Polyhydroxyalkanoate." *Indian Journal of Microbiology* 54 (2): 151–57. https://doi.org/10.1007/s12088-014-0457–9.

Lackner, K. S. 2003. "CLIMATE CHANGE: A Guide to CO_2 Sequestration." *Science* 300 (5626): 1677–78. https://doi.org/10.1126/science.1079033.

Lal, R. 2008. "Sequestration of Atmospheric CO_2 in Global Carbon Pools." *Energy and Environmental Science* 1 (1): 86–100. https://doi.org/10.1039/b809492f.

Lee, Kwangyong, and Choul Gyun Lee. 2002. "Nitrogen Removal from Wastewater by Microalgae without Consuming Organic Carbon Sources." *Journal of Microbiology and Biotechnology* 12(6): 979–85.

León, Rosa, and Francisco Galván. 1999. "Interaction between Saline Stress and Photoinhibition of Photosynthesis in the Freshwater Green Algae *Chlamydomonas reinhardtii*. Implications for Glycerol Photoproduction." *Plant Physiology and Biochemistry* 37 (7): 623–28. https://doi.org/https://doi.org/10.1016/S0981–9428(00)80115–1.

Leone, Gian Paolo, Roberto Balducchi, Sanjeet Mehariya, Maria Martino, Vincenzo Larocca, Giuseppe Di Sanzo, Angela Iovine, et al. 2019. "Selective Extraction of ω-3 Fatty Acids from *Nannochloropsis* sp. Using Supercritical CO_2 Extraction." *Molecules* 24 (13). https://doi.org/10.3390/molecules24132406.

Le Moullec, Y., and M. Kanniche. 2011. "Screening of Flowsheet Modifications for an Efficient Monoethanolamine (MEA) Based Post-Combustion CO_2 Capture." *International Journal of Greenhouse Gas Control* 5 (4): 727–40. https://doi.org/10.1016/j.ijggc.2011.03.004.

Li, Li, Wei Li, Yong-ho Kim, and Yong Woo Lee. 2013. "*Chlorella vulgaris* Extract Ameliorates Carbon Tetrachloride-Induced Acute Hepatic Injury in Mice." *Experimental and Toxicologic Pathology* 65 (1–2): 73–80.

Li, Ming, Xiao-jie Pan, Yi Zou, Xiao-juan Chen, and Cheng-yan Wan. 2012. "Effects of Hg~(2+) Exposure on the Growth and Chlorophyll Fluorescence of Three Microalgal Strains." *Journal of Hydroecology* 2.

Liang, Yanna, Nicolas Sarkany, and Yi Cui. 2009. "Biomass and Lipid Productivities of *Chlorella Vulgaris* under Autotrophic, Heterotrophic and Mixotrophic Growth Conditions." *Biotechnology Letters* 31 (7): 1043–49. https://doi.org/10.1007/s10529-009-9975-7.

Lowrey, Joshua, Marianne S. Brooks, and Patrick J. McGinn. 2015. "Heterotrophic and Mixotrophic Cultivation of Microalgae for Biodiesel Production in Agricultural Wastewaters and Associated Challenges—a Critical Review." *Journal of Applied Phycology* 27 (4): 1485–98. https://doi.org/10.1007/s10811-014-0459-3.

Maeda, K, M Owada, N Kimura, K Omata, and I Karube. 1995. "CO_2 Fixation from the Flue Gas on Coal-Fired Thermal Power Plant by Microalgae." *Energy Conversion and Management* 36 (6–9): 717–20.

Mandalam, Ramkumar K, and Bernhard Palsson. 1998. "Elemental Balancing of Biomass and Medium Composition Enhances Growth Capacity in High-density *Chlorella vulgaris* Cultures." *Biotechnology and Bioengineering* 59 (5): 605–11.

Matsukawa, R, M Hotta, Y Masuda, M Chihara, and I Karube. 2000. "Antioxidants from Carbon Dioxide Fixing *Chlorella sorokiniana*." *Journal of Applied Phycology* 12 (3–5): 263–67.

Matsumoto, Hiroyo, Akihiro Hamasaki, Norio Sioji, and Yosiaki Ikuta. 1997. "Influence of CO_2, SO_2 and NO in Flue Gas on Microalgae Productivity." *Journal of Chemical Engineering of Japan*. https://doi.org/10.1252/jcej.30.620.

Matsumoto, Hiroyo, Norio Shioji, Akihiro Hamasaki, Yoshiaki Ikuta, Yoshinori Fukuda, Minoru Sato, Noriyoshi Endo, and Toshiaki Tsukamoto. 1995. "Carbon Dioxide Fixation by Microalgae Photosynthesis Using Actual Flue Gas Discharged from a Boiler." *Applied Biochemistry and Biotechnology* 51 (1): 681.

Mehariya, Sanjeet, Angela Iovine, Patrizia Casella, Dino Musmarra, Alberto Figoli, Tiziana Marino, Neeta Sharma, and Antonio Molino. 2020. "Chapter 7 - Fischer–Tropsch Synthesis of Syngas to Liquid Hydrocarbons." In *Lignocellulosic Biomass to Liquid Biofuels*, edited by Abu Yousuf, Domenico Pirozzi, and Filomena Sannino, 217–48. Academic Press. https://doi.org/10.1016/B978-0-12-815936-1.00007-1.

Mehariya, Sanjeet, Angela Iovine, Giuseppe Di Sanzo, Vincenzo Larocca, Maria Martino, Gian Paolo Leone, Patrizia Casella, Despina Karatza, Tiziana Marino, and Dino Musmarra. 2019a. "Supercritical Fluid Extraction of Lutein from *Scenedesmus almeriensis*." *Molecules* 24 (7): 1324.

Mehariya, Sanjeet, Angela Iovine, Patrizia Casella, Dino Musmarra, Simeone Chianese, Tiziana Marino, Alberto Figoli, Neeta Sharma, and Antonio Molino. 2019b. "Chapter 12 - Bio-Based and Agriculture Resources for Production of Bioproducts." In *Current Trends and Future Developments on (Bio-) membranes*, edited by Alberto Figoli, Yongdan Li, and Angelo Basile, 263–82. Academic Press. ISBN: 978-0-12-816778-6.

Mehariya, Sanjeet, Anil Kumar Patel, Parthiba Karthikeyan Obulisamy, Elumalai Punniyakotti, and Jonathan W C Wong. 2018. "Co-Digestion of Food Waste and Sewage Sludge for Methane Production: Current Status and Perspective." *Bioresource Technology* 265: 519–31.https://doi.org/10.1016/j.biortech.2018.04.030.

Milano, Jassinnee, Hwai Chyuan Ong, H. H. Masjuki, W. T. Chong, Man Kee Lam, Ping Kwan Loh, and Viknes Vellayan. 2016. "Microalgae Biofuels as an Alternative to Fossil Fuel for Power Generation." *Renewable and Sustainable Energy Reviews* 58: 180–97. https://doi.org/10.1016/j.rser.2015.12.150.

Miller, Anthony G, and Brian Colman. 1980. "Evidence for $HCO3^-$ Transport by the Blue-Green Alga (Cyanobacterium) *Coccochloris peniocystis*." *Plant Physiology* 65 (2): 397–402.

Miura, Y, W Yamada, K Hirata, K Miyamoto, and M Kiyohara. 1993. "Stimulation of Hydrogen Production in Algal Cells Grown under High CO_2 Concentration and Low Temperature." *Applied Biochemistry and Biotechnology* 39 (1): 753.

Miyairi, Sachio. 1995. "CO_2 Assimilation in a Thermophilic Cyanobacterium." *Energy Conversion and Management* 36 (6–9): 763–66.

Miyasaka, Hitoshi, Yosuke Ohnishi, Toru Akano, Kiyomi Fukatsu, Tadashi Mizoguchi, Kiyohito Yagi, Isamu Maeda, et al. 1998. "Excretion of Glycerol by the Marine Chlamydomonas Sp. Strain W-80 in High CO_2 Cultures." *Journal of Fermentation and Bioengineering* 85 (1): 122–24. https://doi.org/10.1016/S0922-338X(97)80367–4.

Molino, Antonio, Maria Martino, Vincenzo Larocca, Giuseppe Di Sanzo, Anna Spagnoletta, Tiziana Marino, Despina Karatza, Angela Iovine, Sanjeet Mehariya, and Dino Musmarra. 2019. "Eicosapentaenoic Acid Extraction from *Nannochloropsis gaditana* Using Carbon di oxide at Supercritical Conditions." *Marine Drugs* 17 (2). https://doi.org/10.3390/md17020132.

Molino, Antonio, Sanjeet Mehariya, Angela Iovine, Vincenzo Larocca, Giuseppe Di Sanzo, Maria Martino, Patrizia Casella, Simeone Chianese, and Dino Musmarra. 2018. "Extraction of Astaxanthin and Lutein from Microalga *Haematococcus pluvialis* in the Red Phase Using CO_2 Supercritical Fluid Extraction Technology with Ethanol as Co-Solvent." *Marine Drugs* 16 (11). https://doi.org/10.3390/md16110432.

Molino, Antonio, Sanjeet Mehariya, Despina Karatza, Simeone Chianese, Angela Iovine, Patrizia Casella, Tiziana Marino, and Dino Musmarra. 2019. "Bench-Scale Cultivation of Microalgae Scenedesmus Almeriensis for CO_2 Capture and Lutein Production." *Energies* 12 (14). https://doi.org/10.3390/en12142806.

Molino, Antonio, Sanjeet Mehariya, Giuseppe Di Sanzo, Vincenzo Larocca, Maria Martino, Gian Paolo Leone, Tiziana Marino, Simeone Chianese, Roberto Balducchi, and Dino Musmarra. 2020. "Recent Developments in Supercritical Fluid Extraction of Bioactive Compounds from Microalgae: Role of Key Parameters, Technological Achievements and Challenges." *Journal of CO_2 Utilization* 36: 196–209. https://doi.org/10.1016/j.jcou.2019.11.014.

Molino, Antonio, Juri Rimauro, Patrizia Casella, Antonietta Cerbone, Vincenzo Larocca, Simeone Chianese, Despina Karatza, et al. 2018. "Extraction of Astaxanthin from Microalga *Haematococcus pluvialis* in Red Phase by Using Generally Recognized as Safe Solvents and Accelerated Extraction." *Journal of Biotechnology* 283: 51–61. https://doi.org/10.1016/j.jbiotec.2018.07.010.

de Morais, Michele Greque, and Jorge Alberto Vieira Costa. 2007. “Isolation and Selection of Microalgae from Coal Fired Thermoelectric Power Plant for Biofixation of Carbon di oxide.” *Energy Conversion and Management* 48 (7): 2169–73. https://doi.org/10.1016/j.enconman.2006.12.011.

Morales-Sánchez, Daniela, Oscar A. Martinez-Rodriguez, John Kyndt, and Alfredo Martinez. 2015. “Heterotrophic Growth of Microalgae: Metabolic Aspects.” *World Journal of Microbiology & Biotechnology* 31 (1): 1–9. https://doi.org/10.1007/s11274-014-1773–2.

Moreira, Diana, and José C.M. Pires. 2016. “Atmospheric CO_2 Capture by Algae: Negative Carbon Dioxide Emission Path.” *Bioresource Technology* 215 (2016): 371–79. https://doi.org/10.1016/j.biortech.2016.03.060.

Moroney, J V, S G Bartlett, and Göran Samuelsson. 2001. “Carbonic Anhydrases in Plants and Algae.” *Plant, Cell & Environment* 24 (2): 141–53.https://doi.org/10.1111/j.1365–3040.2001.00669.x

Nagase, Hiroyasu, Kaoru Eguchi, Ken-Ichi Yoshihara, Kazumasa Hirata, and Kazuhisa Miyamoto. 1998. “Improvement of Microalgal NOx Removal in Bubble Column and Airlift Reactors.” *Journal of Fermentation and Bioengineering* 86 (4): 421–23.

Nakano, Y, K Miyatake, H Okuno, K Hamazaki, S Takenaka, N Honami, M Kiyota, I Aiga, and J Kondo. 1996. “Growth of Photosynthetic Algae Euglena in High CO_2 Conditions and Its Photosynthetic Characteristics.” In *International Symposium on Plant Production in Closed Ecosystems 440*, 49–54.

Ndimba, Bongani Kaiser, Roya Janeen Ndimba, T Sudhakar Johnson, Rungaroon Waditee-Sirisattha, Masato Baba, Sophon Sirisattha, Yoshihiro Shiraiwa, Ganesh Kumar Agrawal, and Randeep Rakwal. 2013. “Biofuels as a Sustainable Energy Source: An Update of the Applications of Proteomics in Bioenergy Crops and Algae.” *Journal of Proteomics* 93: 234–44. https://doi.org/10.1016/j.jprot.2013.05.041.

Olajire, Abass A. 2013. “A Review of Mineral Carbonation Technology in Sequestration of CO_2.” *Journal of Petroleum Science and Engineering*. Elsevier B.V. https://doi.org/10.1016/j.petrol.2013.03.013.

Patil, Vishwanath, Torsten Källqvist, Elisabeth Olsen, Gjermund Vogt, and Hans R Gislerød. 2007. “Fatty Acid Composition of 12 Microalgae for Possible Use in Aquaculture Feed.” *Aquaculture International* 15 (1): 1–9. https://doi.org/10.1007/s10499-006-9060–3.

Perez-Garcia, Octavio, Froylan M.E. Escalante, Luz E. de-Bashan, and Yoav Bashan. 2011. “Heterotrophic Cultures of Microalgae: Metabolism and Potential Products.” *Water Research* 45 (1): 11–36. https://doi.org/10.1016/j.watres.2010.08.037.

Pires, José C.M. 2017. “COP21: The Algae Opportunity?” *Renewable and Sustainable Energy Reviews* 79 (February): 867–77. https://doi.org/10.1016/j.rser.2017.05.197.

Pulz, Otto, and Karl Scheibenbogen. 2007. “Photobioreactors: Design and Performance with Respect to Light Energy Input.” *Bioprocess and Algae Reactor Technology, Apoptosis* 59: 123–52. https://doi.org/10.1007/bfb0102298.

Rahaman, Muhammad Syukri Abd, Li Hua Cheng, Xin Hua Xu, Lin Zhang, and Huan Lin Chen. 2011. “A Review of Carbon di oxide Capture and Utilization by Membrane Integrated Microalgal Cultivation Processes.” *Renewable and Sustainable Energy Reviews* 15 (8): 4002–12. https://doi.org/10.1016/j.rser.2011.07.031.

Raja, R, S Hemaiswarya, N Ashok Kumar, S Sridhar, and R Rengasamy. 2008. “A Perspective on the Biotechnological Potential of Microalgae.” *Critical Reviews in Microbiology* 34 (2): 77–88.

Raja, R, S Hemaiswarya, and R Rengasamy. 2007. “Exploitation of *Dunaliella* for β-Carotene Production.” *Applied Microbiology and Biotechnology* 74 (3): 517–23.

Razzak, Shaikh Abdur, Saad Aldin M Ali, Mohammad Mozahar Hossain, and Hugo deLasa. 2017. “Biological CO_2 Fixation with Production of Microalgae in Wastewater – A Review.” *Renewable and Sustainable Energy Reviews* 76: 379–90. https://doi.org/10.1016/j.rser.2017.02.038.

Ritchie, Hannah, and Max Roser. 2018. "CO_2 and Other Greenhouse Gas Emissions." *Online, Accessed* 7.

Salek, Shiva S., Robbert Kleerebezem, Henk M. Jonkers, Geert jan Witkamp, and Mark C.M. Van Loosdrecht. 2013. "Mineral CO_2 Sequestration by Environmental Biotechnological Processes." *Trends in Biotechnology*. https://doi.org/10.1016/j.tibtech.2013.01.005.

Salih, Fadhil M. 2011. "Microalgae Tolerance to High Concentrations of Carbon di oxide: A Review." *Journal of Environmental Protection* 2 (05): 648.

Sánchez, J F, J M Fernández-Sevilla, F G Acién, M C Cerón, J Pérez-Parra, and E Molina-Grima. 2008. "Biomass and Lutein Productivity of *Scenedesmus almeriensis*: Influence of Irradiance, Dilution Rate and Temperature." *Applied Microbiology and Biotechnology* 79 (5): 719–29.

Sanzo, Giuseppe Di, Sanjeet Mehariya, Maria Martino, Vincenzo Larocca, Patrizia Casella, Simeone Chianese, Dino Musmarra, Roberto Balducchi, and Antonio Molino. 2018. "Supercritical Carbon di oxide Extraction of Astaxanthin, Lutein, and Fatty Acids from *Haematococcus pluvialis* Microalgae." *Marine Drugs* 16 (9). https://doi.org/10.3390/md16090334.

Sathasivam, Ramaraj, Ramalingam Radhakrishnan, Abeer Hashem, and Elsayed F. Abd_Allah. 2019. "Microalgae Metabolites: A Rich Source for Food and Medicine." *Saudi Journal of Biological Sciences* 26 (4): 709–22. https://doi.org/10.1016/j.sjbs.2017.11.003.

Seckbach, J, H Gross, and M B Nathan. 1971. "Growth and Photosynthesis of *Cyanidium caldarium* Cultured under Pure CO_2." *Israel Journal of Botany* 20: 84–90.

Shi, Xian-Ming, Hui-Jun Liu, Xue-Wu Zhang, and Feng Chen. 1999. "Production of Biomass and Lutein by *Chlorella protothecoides* at Various Glucose Concentrations in Heterotrophic Cultures." *Process Biochemistry* 34 (4): 341–47. https://doi.org/10.1016/S0032–9592(98)00101–0.

Siciliano, Alessio, Carlo Limonti, Sanjeet Mehariya, Antonio Molino, and Vincenza Calabrò. 2018. "Biofuel Production and Phosphorus Recovery through an Integrated Treatment of Agro-Industrial Waste." *Sustainability* 11 (1). https://doi.org/10.3390/su11010052.

Silva, G. P.D. De, P. G. Ranjith, and M. S.A. Perera. 2015. "Geochemical Aspects of CO_2 Sequestration in Deep Saline Aquifers: A Review." *Fuel*. Elsevier Ltd. https://doi.org/10.1016/j.fuel.2015.03.045.

Slade, Raphael, and Ausilio Bauen. 2013. "Micro-Algae Cultivation for Biofuels: Cost, Energy Balance, Environmental Impacts and Future Prospects." *Biomass and Bioenergy* 53: 29–38. https://doi.org/10.1016/j.biombioe.2012.12.019.

Sung, K. D., J. S. Lee, C. S. Shin, and S. C. Park. 1999. "Isolation of a New Highly CO_2 Tolerant Fresh Water Microalga *Chlorella* sp. KR-1." *Renewable Energy* 16 (1–4): 1019–22. https://doi.org/10.1016/s0960-1481(98)00362–0.

Sung, Ki-Don, Jin-Suk Lee, Chul-Seung Shin, and Soon-Chul Park. 1998. "Isolation of a New Highly CO_2 Tolerant Fresh Water Microalga *Chlorella* sp. KR-1." *Korean Journal of Chemical Engineering* 15 (4): 449–50. https://doi.org/10.1007/BF02697138.

Stepan Daniel J., Richard E. Shockey, Thomas A. Moe, and Ryan Dorn. 2002. "Carbon di oxide Sequestering Using Microalgal Systems." https://doi.org/10.2172/882000.

Tang, Dahai, Wei Han, Penglin Li, Xiaoling Miao, and Jianjiang Zhong. 2011. "CO_2 Biofixation and Fatty Acid Composition of *Scenedesmus obliquus* and *Chlorella pyrenoidosa* in Response to Different CO_2 Levels." *Bioresource Technology* 102 (3): 3071–76. https://doi.org/10.1016/j.biortech.2010.10.047.

Tang, Yong, Ruizhi Yang, and Xiaoqiang Bian. 2014. "A Review of CO_2 Sequestration Projects and Application in China." *Scientific World Journal*. Hindawi Publishing Corporation. https://doi.org/10.1155/2014/381854.

Usui, Naoto, and Masahiro Ikenouchi. 1997. "The Biological CO_2 Fixation and Utilization Project by RITE(1) Highly-Effective Photobioreactor System." *Energy Conversion and Management* 38: S487–92. https://doi.org/10.1016/S0196–8904(96)00315–9.

Ummalyma, Sabeela Beevi, and Rajeev K. Sukumaran. 2014. "Cultivation of Microalgae in Dairy Effluent for Oil Production and Removal of Organic Pollution Load." *Bioresource Technology* 165 (C): 295–301. https://doi.org/10.1016/j.biortech.2014.03.028.

White, Curt M., Brian R. Strazisar, Evan J. Granite, James S. Hoffman, and Henry W. Pennline. 2003. "Separation and Capture of CO_2 from Large Stationary Sources and Sequestration in Geological Formations Coalbeds and Deep Saline Aquifers." *Journal of the Air and Waste Management Association* 53 (6): 645–715. https://doi.org/10.1080/10473289.2003.10466206.

Wilbanks, Thomas J., and Steven J. Fernandez. 2014. *Climate Change and Infrastructure, Urban Systems, and Vulnerabilities: Technical Report for the U.S. Department of Energy in Support of the National Climate Assessment. Climate Change and Infrastructure, Urban Systems, and Vulnerabilities: Technical Report for the U.S. Department of Energy in Support of the National Climate Assessment.* Island Press-Center for Resource Economics. https://doi.org/10.5822/978-1-61091-556-4.

Xiong, Fusheng, Jiri Kopecky, and Ladislav Nedbal. 1999. "The Occurrence of UV-B Absorbing Mycosporine-like Amino Acids in Freshwater and Terrestrial Microalgae (*Chlorophyta*)." *Aquatic Botany* 63 (1): 37–49. https://doi.org/10.1016/S0304-3770(98)00106-5.

Xu, Nianjun, Xuecheng Zhang, Xiao Fan, Lijun Han, and Chengkui Zeng. 2001. "Effects of Nitrogen Source and Concentration on Growth Rate and Fatty Acid Composition of *Ellipsoidion* sp. (Eustigmatophyta)." *Journal of Applied Phycology* 13 (6): 463–69. https://doi.org/10.1023/A:1012537219198.

Yongmanitchai, Wichien, and Owen P Ward. 1991. "Growth of and Omega-3 Fatty Acid Production by *Phaeodactylum tricornutum* under Different Culture Conditions." *Appl. Environ. Microbiol.* 57(2): 419–25.

Yoshihara, Ken-Ichi, Hiroyasu Nagase, Kaoru Eguchi, Kazumasa Hirata, and Kazuhisa Miyamoto. 1996. "Biological Elimination of Nitric Oxide and Carbon di oxide from Flue Gas by Marine Microalga NOA-113 Cultivated in a Long Tubular Photobioreactor." *Journal of Fermentation and Bioengineering* 82 (4): 351–54.

Yue, Lihong, and Weigong Chen. 2005. "Isolation and Determination of Cultural Characteristics of a New Highly CO_2 Tolerant Fresh Water Microalgae." *Energy Conversion and Management* 46 (11): 1868–76. https://doi.org/https://doi.org/10.1016/j.enconman.2004.10.010.

Zeng, Xianhai, Michael K. Danquah, Xiao Dong Chen, and Yinghua Lu. 2011. "Microalgae Bioengineering: From CO_2 Fixation to Biofuel Production." *Renewable and Sustainable Energy Reviews* 15(6): 3252–60.

Zhan, Jiao, Junfeng Rong, and Qiang Wang. 2017. "Mixotrophic Cultivation, a Preferable Microalgae Cultivation Mode for Biomass/Bioenergy Production, and Bioremediation, Advances and Prospect." *International Journal of Hydrogen Energy* 42 (12): 8505–17. https://doi.org/10.1016/j.ijhydene.2016.12.021.

Zhao, Bingtao, and Yaxin Su. 2014. "Process Effect of Microalgal-Carbon di oxide Fixation and Biomass Production: A Review." *Renewable and Sustainable Energy Reviews* 31: 121–32. https://doi.org/10.1016/j.rser.2013.11.054.

Zhou, Wenguang, Jinghan Wang, Paul Chen, Chengcheng Ji, Qiuyun Kang, Bei Lu, Kun Li, Jin Liu, and Roger Ruan. 2017. "Bio-Mitigation of Carbon di oxide Using Microalgal Systems: Advances and Perspectives." *Renewable and Sustainable Energy Reviews* 76 (February): 1163–75. https://doi.org/10.1016/j.rser.2017.03.065.

12 Mechanisms for Carbon Assimilation and Utilization in Microalgae and Their Metabolites for Value-Added Products

Varsha S.S. Vuppaladadiyam, Zenab T. Baig, Abdul F. Soomro, and Arun K. Vuppaladadiyam

CONTENTS

12.1 INTRODUCTION

Microalgae has gained much attention in the past few decades and has emerged as an attractive bioenergy feedstock as compared with traditional energy crops. The main driving force to commercially cultivate microalgae is to yield its metabolic products (Perez-Garcia, Bashan, and Esther Puente 2011). Microalgae, light-driven cell factories, convert CO_2 into the value-added molecules and carbon-rich lipids, which are in turn transformed into biofuels and other products of commercial value.

Adopting the photosynthesis route, microalgae can transform CO_2 and water into organic materials without releasing pollutants and requiring energy beyond that procured via sunlight (Klinthong et al. 2015). Microalgae does not require high-purity CO_2, it does not need to be grown in freshwater, and harmful combustion products such as NO_X and SO_X can potentially be used as nutrients during cultivation (Klinthong et al. 2015, Zhang et al. 2016, Camerini et al. 2016, Thomas, Mechery, and Paulose 2016, Zhu et al. 2016). Aquatic microalgal species are ideal for biological carbon capture because of their rapid growth, high biomass yields, and a range of potential value-added products (Gao et al. 2012, Sing et al. 2013). Furthermore, microalgae do not compete directly with food crops for land or water (Global CCS Institute), can fix CO_2 10–50 times more efficiently than terrestrial higher plants, and do not require a vascular system for transporting nutrients (Cuellar-Bermudez, Garcia-Perez, et al. 2015, Lam, Lee, and Mohamed 2012, Khan et al. 2009).

The formation of metabolic products depends upon many factors such as light harvesting mechanism; acquisition and assimilation of carbon, nitrogen, and sulphur; and generation of unique secondary metabolites (Beardall and Raven 2001). Several marine and freshwater microalgal species have developed resilience strategies from different metabolic pathways to survive in competitive environments. Furthermore, microalgae provides vital compounds for human nutrition and novel biologically active materials. Therefore, the need of algal extracts or pure compounds from algae demands the study of both primary and secondary metabolisms to commercially explore microalgal technologies (Cardozo et al. 2007). The microalgal lipid fraction mainly contains fatty acids (FAs), sterols, waxes, ketones, hydrocarbons, and pigments (Adarme-Vega et al. 2012). Few other metabolites that are much explored include phycocolloids, carotenoids, lectins, mycosporine, halogenated compounds, and polyketides. In this chapter, an attempt has been made to critically discuss the vital concepts of microalgae and its importance as a carbon sink. Also, the most important carbon partitioning strategies that a microalgae can adopt to increase the yields of a specific metabolic product are critically discussed. In the last part, a wide range of microalgal metabolites are discussed in detail. The information presented in this chapter can enhance the basic knowledge of readers in view of microalgae and its application in carbon capture and utilization.

12.2 CO_2 ASSIMILATION AND METABOLISM

Understanding the mechanisms of carbon uptake and utilization is crucial to maximize the efficiency of microalgae to convert carbon and energy into biomass. There are three different strategies in which microalgae can assimilate inorganic carbon: (i) conversion of bicarbonates into CO_2 using extracellular carbonic anhydrase (CA) that can freely diffuse into the cells, (ii) direct assimilation of CO_2 through the plasmatic membrane, and (iii) uptake of bicarbonates directly into the cells via active transporters in the membrane (Spalding 2008). The strategies of CO_2 uptake by few microalgal species reported in the literature are listed in Table 12.1, which shows that assimilation of inorganic carbon using all the three strategies is not possible with every microalga. Few microalgal species can use all the three strategies to assimilate inorganic carbon, and few can use only one strategy. The inorganic carbon

TABLE 12.1
Different Inorganic Carbon-Uptake Strategies Adopted by Microalgal Species

Species	Carbonic Anhydrase	CO_2 Direct	HCO_3^-	Refs.
		Bacillariophyceae		
Chaetoceros calcitrans	NA	AV	AV	Korb et al. (1997)
Navicula pelliculosa	NA	AV	AV	Rotatore and Colman (1992)
Phaeodactylum tricornutum	AV	AV	AV	Colman and Rotatore (1996)
		Chlorophyceae		
Chlamydomonas reinhardtii	AV	AV	AV	Sültemeyer et al. (1989)
Chlamydomonas noctigama	AV	AV	DUN	van Hunnik et al. (2001)
Dunaliella tertiolecta	AV	AV	AV	Amoroso et al. (1998)
Scenedesmus obliquus	AV	AV	AV	Palmqvist, Yu, and Badger (1994)
Spirulina platensis	AV	AV	AV	Camerini et al. (2016)
Tetraedron minimum	AV	AV	DUN	van Hunnik et al. (2001)
Nannochloris atomus	NA	AV	NA	Huertas, Colman, et al. (2000)
Nannochloris maculata	NA	AV	NA	Huertas, Colman, et al. (2000)
		Chrysophyceae		
Vischeria stellate	NA	AV	AV	Huertas, Colman, and Espie (2002)
		Coscinodiscophyceae		
Ditylum brightwellii	NA	AV	AV	Korb et al. (1997)
Skeletonema costatum	NA	AV	AV	Korb et al. (1997)
Thalassiosira punctigera	NA	AV	N/A	Elzenga, Prins, and Stefels (2000)
Thalassiosira pseudonana	AV	N/A	AV	Elzenga, Prins, and Stefels (2000), Armbrust et al. (2004)
		Cyanophyceae		
Synechococcus PCC7942	AV	DUN	DUN	Price, Coleman, and Badger (1992)
Nostoc calcicola	AV	AV	DUN	Jaiswal, Prasanna, and Kashyap (2005)
Prochlorococcus sp.	AV	DUN	DUN	So et al. (2004)
		Dinophyceae		
Amphidinium carterae	NA	AV	NA	Colman et al. (2002)

(*Continued*)

TABLE 12.1 (*Continued*)
Different Inorganic Carbon-Uptake Strategies Adopted by Microalgal Species

Species	Carbonic Anhydrase	CO_2 Direct	HCO_3^-	Refs.
		Eustigmatophyceae		
Monodus subterraneus	NA	AV	NA	Huertas, Espie, et al. (2000)
Nannochloropsis gaditana	AV	NA	AV	Huertas, Espie, et al. (2000), Radakovits et al. (2012)
Nannochloropsis oculata	NA	NA	AV	Huertas, Espie, et al. (2000)
		Mediophyceae		
Cyclotella sp.	AV	AV	AV	Rotatore, Colman, and Kuzma (1995)
		Porphyridiophyceae		
Porphyridium cruentum	AV	AV	AV	Colman and Gehl (1983)
		Prymnesiophyceae		
Emiliania huxleyi	AV	AV	AV	Elzenga, Prins, and Stefels (2000)
Dicrateria inornata	AV	AV	AV	Colman et al. (2002)
Isochrysis galbana	AV	AV	AV	Colman et al. (2002)
Phaeocystis globosa	AV	AV	DUN	Elzenga, Prins, and Stefels (2000)
		Trebouxiophyceae		
Chlorella ellipsoidea	NA	AV	AV	Rotatore and Colman (1991)
Chlorella saccharophila	AV	AV	AV	Rotatore and Colman (1991)
Chlorella kessleri	NA	AV	AV	Bozzo, Colman, and Matsuda (2000)
Chlorella sp.	AV	AV	NA	Badger et al. (1998)
Coccomyxa	AV	DUN	AV	Palmqvist et al. (1995)
Eremosphaera viridis	NA	AV	NA	Rotatore and Colman (1992)

Source: Adapted from Vuppaladadiyam et al. (2018).
AV, present; NA, absent; DUN, data unavailable in the literature.

transport system is of prime importance as it enables the microalgae to acquire the inorganic carbon from the external environment. Once the transport process is fully stimulated, the cell attains the ability to utilize the inorganic carbon in the form of both CO_2 and HCO_3^- (Sültemeyer et al. 1989). However, it is important to note that the enzyme RuBisCo has a low affinity towards CO_2, and this property of the enzyme is considered as a serious challenge to the microalgal species (Badger and Price 1992). At atmospheric levels of CO_2, RuBisCo can perform up to 25% of its catalytic activity because Michaelis constant (K_m) is higher than the dissolved CO_2

concentration. With elevated levels of O_2, microalgae can experience fluctuations in the levels of CO_2 and HCO_3^-, affecting the pH of the environment, which in turn impacts the availability of inorganic compounds for photosynthesis process (Mondal et al. 2016). The CA is available in animals, plants, and microorganisms and plays a role in CO_2 acquisition, respiration, ion exchange and transport, and photosynthesis. The CA is grouped into five families, which are named as α-CA, β-CA, γ-CA, δ-CA, and ε-CA (15). A wide range of CA enzymes are reported in microalgae, such as *Chlamydomonas reinhardtii*, *Phaeodactylum tricornutum*, *Symbiodinium* sp., and *Porphyridium purpurea* (Moroney et al. 2001, Badger and Price 1994).

The CO_2 uptake by microalgae is then fixed into organic carbon through three pathways: autotrophic, heterotrophic, and mixotrophic metabolism (Van Den Hende, Vervaeren, and Boon 2012). The metabolic mechanisms of these three pathways are discussed in the next three subsections.

12.2.1 Autotrophic Metabolism

The two subcategories that are included in autotrophic metabolism are photoautotrophic and chemoautotrophic (Min et al. 2012). In photoautotrophic metabolism (Figure 12.1), light is used as a source of energy to convert inorganic CO_2 into carbohydrates via the process of photosynthesis (which is shown in Eq 12.1); algae and cyanobacteria come under photoautotrophs.

$$H_2O + CO_2 + \text{ Photons}(\text{light}) \rightarrow [CH_2O]_n + O_2 \quad (12.1)$$

The transport molecules and energy-storingmolecules such as adenosine triphosphate (ATP) and nicotinamide adenine dinucleotide phosphate (NADPH) are formed during the first stage of photosynthesis through the light-dependent reaction, which is expressed in the following equation (Carvalho et al. 2011):

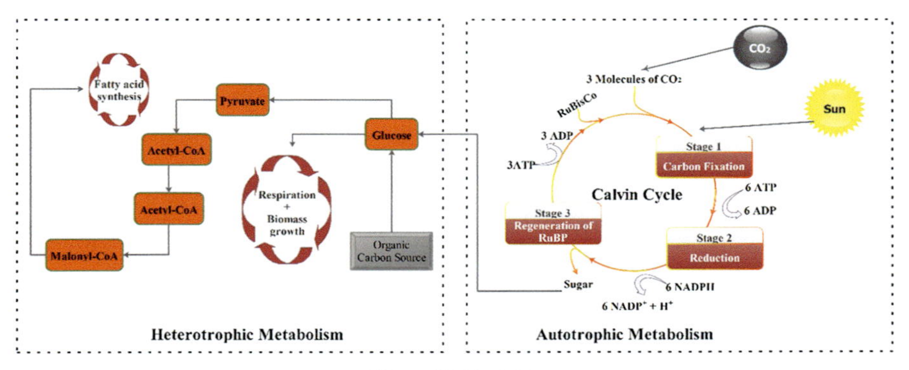

FIGURE 12.1 Various metabolisms for carbon assimilation in microalgal cultivation (Mohan et al. 2015).

$$2H_2O + 2NADP^+ + 3ADP + 3P + light \rightarrow 2NADPH + 2H^+ + 3ATP + O_2 \quad (12.2)$$

where NADP, ADP, and P are nicotinamide adenine dinucleotide phosphate, adenosine diphosphate, and phosphorus, respectively.

Later, glucose is produced by fixing CO_2 using the ATP and NADPH via light-independent reaction (Eq. 12.3). As part of the Calvin–Benson (CB) cycle, enzyme ribulose bisphosphate carboxylase/oxygenase (RuBisCo) catalyses the following reaction (Williams and Laurens 2010):

$$3CO_2 + 9ATP + 6NADPH + 6H^+ \rightarrow C_3H_6O_3 - \text{phosphate} + 9ADP + 8P + 6NADP^+ + 3H_2O \quad (12.3)$$

The CB cycle or the reductive pentose phosphate cycle consists of an irreversible reaction catalysed by RuBisCo, where one molecule of ribulose 1,5-bisphosphate (RuBP), water, and CO_2 combine to produce two molecules of 3-phospoglycerate (3-PGA). Five out of six molecules of 3-PGA, which are produced from three molecules of CO_2 and RuBP, are used to bring back RuBP, and the excess molecule of 3-PGA is consumed to build cell material. The RuBP is then regenerated. Together with the light-independent reaction, through photorespiration, the oxygenation of RuBP may be catalysed by the RuBisCo enzyme. This uses the energy from photosynthesis to produce phosphoglycolate that cannot be used in the CB cycle and leads to a reduction in the efficiency of carbon fixation via photosynthesis by 20%–30% (Sayre 2010, Zhu, Long, and Ort 2008). A few eukaryotic algae and cyanobacteria possess carbon-concentrating mechanisms (CCMs), which minimizes this problem (Hügler and Sievert 2011). For instance, in cyanobacteria, bicarbonates are initially accumulated and then transported to carboxysome, where they are converted into CO_2 by CA to create a microenvironment rich in CO_2 (Moroney et al. 2013).

12.2.2 Heterotrophic Metabolism

Heterotrophic metabolism requires organic nutrients instead of CO_2 as the raw material. Many species of algae live in extreme conditions which are mostly deprived of sunlight, and so, photoautotrophic metabolism is not possible. Such species are forced to rely on heterotrophic metabolism (Tuchman et al. 2006, Lowrey, Armenta, and Brooks 2016, Morales-Sánchez et al. 2015, Liu, Sun, and Chen 2014), and a large number of studies reported that algae can metabolize a wide range of organic carbon sources (chemo)heterotrophically under light-deprived conditions (Tuchman et al. 2006, Lowrey, Armenta, and Brooks 2016, Morales-Sánchez et al. 2015, Liu, Sun, and Chen 2014). Microalgae is considered suitable for heterotrophic cultivation because it can able to: (i) metabolize and grow in the absence of light, (ii) rapidly grow in sterilized culture media, (iii) quickly adapt to changing environment, and (iv) metabolize a wide range of organic carbon sources such as pyruvate, acetate, saturated FAs, lactate, ethanol, C6 sugars, glycolate, glycerol, C5 monosaccharides, disaccharides, and amino acids (Morales-Sánchez et al. 2015).

Several studies reported the application of microalgal strains for biomass and other high-value products generation under heterotrophic conditions using glucose as their carbon source (Im et al. 2012, Pleissner et al. 2013, Yan et al. 2011). In comparison with other carbon substrates, such as organic acids, sugar alcohols, monohydric alcohols, and other sugars, glucose as a carbon source resulted in higher growth rates (Perez-Garcia et al. 2011). Embden–Meyerhof–Parnas (EMP) pathway is responsible for carbohydrate metabolism in autotrophic and mixotrophic metabolisms, whereas the pentose phosphate pathway (PPP) plays a major role in heterotrophic growth (Perez-Garcia et al. 2011). Low quantities of enzyme, like lactate dehydrogenase, and inadequate energy released during the dissimilation of glucose are the two constrains that limit the glucose metabolism under dark and anaerobic conditions (Droop 1974). More than 85% of the total glucose consumed by microalgae is assimilated and transformed into oligo- and polysaccharides; however, only 1% of it remains as free glucose (Tanner 2000). The heterotrophic metabolism pathway is explained in Figure 12.1.

12.2.3 Mixotrophic Metabolism

Mixotrophic metabolism allows the flexibility to simultaneously use both photoautotrophic and heterotrophic metabolic strategies (e.g. light and/or organic chemical energy sources) along with the possibility to switch between them. As shown in Figure 12.1, in the mixotrophic metabolism, the acetyl-CoA pool is sustained via the combined use of extracellular organic carbon and CO_2 (Mohan et al. 2015). Organic carbon and inorganic carbon are assimilated through aerobic respiration and photosynthesis, respectively (Hu et al. 2012). This arrangement to switch between strategies enables a higher biomass productivity (Wang, Yang, and Wang 2014).

In the mixotrophic cultivation, microalgae do not rely heavily on irradiance as they have a potential to use a wide range of organic compounds, thus avoiding the chances of photoinhibition and photolimitation (Fernández, Sevilla, and Grima 2013). Enhancement of biomass productivity, improved growth rates, negligible loss of biomass during dark hours, and short growth cycles are few advantages of mixotrophic mode of cultivation over the other methods (Park et al. 2012, Kong et al. 2012, Bassi, Saxena, and Aguirre 2014). However, mixotrophic mode of cultivation is cost-intensive and poses a high risk of invasion from heterotrophic bacteria that compete for organic carbon source (Capital Energy). Additionally, there is always a challenge in maintaining an optimum balance between two metabolic pathways.

12.3 CARBON PARTITIONING STRATEGY OF MICROALGAE

Carbon partitioning can be explained as the distribution of photoassimilates (products of photosynthesis process) such as water-soluble polysaccharides, starch, and triacylglycerols (TAGs) (Guarnieri et al. 2011). As these organic carbon products often demand the same precursors, manipulating the biosynthetic pathway through omic-based strategies, genetic modification, and microalgal cultivation conditions are crucial to obtaining desired products (Jia et al. 2015, Guarnieri et al. 2011, Sivaramakrishnan and Incharoensakdi 2017, Xue and Jiang 2017, de Jesus and Maciel Filho 2017, Patrinos 2017). There are two ways in which a microalgae responds to variations in the environmental conditions: acclimation response and homeostatic response. The cells modulate and change their metabolic pathways and cell composition under acclimation response, resulting in a significant variation in their biochemical and elemental composition. Under such response, they often deviate away from canonical Redfield ratio (Jeyasingh et al. 2017). On the other hand, microalgae can maintain a balanced cell composition irrespective of variations in the physicochemical composition of the external environment, and this can be considered as homeostatic response (Montechiaro et al. 2006).

Most of the microalgal species possess a specific carbon partitioning strategy under specific culture conditions; for instance, *Dunaliella salina*, *Botryococcus braunii*, and *Chlorella* sp., can produce 30%–50% protein, 8%–15% lipids, and 20%–40% carbohydrate when cultivated under favourable conditions irrespective of the strains considered. However, when cultivated under stress-induced conditions, these strains can produce 40% of glycerol, 80% of hydrocarbons, and 80% of FAs, respectively, via acclimation response (Richmond 2008). In view of the advantages presented by microalgae under stress-induced conditions, acclimation response has been considered as an approach to enhance the yields of a particular carbon product. When microalgal cultivation is considered for biofuel production, improving the yield of FAs and TAG synthesis is of prime interest. Jiang et al.(2012) studied the impact of shifting from nitrogen-rich medium to nitrogen (N)-depleted conditions on lipid accumulation in two marine microalgae *Dunaliella tertiolecta* and *Thalassiosira pseudonana* and found that the lipid accumulation increased in the N-depleted environment from 2- to 2.5-fold after 10 days (5 days for *T. pseudonana*) before it declined. Moreover, the study indicated that continuously depleting N from the medium will not increase the accumulation of the lipids once the maximum oil formation is reached. However, depleting N from the medium negatively impacted the specific growth rates of the two strains. The growth rates reported were noted to drop from 0.84 day^{-1} and 1.21 day^{-1} in the N-repleted medium to −0.1 and 0.1 day^{-1} in the N-depleted medium for the two strains, respectively (Jiang, Yoshida, and Quigg 2012). It was observed that the carbon partitioning strategies of the two microalgal species were also different, under N-stressed conditions. *D. tertiolecta* fixed carbon into carbohydrates, whereas *T. pseudonana* accumulated carbon as lipids. In a study done by Guarnieri et al. (2011) on an oleaginous microalgal strain *C. vulgaris* to understand the TAG synthesis pathway, it was identified that similar growth rates were recorded in both N-depleted and N-repleted conditions for the first 24 hours; however, the growth rates eventually declined later under N-free environment (Guarnieri et al. 2011). An enhancement in the FA content from 10% to 60% was recorded in the N-free cultures. Ho et al. (2010) cultivated microalgae *S. obliquus* in two stages using flue gas containing 10 vol.% CO_2 for biodiesel production. In the first stage, the microalgae was cultivated in an N-rich environment for a period of 12 days and later was shifted to an N-depleted environment (Ho, Chen, and Chang 2010). They reported a lipid productivity, CO_2 consumption rate, and biomass productivity of ~78, ~379, and ~201 mg $L^{-1}day^{-1}$ after 12 days, respectively. Although they reported a twofold increase in the lipid productivity at the end of the study, there was a drop in the CO_2 consumption and biomass productivity by approximately 30%

over the next 12-day period. These findings indicate that even though N-depleted conditions significantly enhance the lipid production, there is an undesirable effect on the microalgal growth rates. In a mathematical model developed by Packer et al. (2011), an attempt was made to understand the impact of nitrogen starvation on lipid accumulation and biomass production. They reported that the possible reason for high lipid production would be decoupling photosynthesis from cellular growth. Additionally, it was also reported that the measuring N:C ratio would explain the threshold of N quota during lipid synthesis (Packer et al. 2011). Thus, pilot studies that examine optimal configurations for the two-stage systems and optimization of N supply rate, to balance lipid content and growth rate, are crucial to make these strategies sustainable. Tailoring the cultivation conditions to different microalgal strains will also be a technical challenge.

Alternatively, genetic engineering can also be considered as a potential approach to enhance the lipid contents in microalgae. In the early studies on carbon partitioning, modification of one of the genes in the lipid synthesis pathway was considered as a way to increase lipid yields. Ibáñez-Salazar et al. (2014) reported a twofold increase in the lipid production by performing an overexpression of DOF-type transcription factor genes. Through overexpression of a BHLH transcription factor in *Nannochloropsis salina* under N-limited conditions for 8 days, a 33% increase in the yields of lipids was reported in the literature (Kang et al. 2015). However, the specific growth rate in such conditions was 60% slower than that of unmodified strain (Ibáñez-Salazar et al. 2014). A twofold increase in the yields of lipids and a 15% drop in the biomass productivity were reported by Dinamarca et al. (2017) during overexpression of diacylglycerol acyltransferase (DGAT) gene in *P. tricornutum* (Dinamarca et al. 2017). Instead of overexpression of a specific gene, Li et al. (2010) in their study on *C. reinhardtii* inactivated the ADP-glucose pyrophosphorylase by shunting the carbon precursors from the starch synthesis pathway (Li et al. 2010). As both TAG and starch synthesis pathways share the same precursor, the yields of TAG increased by 10-fold. However, there was a decrease in the biomass growth rates by *ca.* 12%–30% over the study duration of 6 days. Although the metabolic engineering strategies appear promising, the results of these studies suggest that an increase in lipid synthesis heavily relies on N-deficient conditions. This means that they share the similar barriers to scale-up as their unmodified counterparts. A more thorough economic and technical analysis should be conducted to justify whether the improvements in lipid content are worth the additional effort of metabolic engineering.

12.4 METABOLITES

Microalgae can convert CO_2 and carbon-containing organic substrates into primary metabolites, which include lipids, carbohydrates, and proteins, and into a wide range of secondary metabolites, e.g. waxes, sterols, ketones, vitamins, hydrocarbons, pigments, carotenoids, phycocolloids, lectins, polyketides, and mycosporine (Vuppaladadiyam et al. 2018a). Figure 12.2 shows the chemical composition of primary metabolites in different microalgal species. In general, proteins are present in higher percentage, whereas liquids the lowest percentage. Proteins are biosynthesized during the growth phase, whereas lipids and carbohydrates are biosynthesized

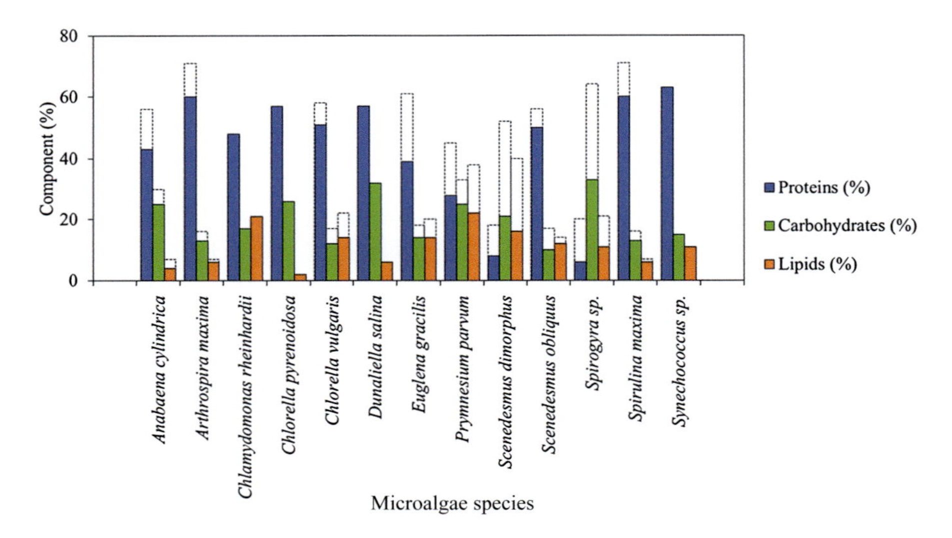

FIGURE 12.2 Variation in the primary metabolites for different microalgal species. (Adapted from Klinthong et al., 2015.)

during nutrient deficiency or light-conditioning stage, mainly as energy storage molecules (Barry et al. 2016). Cyanobacteria generally accumulates carbohydrates and a minor amount of lipids, mainly in the forms of polar membrane glycerolipids (few contain up to 14%) (Cuellar-Bermudez, Aguilar-Hernandez, et al. 2015).

12.4.1 Fatty Acids (FA) and Triacylglycerols (TAG)

Lipophilic compounds in algae belong to both primary and secondary metabolites. Under photo-oxidative stress, TAG synthesis can also serve as an electron sink. Recent studies showed that under N-depleted conditions, increased *de novo* TAG synthesis from acyl-CoA and increased carbon flux towards acyl-CoA and glycerol-3-phosphate for FA synthesis are responsible for the lipid accumulation in green algae (Miller et al. 2010, Fan et al. 2012). The highest lipid contents in most microalgae are reported when the microalgae are cultivated in stressed or unfavourable conditions. Sharma et al. (2013) used UV-C irradiation at small doses to increase the lipid contents and promote rapid settling on further increasing the doses (Sharma et al. 2013). This procedure decouples biomass growth from harvesting and lipid accumulation, and is successfully demonstrated in pilot-scale outdoor cultivation for several microalgal strains.

Lipids from algae are mainly classified into polar lipids, which mainly consist of amphipathic phospholipids and glycolipids, and neutral lipids, which are mainly composed of TAG. Glycolipids and phospholipids and other lipids are dominant categories in chloroplast membranes and extra-chloroplast membranes, respectively (Haneklaus et al. 2005). When the TAGs are generated from membrane lipids in storage bodies under stress conditions, it is possible to interconvert neutral lipids and polar lipids. In a study done on *P. tricornutum* in N-depleted conditions, it was noticed that only 60% of TAG was synthesized from *de novo* carbon fixation and the remaining 40% was generated from the transformation of pigment, protein, carbohydrate, and other membrane components (Burrows et al. 2012). In a review reported by Williams and Laurens (2010), different microalgal and cyanobacterial species were selected and the distribution of three classes of lipids, namely simple lipids, glycolipids, and phospholipids, was critically discussed. They considered four species of microalgae, namely diatoms, green algae, blue-green algae, and other species, and noticed that diatoms were dominated by simple lipids followed by glycolipids and phospholipids (Williams and Laurens 2010). A clear domination of a particular class of lipids was not seen in other species of microalgae. A number of factors give rise to a variation in the distribution of lipid classes. As the rate of metabolism slows down, storage lipids may gain a large proportion of the lipid fraction, and for the same reason, a shift in the lipid composition may happen throughout the growth phase.

In another review on understanding the microalgal cultivation and the production of metabolites, Vuppaladadiyam et al. (2018a) critically discussed the importance of chain length distribution of FAs in microalgae and compared them against the same in higher plants. It is important to note that the fuel characteristics highly depend on FA composition, which varies with culture conditions. The distribution of FAs in microalgae is more diverse when compared to higher plants. This quality makes certain species of microalgae suitable for generating biomass with selected fuel properties. For instance, certain microalgal strains accumulate short-chain FAs ($<C_{16}$), making them suitable for the production of jet fuels. Also, there are few strains that accumulate long-chain FAs ($>C_{20}$), making them suitable for producing lubricants.

12.4.2 Poly-unsaturated Fatty Acids

Long-chain poly-unsaturated fatty acids (PUFAs) are of special interest mainly because of their nutritional values and their potential in the prevention and treatment of various diseases (Morales-Sánchez et al. 2015, Fraeye et al. 2012, Ryckebosch et al. 2012). Eicosapentaenoic acid (EPA 20:5 ω-3), α-linolenic acid (ALA 18:3 ω-3), docosahexaenoic acid (DHA 22:6 ω-3), arachidonic acid (ARA 20:6 ω-6), and γ-linolenic acid (GLA 18:3 ω-6) are the most prominent PUFAs reported in the literature. Fish and fish oil are widely recognized as the source of PUFAs; however, sustainability of this source is a serious concern and algae-derived PUFAs appear as a promising alternative (Perez-Garcia et al. 2011). Table 12.2 lists the microalgal species containing significant amounts of EPA and DHA. Interestingly, in a study done on *Thraustochytrium* sp., the synthesis of DHA was stimulated by N-depletion after a high supply of monosodium glutamate and yeast extract, was reported to be 35% of the total FA content (Lenihan-Geels, Bishop, and Ferguson 2013). Large amount of ARA can be produced from few green algae species, like *Parietochloris incisa*, under N-limiting conditions, where a cell can contain up to 20% ARA of the biomass (Khozin-Goldberg et al. 2002). Additionally, the yields can be increased by decreasing the operational temperature (Bigogno, Khozin-Goldberg, and Cohen 2002). In a study done on *P. incisa*, it was noted that the ARA-rich TAG demonstrated medicinal properties and helped faster recovery from infection

TABLE 12.2
Prominent Microalgal Strains Rich in PUFA

	EPA		DHA	EPA + DHA	
Species	%	mg g^{-1}C	mg g^{-1}C	%	**Refs.**
C. vulgaris (green)		0.2	0.2		Batista et al. (2013)
Chlorella vulgaris (orange)		0.4	0.8		Ryckebosch et al. (2014)
Diacronema vlkianum		32.1	8.4		Batista et al. (2013)
Haematococcus . pluvialis		5.8	11.6		Batista et al. (2013)
Phaeodactylum tricornutum	9.3 (total FA)				Ryckebosch et al. (2014)
Nannochloropsis limnetica		100.9	0.0		Martin-Creuzburg, Wacker, and Basen (2010)
Cryptomonas sp.		29.9	5.3		Martin-Creuzburg, Wacker, and Basen (2010)
Rhodomonas lacustris		44.5	3.0		Wenzel et al. (2012)
Rhodomonas salina		20.7	13.3		Chen, Wakeham, and Fisher (2011)
Rhodomonas sp.		5.8	2.9		Chen, Liu, and Chen (2012)
Thalassiosira oceanica		45.9	7.5		Jiménez et al. (2003)
Thalassiosira weissflogii		18.6	3.2		Chen, Liu, and Chen (2012)
Prorocentrum dentatum		0.5	6.2		Chen, Liu, and Chen (2012)
Nannochloropsis salina	28 (total lipids)				Van Wagenen et al. (2012)
Pinguiococcus pyrenoidosus				22.0 (total lipids)	Sang et al. (2012)
Thraustochytrium sp.				45.1 (total lipids)	Scott et al. (2011)
Dunaliella salina	21.4 (total lipids)				Bhosale, Rajabhoj, and Chaugule (2010)
Isochrysis galbana				411.5 (total lipids)	Yago et al. (2010)

(Khozin-Goldberg et al. 2006). However, the risk, like toxicity, associated with the overdosage of PUFAs in nutritional products cannot be overlooked (Ikawa 2004).

12.4.3 Sterols

Sterols are bioactive compounds and are presented in eukaryotes, which can significantly affect the physicochemical properties of plasma membrane

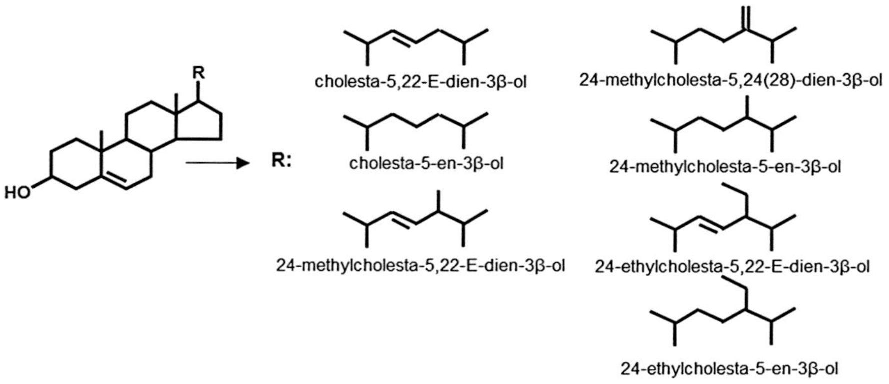

R

HO

R:

cholesta-5,22-E-dien-3β-ol

cholesta-5-en-3β-ol

24-methylcholesta-5,22-E-dien-3β-ol

24-methylcholesta-5,24(28)-dien-3β-ol

24-methylcholesta-5-en-3β-ol

24-ethylcholesta-5,22-E-dien-3β-ol

24-ethylcholesta-5-en-3β-ol

FIGURE 12.3 Sterols isolated from fresh and marine microalgae. (Adapted from Ponomarenko et al., 2004.)

(Volkman 2016). The sterol content in the plasma membrane can reach up to 20%–30% in the eukaryotes and is highly diverse when compared to sterols in animals (Abdul et al. 2016, Martin-Creuzburg and Merkel 2016). Sterols derived from microalgae include sterols, fucosterols, and phytosterols and found their application in health and food industry (Volkman 2016). Structures of common sterols found in marine microalgae are shown in Figure 12.3. Sterols are reported to lower the levels of cholesterol in blood and are also considered as anticarcinogenic (Leblond et al. 2011, Tang and Suter 2011), anti-inflammatory (Caroprese et al. 2012, Jung et al. 2013), immunomodulatory (Caroprese et al. 2012), antioxidative (Abdul et al. 2016, Lee et al. 2003), hepatoprotective (Jung et al. 2013, Demirbas and Fatih Demirbas 2011), and antihypercholesterolaemic (Chen et al. 2014). Additionally, they act as precursors to other bioactive compounds. Recently, in a study done on sterol extracts from *Schizochytrium*, Adarme-Vega et al. (2012) reported that the intestinal gene expression ACAT2283 was found to be responsible for the absorption of cholesterol in intestine.

12.4.4 Carbohydrates

Carbohydrate metabolism is the basis for the carbon and energy balances of a cell and plays an important role in understanding and improving the overall yield of different microalgal-derived bioproducts (Markou, Angelidaki, and Georgakakis 2012, Chen et al. 2013). Popper et al. (2011) presented a detailed review on the composition and characteristics of carbohydrate and structural components in microalgae. It was reported that many green algae species possess rigid cell walls, as seen in plants, mainly with cellulose and/or other biopolymers. Next to structural carbohydrates, few microalgal species, for instance red algae and dinoflagellates, accumulate significant amounts of carbohydrates in the form of energy storage molecules like starch. While most cyanobacteria stocks energy in the form of glycogen, species from brown algae and diatoms produce alginate, laminaran, mannitol, and fucoidin as food reserves (Yoo, Spalding, and Jane 2002). Similar to gram-negative bacteria, most cyanobacteria have a peptidoglycan layer and a cell envelope that is encased in a polysaccharide (Hoiczyk and Hansel 2000). Carbohydrates from microalgae show widely varying structures and biological activities and find commercial applications in cosmetics and skin-care products (Richmond and Hu 2013, Shrestha, St. Clair, and O'Neill 2015), emulsifiers, food industry, and medicine (Al-Fawwaz and Abdullah 2016). Few sulphated polysaccharides are reported to have remarkable medical applications (Berri et al. 2016, Brito et al. 2016).

The carbon partitioning explains the competition for 3-glyceraldehyde and pyruvate in carbohydrate and lipid syntheses along with the biosynthesis route to starch production. In the process of starch biosynthesis, active protein initiates complexation and ATP transports bicarbonate into the chloroplast stroma and then to the thylakoid lumen. The conversion of bicarbonate to CO_2 is catalysed in the lumen by CAs that are localized in the stroma. The oxygenase reaction of RuBisCo, which regulates the photorespiration, is inhibited by the internal accumulation of CO_2, resulting in the loss of CO_2 during the decarboxylation of serine in the mitochondria, which reduces the overall photosynthesis efficiency (Vuppaladadiyam et al. 2018a). It is important to note that RuBisCo is comparatively slow among natural enzymes and has a turnover rate of 2–10 CO_2 molecules s^{-1} (Sharwood, Ghannoum, and Whitney 2016). Its abundance in the cell is *ca.* 1000 times higher than reactive centres in PSI and PSII, which is one of the reasons why the overall rates of CO_2 fixation reach the maximum rates of electron transfer. In the past, the evolution of the O_2 release was considered as a measure of photosynthetic activity; however, higher photosynthetic activities did not show higher biomass productivity. This limitation is connected to the transportation of electrons and protons as well as the phosphorylation cycle. Due to photorespiration, it has been estimated that there has been a drop in the CO_2 fixation by *ca.* 25% of the maximum carboxylation efficiency. Various active strategies have been adopted by algae and higher terrestrial plants to increase the concentration of CO_2 near the active site of RuBisCo enzyme, e.g. carboxysomes and transporter complexes. However, they require additional ATP, and therefore, for active bicarbonate usage, the cyclic photophosphorylation is also an essential process, at least in green algae (Spreitzer and Salvucci 2002, Jia et al. 2008). A study on *Chlamydomonas* demonstrated that the accumulation of most proteins, such as NADH dehydrogenase, ATP synthases, and cytochrome c oxidase, related to oxidative phosphorylation was activated by nitrogen depletion (Baker and Ort 1992). This approach of nitrogen starvation appears to significantly impact on starch accumulation instead of TAG accumulation; however, it depends largely on the C:N ratio of the available nutrients in the cytosol of the algal cell.

As both carbohydrate biosynthesis and lipid biosynthesis compete for the same bulk of CO_2 fixed in the cell, Kumar et al. (2016) compared the kinetics and energy balance of metabolites. This study addressed the key question of which primary metabolite would be the preferred one in terms of carbon and energy balances for biofuel production. However, for accurate kinetic studies, enzymatic data is lacking and relying on the thermodynamics. They concluded that, for starch synthesis when compared to TAG synthesis per carbon atom fixed, the requirements of ATP and NAD(P)H are, respectively, 50% and 45% lower. This indicates that when compared to glucose-derived carbohydrates, TAG synthesis requires 53% higher energy input (Vuppaladadiyam et al. 2018). Therefore, an important question that arises here is: under stress conditions, why would microalgae prefer to store TAG rather than carbohydrates? The answer is evident when considering the energy recovered from these metabolites upon oxidation. Through both β-oxidation and the citric acid cycle oxidation pathways, saturated FA can generate 6.6 ATP equivalents per carbon, whereas through glucose oxidation via glycolysis and citric acid cycle, saturated FA can generate only 5 ATP equivalents. Additionally, regarding the energy density, the energy content per carbon of TAG is 41% higher than for starch (Vuppaladadiyam et al. 2018).

12.4.5 Polysaccharides

Polysaccharides are defined as glycosidic bond-linked polymers of monosaccharides (Rossi and De Philippis 2016). Polysaccharides from the microalgae show a strong structural variability and biological activities. The complexity associated with the monosaccharides, glycosidic linkages, and non-sugar substituents makes the structural investigation into polysaccharides more complicated. Exopolysaccharides, microalgal-produced biopolymers, are embedded in the cell walls (e.g. as in *Porphyridium* sp.), and it is often complicated to differentiate them from the cell wall (Arad and Levy-Ontman 2010). Although their functional role is unclear, few others hypothesized that these metabolic products avoid desiccation (Pignolet et al. 2013). Moreover, it is also believed that exopolysaccharides are responsible for mechanical protection, offering the cells a flexibility to grow in a wide range of environments (Arad and Levy-Ontman 2010). It was also identified that the formation of an algal colony is connected with the production of polysaccharides and is also catalysed by other factors like C:N ratio. A recent study on *S. obliquus* identified that the production of exopolysaccharides increased with the supplementation of glyoxylate (Liu et al. 2010, Shrestha, St. Clair, and O'Neill 2015).

Few microalgal species like *Porphyridium* sp., accumulate carbohydrates in large quantities, which make them commercially attractive. Since the cultivation of *Porphyridium* sp., is easy, and as they belong to a class of fastest-growing microalgae, they have attracted significant research interest. The polysaccharides produced from microalgae find their commercial applications in cosmetics and skin-care products (Richmond and Hu 2013). The medical activities and biological functions of the carbohydrates produced from microalgae have been intensively studied, and the findings like antiviral activities of carbohydrates guided the development of emollients to treat herpes virus (Richmond and Hu 2013, Arad and Levy-Ontman 2010, Shrestha, St. Clair, and O'Neill 2015). However, the high cost of microalgal cultivation limited the economic exploitation of these commercial values (Hardouin et al. 2016). The other applications of microalgal polysaccharides include their usage in food industries, fabrics, emulsifiers, stabilizers, and medicines (Al-Fawwaz and Abdullah 2016). Sulphated polysaccharides are polysaccharides with sulphate esters, which are produced in microalgae, and are used for unique medical applications. *C. vulgaris*, *Porphyridium* sp., and *Scenedesmus quadricauda* are few examples of microalgae that are capable of producing sulphated polysaccharides (Berri et al. 2016, Brito et al. 2016).

In view of human health, polysaccharides play a crucial role. It is to be noted that microalgal polysaccharides have antiviral and antimicrobial activities. Acidic extracellular sulphated polysaccharides produced by microalgae can be used as a therapeutic agent (de Jesus Raposo, de Morais, and de Morais 2014). It has been demonstrated that the higher the sulphate content, the higher the antiviral activities (de Jesus Raposo, de Morais, and de Morais 2014, Arad and Levy-Ontman 2010). The sulphated polysaccharides produced from *Spirulina* have already been identified as antiviral agents (de Jesus Raposo, de Morais, and de Morais 2013). Additionally, the antibacterial potential of the same species was established by many researchers (Medina-Jaritz, Carmona-Ugalde, and Lopez-Cedillo 2013). Polysaccharides were considered in their applications in the reconstruction of tissues and organs (Steffens et al. 2014, de Morais et al. 2010). They have a unique ability to activate macrophage function and thus modulate the immune system (de Morais et al. 2015, Sun et al. 2016). The anticancer potential of microalgal polysaccharides has been widely studied and reported (Rawat et al. 2016, Ruocco et al. 2016, Griffiths et al. 2016). It has been identified that an extracellular polysaccharide (GA3P) from toxic marine microalgae (*Gymnodinium* sp. A3) inhibited the growth of different cancer cells (Talero et al. 2015).

12.4.6 Minor Compounds

Carotenoids, chlorophylls, and phycobiliproteins are the pigments found in microalgae. They help microalgae in harvesting light energy and serve as antioxidants to repair the photosynthetic apparatus (Mohan et al. 2015). The photosystems capture the light spectrum and the pigments that are bind to the light-harvesting complex (LHC) expand the captured spectra to disperse the extra excitation energy from chlorophyll through non-photochemical quenching. The visible and UV irradiations are absorbed by the cells with the help of π-electron system, which is present in their central C40 backbone. Variations in carotenoid structures occur in the oxygen cyclic, and conjugated double-bond moieties (primary and secondary carotenoids). The oxygenic and cyclic modifications along with the number of conjugated double bonds are responsible for the production of a variety of carotenoids (Varela et al. 2015). Carotenoids protect the photosynthetic apparatus of the system against photo-oxidative damage. Few microalgae can accumulate these pigments in lipid globules under unfavourable conditions as seen in green algae *Haematococcus* and *Dunaliella* species in accumulating astaxanthin and β-carotene (Davidi et al. 2014). Adding to their importance in stabilizing photosystem and light harvesting, carotenoids also play a valuable role in plants and animals. In plants, carotenoids are responsible for attracting pollinators, and in animals, reproduction success depends on the intake of carotenoid-derived metabolites like vitamin A (Dugas, Yeager, and Richards-Zawacki 2013, Marri and Richner 2014). Furthermore, carotenoids are further classified into carotenes (unsaturated hydrocarbons) and xanthophylls (containing one or more oxygen-containing functional group) (Varela et al. 2015).

Carotenoids are used in food industry due to their potential as natural colour enhancers. Few of them, including lutein, β-carotene, astaxanthin, and canthaxanthin, are already commercially exploited. The demand for natural pigments has been increasing because it is suspected that synthetic colours may lead to carcinogenesis and retinal and liver toxicity (Vílchez et al. 2011). Furthermore, a combination of *trans* and *cis* isomers, which may have anticancer activity, is available with natural carotenoids. In view of their antioxidant properties, carotenoids (mainly β-carotene) can be used for the production of healthy foods. In addition, carotenoids can be applied in animal and aquaculture feed (Guedes, Amaro, and Malcata 2011). For human nutrition, few carotenoids have provitamin-A activity (Guiheneuf, Khan, and Phan Tran 2016). Along with flavonoids, carotenoids protect the low melanin-containing skin of humans from UV-induced sunburns (Varela et al. 2015, Marri and Richner 2014). Both retinal-containing rhodopsins and photoprotective carotenoids are crucial for vision in humans. As humans and most animals cannot synthesize carotenoids, and as carotenoids play a vital role in signal transduction and/or photoreception, low intake of dietary carotenoids may result in night blindness and xerophthalmia, and may lead to keratinization of the conjunctiva and cornea (Morales-Sánchez et al. 2015). The antioxidant activity and the associated health benefits of carotenoids have been confirmed by many researchers. In humans and animals, dietary carotenoids protect the lipophilic part from lipid peroxidation by scavenging reactive oxygen species (ROS) (Zhong et al. 2011, Christaki et al. 2013).

The other important bioactive compounds that possess commercial value and health benefits include lutein (Talero et al. 2015), vitamins (de Jesus Raposo, de Morais, and de Morais 2013), proteins and peptides (Talero et al. 2015, Kang and Kim 2013), polyketides (Sasso et al. 2012), phycobilins (Borowitzka 2013), stable isotope biochemicals, polyhydroxyalkanoates (Bhati and Mallick 2012), and phenolic compounds (de Jesus Raposo and de Morais 2015). Under stress conditions, lutein protects the cells of microalgae against ROS. Industrial applications of lutein are numerous; among them, its uses as feed additive and food colourant are the most important (Fernández-Sevilla, Fernández, and Grima 2010). Lutein is also believed to protect humans against age-related macular degeneration (Kula et al. 2014). High lutein content in microalgae (0.5%–1.2% dry weight) can be considered as the main reason to commercially exploit microalgae for lutein. Other beneficial effects of lutein include antioxidative (Fu et al. 2013, Lin, Lee, and Chang 2015, Nidhi et al. 2015), anti-inflammatory (Bian et al. 2012), and anticancer (Reynoso-Camacho et al. 2011) activities. Lutein along with zeaxanthin (β, β-carotene-3,3′-diol) accumulates in the central retina, and the studies identified that macular degeneration may result because of insufficient dietary intake of lutein and zeaxanthin or low serum zeaxanthin levels (Chen et al. 2012, Okuyama et al. 2014). Microalgal biomass is a rich source of vitamins such as vitamin A1, vitamin B1, vitamin B2, vitamin B6, vitamin B12, vitamin C, vitamin E, biotin, nicotinate, pantothenic acid, and folic acid (Vuppaladadiyam et al. 2018). Burja et al. (2001) reviewed the antibiotic and antimycotic activities of microalgal compounds. It was identified that vitamins B6 and B12 can minimize the risk of heart stroke and vitamin B12 and vitamin D may reduce the incidence of thrombosis and heart stroke (Burja et al. 2001). The reason behind this behaviour of these bioactive compounds is that they possess the potential to lower the levels of homocysteine, a compound associated with cardiovascular diseases and heart stroke. Vitamin D also helps in preventing cerebrovascular disease, and reduces hypertension and atherosclerosis (Foroughi et al. 2013). Vitamin E is an antioxidant in nature, and protects and repairs against atherosclerosis damage caused by free radicals (Apostolopoulou et al. 2012).

Some unique properties of proteins from marine microalgae include foam and film-forming capacity, antimicrobial activity, and gel-forming ability. Alongside these, other beneficial effects of proteins from microalgae include hepatoprotective, immunomodulating, anti-inflammatory, anticancer, and antioxidant properties (Vuppaladadiyam et al. 2018). Microalgal-derived peptides also possess few unique biological activities; anticancer (Razzak et al. 2013) and anti-inflammatory (Kim and Kim 2013) activities are the most important. Polyketides, due to their high commercial value, are recognized as the most important bioactive compounds produced by microalgal family that has a huge contribution to the pharmaceutical industry (Cardozo et al. 2007). These bioactive products find their applications as antibiotic, antifungal, and anti-coccidiosis agents. Furthermore, it was also identified that a majority of polyketides stored in the form of open-chain polyketides and polycyclic ether molecules are potentially toxic and are not suitable for human therapy (Kellmann et al. 2010). Phycobilins, mainly phycoerythrin and phycocyanin, are unique photosynthetic pigments because of their bond with water-soluble proteins to build phycobiliproteins. The most important function of these pigments is to forward the harvested light energy to chlorophylls for photosynthesis (Koller, Muhr, and Braunegg 2014). They are used as chemical tags in research because they bind phycobiliproteins to antibodies. Such microalgal-derived phycobiliproteins by far possess the highest market value (Parmar et al. 2011). Furthermore, these pigments found their applications at industrial level in cosmetics, dairy products, and as food colourants because of their high colouring effects (Arad and Yaron 1992).

Microalgae are the best possible source for stable isotope compounds (Vuppaladadiyam et al. 2018). Formation of the basis of culture media for mammalian, yeast, and bacteria cells is one of the major applications of the microalgal-derived organic compound, which is complex and isotopically stable (Suganya et al. 2016). Stable isotope compounds derived from microalgae have been used to explain various metabolic pathways (Cunnane and Likhodii 1996). Breath tests, to diagnose the dysfunction of organ and disease, can be done with the help of microalgal-derived stable isotopes. ^{13}C-galactose has been successfully demonstrated to test the functioning of liver (Suganya et al. 2016). The market value for stable isotopes derived from microalgae is assumed to be higher than US$ 13 million per annum (Adarme-Vega et al. 2012, Spolaore et al. 2006, Vílchez et al. 2011). Polyhydroxyalkanoates like poly-3-hydroxybutyrate (P3HB) are mainly produced in cyanobacteria such as *Synechocystis* and *Spirulina*, but considerably at a less quantity than those found in other bacterial species. Although the quantity of P3HB produced is less, cyanobacteria still play a major role in the production of biodegradable plastic. This can be done by implementing techniques such as mutagenesis and genetic engineering (Borowitzka 2013). According to Goiris et al. (2012), phenols are a category of phytochemicals that contribute to the antioxidant property of microalgae (Goiris et al. 2012). Flavonoids along with other phenol compounds participate in inhibiting the lipid oxidation. Phenolic acids such as caffeic and chlorogenic acids extracted from *Arthrospira* exhibit higher antioxidant activity in nature than other phenolic acids (de Jesus Raposo and de Morais 2015). Many researchers suggested that the occurrence of degenerative diseases can be minimized with the increase in the intake of phenols (Milledge, Nielsen, and Bailey 2016). However, the clinical evidence of humans who received phenolic compounds of microalgal origin as dietary supplementation is much inferior (de Jesus Raposo and de Morais 2015, Ku et al. 2013). Table 12.3 illustrates the impact of culture conditions and growth conditions on the production of various secondary metabolites.

12.5 FUTURE PROSPECTS

Microalgae can serve as economic and effective bioreactors for producing high value-added novel biologically active compounds and require inexpensive substrates for their growth. Moreover, the demand for nutritive food and health products increases with an increase in the population growth. The fast growth rate and a wide range of health benefits of algae make them a potential competitor and best source to meet the nutritive demands. In terms of usage, only a small portion of microalgal-derived food supplements are considered for humans. Research in investigating the unexplored algal strains will significantly enhance the usage of microalgae biomass to generate pharmaceuticals and nutraceuticals that benefit human health. Positive health effects of few metabolites derived from microalgae, like C-phycocyanin, are still not completely understood. Microalgae is an attractive option for future as a source for the production of complex eukaryotic proteins, at low cost and low risk of human pathogenic contamination, especially immunoglobulin G (IgG) antibodies, which are found to be secreted by algae into the culture medium. In view of commercial applications, the recombinant protein production from microalgae is still in its infancy, elucidating the importance of future research in this area. Furthermore, high cost incurred in clinical screening for commercial benefits of microalgae prevents the investigation and identification of few strains which may have commercial applications. Further research is expected to identify the methods to lower the cost of screening process. Additionally, the anticancer, anti-inflammatory, and antioxidant activities of bioactive molecules reported in microalgae can prevent the skin-related issues, making the scope of research in metabolites of microalgae widely open. Future research is expected to happen in the field of metabolic engineering of algae to generate novel compounds in pharmaceutical industry. According to the literature, the safety regulations for transgenic microalgae for different countries, transgenic algal culture at large scale for metabolic products, processing of algal cells, purification, quality control, and preclinical and clinical trials for ensuring quality and safety are few areas where extensive research is needed.

TABLE 12.3
Secondary Metabolites Production as a Function of Culture Conditions. Adapted from (Vuppaladadiyam et al. 2018a).

Microalgae	Metabolites	Cultivation Type (vol., L)	Intensity of Light ($\mu mol\ m^{-2}s^{-1}$)	Carbon Source	T (°C)	pH	Yields ($mg\ L^{-1}$)	Note
Chlorella vulgaris	Omega-3 FAs	EF (0.5)	72[a]	Glucose	23	-	-	Increase in the light intensity to values beyond 72 $\mu E\ m^{-2}s^{-1}$ did not increase the lipid production significantly.
Chlorococcum sp.	Astaxanthin	EF (0.5)	22[a]	-	25	8.4	7.1 ± 0.4[b]	H_2O_2 treatment induced the formation of *trans*-astaxanthin.
Chlorella zofingiensis	Astaxanthin	CF (0.2)	90	1*	28		1.8 ± 0.2	Total and specific carotenoids altered with light irradiance, temperature, and nutrient limitation.
			460				11.2 ± 1.0	
			920				19.0 ± 1.5	
	Lutein		90		28	-	17.3 ± 1.5	
			460		28		19.3 ± 1.8	
			920		28		24.7 ± 2.1	
Dunaliella salina	β-Carotene	EF (0.25)	70	2.5*	25	7.5	41[b]	High or low light intensities with nitrogen starvation led to increased β-carotene contents.
Dunaliella salina	Carotenoids	EF (0.25)	120	-	22	8.5	8.8[b]	Total carotenoids increased when cultures were treated with 10 $\mu mol\ L^{-1}$ methyl jasmonate.
Dunaliella salina	Zeaxanthin	PB (0.3)	255[a]	2.5*	25	6.5–7.5	0.5 ± 0.1[b]	β-Carotene and lutein accumulation when the red LED was replaced with blue LED.
	Lutein	PB (0.3)	128[a]	2.5*	25	6.5–7.5	5.5 ± 0.8[b]	
	Trans-β-carotene	PB (0.3)	128[a]	2.5*	25	6.5–7.5	4.4 ± 0.5[b]	

(*Continued*)

TABLE 12.3 (*Continued*)
Secondary Metabolites Production as a Function of Culture Conditions. Adapted from (Vuppaladadiyam et al. 2018a).

Microalgae	Metabolites	Cultivation Type (vol., L)	Intensity of Light ($\mu mol\ m^{-2}s^{-1}$)	Carbon Source	T (°C)	pH	Yields ($mg\ L^{-1}$)	Note
Eustigmatos cf. *polyphem*	β-Carotene	PB (0.3)	150/300	1*	23	7.5	450	β-Carotene production higher in bubble column bioreactor than in flat plate bioreactor.
H. pluvialis	Astaxanthin	PB (0.4)	108	6*	25	6.5–8.0	46[b]	Astaxanthin yield changed with high CO_2 concentration and light intensity.
Isochrysis galbana	Fucoxanthin	PB	Sunlight	-	-	-	-	Carotenoid profiles were species specific.
Porphyridium cruentum	Zeaxanthin	PB	Sunlight	-	-	-	-	Carotenoid profiles were species specific.
Pseudokirchneriella subcapitata	Omega-3 FAs	EF (0.5)	72[a]	Glucose	23	-	-	Increase in the light intensity to values beyond 72 $\mu E\ m^{-2}s^{-1}$ did not increase the lipid production significantly.
Phaeodactylum tricornutum	Fucoxanthin	PB	Sunlight	-	-	-	-	Carotenoid profiles were species specific.
Scenedesmus sp.	Lutein	PB (5.0)	300	3.5*	30	7	-	Microwave-assisted binary-phase solvent extraction favoured the recovery of highest lutein.
Spirulina platensis	C-phycocyanin	PB (1.0)	100	2.5*	28	9	12.6[c]	At 700 $\mu mol\ m^{-2}s^{-1}$, CO_2 and nitrogen consumption rates increased.
Spirulina sp.	Phycobiliprotein	EF (0.5)	48	6*	30	9	250[b]	Under nutrient starvation conditions, the potential of *Spirulina* sp. declined.

(*Continued*)

TABLE 12.3 (*Continued*)

Secondary Metabolites Production as a Function of Culture Conditions. Adapted from (Vuppaladadiyam et al. 2018a).

Microalgae	Metabolites	Cultivation Type (vol., L)	Intensity of Light ($\mu mol\ m^{-2}s^{-1}$)	Carbon Source	T (°C)	pH	Yields ($mg\ L^{-1}$)	Note
Tetraselmis suecica	Carotenoids	EF (0.25)	120	-	22	8.5	8.1[b]	Total carotenoids increased when cultures were treated with 70 $\mu mol\ L^{-1}$ salicylic acid.
T. suecica	Lutein, violaxanthin	PB	Sunlight	-	-	-		Carotenoid profiles were species specific.
Vischeria helvetica	Carotenoids, FAs	EF (0.5)	40	-	25	8.0	-	Carotenoids production can be increased by FA inhibitors.
Nannochloropsis gaditana	β-Carotene	PB	Sunlight	-	-	-		Carotenoid profiles were species specific.

Source: Adapted from Vuppaladadiyam et al. (2018).

CF, cylindrical flask; PB, photobioreactor; EF, Erlenmeyer flask.

*CO_2 (v/v %).

[a] $\mu E\ m^{-2}s^{-1}$.

[b] $mg\ g^{-1}$.

[c] dry wt%.

12.6 CONCLUSIONS

Microalgae is considered as important natural resource that currently contains commercial and valuable bioactive compounds. The content of these bioactive compounds within microalgae can be enhanced when they experience stress conditions, including change in temperature, pH, salinity, nutrient limitation, and composition of community. Moreover, as a pristine type of bioreactor, microalgae can generate recombinant products such as antibodies and vaccines. In this chapter, we outline a wide range of bioactive compounds particularly in view of economic and human health interest. The ability of producing bioactive compounds makes it an attractive tool for many sectors of biotechnology. Furthermore, toxins produced by few species, like cyanobacteria, make them potential candidates for extracting anticancer drugs. The capability to handle various diseases with bioactive compounds generated from microalgae is clearly elucidated in this chapter. With the advancements in new techniques for metabolic engineering and manipulations at molecular levels, advancements in the generation of novel products will be inevitably achieved and the potential of these microorganisms in the production of pharmaceuticals will be recognized in the near future.

REFERENCES

Abdul, Qudeer Ahmed, Ran Joo Choi, Hyun Ah Jung, and Jae Sue Choi. "Health benefit of fucosterol from marine algae: A review." *Journal of the Science of Food and Agriculture* 96, 6 (2016):1856–1866.

Aburai, Nobuhiro, and Abe Katsuya. "Metabolic switching: Synergistic induction of carotenogenesis in the aerial microalga, Vischeriahelvetica, under environmental stress conditions by inhibitors of fatty acid biosynthesis." *Journal of Biotechnology Letters* 37, 5 (2015): 1073–1080.

Adarme-Vega, T. Catalina, David K. Y. Lim, Matthew Timmins, Felicitas Vernen, Yan Li, and Peer M. Schenk. "Microalgal biofactories: A promising approach towards sustainable omega-3 fatty acid production." *Microbial Cell Factories* 11, 1 (2012): 1–10.

Ahmed, Faruq, Kent Fanning, Michael Netzel, and Peer M. Schenk. "Induced carotenoid accumulation in Dunaliella salina and Tetraselmis suecica by plant hormones and UV-C radiation." *Applied Microbiology and Biotechnology* 99, 22 (2015): 9407–9416.

Al-Fawwaz, Abdullah T, and Mufida Abdullah. "Decolorization of Methylene Blue and Malachite Green by Immobilized Desmodesmus sp. Isolated from North Jordan." *International Journal of Environmental Science and Development* 7, 2 (2016): 95.

Amoroso, Gabi, Dieter Sültemeyer, Christoph Thyssen, and Heinrich P Fock. "Uptake of HCO3– and CO_2 in cells and chloroplasts from the microalgae *Chlamydomonas reinhardtii* and *Dunaliella tertiolecta*." *Plant Physiology* 116, 1 (1998): 193–201.

Apostolopoulou, Martha, Konstantinos Michalakis, Alexander Miras, Apostolos Hatzitolios, and Christos Savopoulos. "Nutrition in the primary and secondary prevention of stroke." *Maturitas* 72, 2 (2012): 29–34.

Arad, S., and A. Yaron. "Natural pigments from red microalgae for use in foods and cosmetics." *Trends in Food Science & Technology* 3 (1992):92–97.

Arad, Shoshana Malis, and Oshrat Levy-Ontman. "Red microalgal cell-wall polysaccharides: Biotechnological aspects." *Current Opinion in Biotechnology* 21, 3 (2010): 358–364.

Armbrust, E Virginia, John A Berges, Chris Bowler, et al. "The genome of the diatom Thalassiosira pseudonana: Ecology, evolution, and metabolism." *Science* 306, 5693 (2004): 79–86.

Badger, Murray R, T John Andrews, SM Whitney, et al. "The diversity and coevolution of Rubisco, plastids, pyrenoids, and chloroplast-based CO_2-concentrating mechanisms in algae." *Canadian Journal of Botany Price* 76, 6 (1998): 1052–1071.

Badger, Murray R, and G Dean, Price. "The role of carbonic anhydrase in photosynthesis." *Annual Review of Plant Biology* 45, 1 (1994): 369–392.

Badger, Murray R, and G Dean, Price. "The CO_2 concentrating mechanism in cyanobactiria and microalgae." *Journal of Physiologia Plantarum* 84, 4 (1992): 606–615.

Baker, NR, and DR Ort. 1992. "Light and crop photosynthetic performance." *Topics in Photosynthesis* 12: 289–312.

Barry, A, A Wolfe, C English, C Ruddick, and D Lambert. 2016. National Algal Biofuels Technology Review. edited by U.S. Department of Energy. *Office of Energy Efficiency and Renewable Energy*. Bioenergy Technologies Office, 1–212.

Bassi, Amarjeet, Priyanka Saxena, and Ana-Maria Aguirre. 2014. Mixotrophic algae cultivation for energy production and other applications. In *Algal biorefineries*, 177–202. Dordrecht: Springer:.

Batista, Ana Paula, Luísa Gouveia, Narcisa M Bandarra, José M Franco, and Anabela Raymundo. "Comparison of microalgal biomass profiles as novel functional ingredient for food products." *Algal Research* 2, 2 (2013): 164–173.

Beardall, John, and John A. Raven. 2001. *Algal Metabolism*. In eLS. John Wiley & Sons, Ltd.

Berri, Mustapha, Cindy Slugocki, Michel Olivier, et al. "Marine-sulfated polysaccharides extract of Ulva armoricana green algae exhibits an antimicrobial activity and stimulates cytokine expression by intestinal epithelial cells." *Journal of Applied Phycology* (2016): 1–10.

Bhati, Ranjana, and Nirupama Mallick. "Production and characterization of poly (3-hydroxybutyrate-co-3-hydroxyvalerate) co-polymer by a N2-fixing cyanobacterium, *Nostoc muscorum* Agardh." *Journal of Chemical Technology and Biotechnology* 87, 4 (2012): 505–512.

Bhosale, Rahul A, MP Rajabhoj, and BB Chaugule. "*Dunaliella salina* Teod. as a prominent source of eicosapentaenoic acid." *International Journal on Algae* 12, 2 (2010): 185–189.

Bian, Qingning, Shasha Gao, Jilin Zhou, et al. "Lutein and zeaxanthin supplementation reduces photooxidative damage and modulates the expression of inflammation-related genes in retinal pigment epithelial cells." *Free Radical Biology and Medicine* 53, 6 (2012): 1298–1307.

Bigogno, Chiara, Inna Khozin-Goldberg, and Zvi Cohen. "Accumulation of arachidonic acid-rich triacylglycerols in the microalga *Parietochloris incisa* (Trebuxiophyceae, Chlorophyta)." *Phytochemistry* 60, 2 (2002): 135–143.

Borowitzka, Michael A. 2013. "High-value products from microalgae—their development and commercialisation." *Journal of Applied Phycology* 25 (3): 743–756.

Bozzo, Gale G, Brian Colman, and Yusuke Matsuda. "Active transport of CO_2 and bicarbonate is induced in response to external CO_2 concentration in the green alga *Chlorella kessleri*." *Journal of Experimental Botany* 51, 349 (2000): 1341–1348.

Brito, Tarcisio V, Francisco CN Barros, Renan O Silva, et al. "Sulfated polysaccharide from the marine algae *Hypnea musciformis* inhibits TNBS-induced intestinal damage in rats." *Carbohydrate Polymers* 151(2016): 957–964.

Burja, Adam M, Bernard Banaigs, Eliane Abou-Mansour, J Grant Burgess, and Phillip C Wright. "Marine cyanobacteria—a prolific source of natural products." *Tetrahedron* 57, 46 (2001): 9347–9377.

Burrows, Elizabeth H, Nicholas B Bennette, Damian Carrieri et al. "Dynamics of lipid biosynthesis and redistribution in the marine diatom *Phaeodactylum tricornutum* under nitrate deprivation." *BioEnergy Research* 5, 4 (2012): 876–885.

Camerini, Felipe, Michele Greque de Morais, Bruna da Silva Vaz, Etiele Greque de Morais, and Jorge Alberto Vieira Costa. "Biofixation of CO_2 on a pilot scale: Scaling of the process for industrial application." *African Journal of Microbiology Research* 10, 21 (2016): 768–774.

CapitalEnergy. http://capitalenergy.biz/?p=16989, (accessed July 2019).

Cardozo, Karina HM, Thais Guaratini, Marcelo P Barros, Vanessa R Falcão, Angela P Tonon, Norberto P Lopes, Sara Campos, Moacir A Torres, Anderson O Souza, and Pio Colepicolo. "Metabolites from algae with economical impact." *Comparative Biochemistry and Physiology Part C: Toxicology & Pharmacology* 146, 1 (2007): 60–78.

Caroprese, Mariangela, Marzia Albenzio, Maria Giovanna Ciliberti, Matteo Francavilla, and Agostino Sevi. "A mixture of phytosterols from *Dunaliella tertiolecta* affects proliferation of peripheral blood mononuclear cells and cytokine production in sheep." *Veterinary immunology and immunopathology* 150, 1 (2002): 27–35.

Carvalho, Ana P, Susana O Silva, José M Baptista, and F Xavier Malcata. "Light requirements in microalgal photobioreactors: An overview of biophotonic aspects." *Applied Microbiology and Biotechnology* 89, 5 (2011): 1275–1288.

Chen, Chao-Rui, Siang-En Hong, Yuan-Chuen Wang, Shih-Lan Hsu, Daina Hsiang, and Chieh-Ming J Chang. "Preparation of highly pure zeaxanthin particles from sea water-cultivated microalgae using supercritical anti-solvent recrystallization." *Bioresource Technology* 104 (2012): 828–831.

Chen, Chun-Yen, Xin-Qing Zhao, Hong-Wei Yen, et al. "Microalgae-based carbohydrates for biofuel production." *Biochemical Engineering Journal* 78 (2013): 1–10.

Chen, Jingnan, Rui Jiao, Yue Jiang, Yanlan Bi, and Zhen-Yu Chen. "Algal Sterols are as effective as β-sitosterol in reducing plasma cholesterol concentration." *Journal of Agricultural and Food Chemistry* 62, 3 (2014): 675–681.

Chen, Mianrun, Hongbin Liu, and Bingzhang Chen. "Effects of dietary essential fatty acids on reproduction rates of a subtropical calanoid copepod, *Acartia erythraea*." *Marine Ecology Progress Series* 455 (2012): 95–110.

Chen, Xi, Stuart G Wakeham, and Nicholas S Fisher. "Influence of iron on fatty acid and sterol composition of marine phytoplankton and copepod consumers." *Limnology and Oceanography* 56, 2 (2011): 716.

Christaki, Efterpi, Eleftherios Bonos, Ilias Giannenas, and Panagiota Florou-Paneri. "Functional properties of carotenoids originating from algae." *Journal of the Science of Food and Agriculture* 93, 1 (2013): 5–11. doi: 10.1002/jsfa.5902.

Colman, B, and C Rotatore. "Photosynthetic inorganic carbon uptake and accumulation in two marine diatoms." *Oceanographic Literature Review* 2, 43 (1996): 131–132.

Colman, Brian, and Katharina A Gehl. "Physiological characteristics of photosynthesis in *Porphyridium cruentum*: Evidence for bicarbonate transport in a unicellular red algal." *Journal of Phycology* 19, 2 (1993): 216–219.

Colman, Brian, I Emma Huertas, Shabana Bhatti, and Jeffrey S Dason. "The diversity of inorganic carbon acquisition mechanisms in eukaryotic microalgae." *Functional Plant Biology* 29, 3 (2002): 261–270.

Cuellar-Bermudez, S. P., I. Aguilar-Hernandez, D. L. Cardenas-Chavez, N. Ornelas-Soto, M. A. Romero-Ogawa, and R. Parra-Saldivar. "Extraction and purification of high-value metabolites from microalgae: Essential lipids, astaxanthin and phycobiliproteins." *Microbial Biotechnology* 8, 2 (2015): 190–209. doi: 10.1111/1751-7915.12167.

Cuellar-Bermudez, Sara P., Jonathan S. Garcia-Perez, Bruce E. Rittmann, and Roberto Parra-Saldivar. "Photosynthetic bioenergy utilizing CO_2: An approach on flue gases utilization for third generation biofuels." *Journal of Cleaner Production* 98 (2015): 53–65. doi: http://dx.doi.org/10.1016/j.jclepro.2014.03.034.

Cunnane, Stephen C, and Sergei S Likhodii. "13C NMR spectroscopy and gas chromatograph-combustion-isotope ratio mass spectrometry: Complementary applications in monitoring the metabolism of 13C-labelled polyunsaturated fatty acids." *Canadian Journal of Physiology and Pharmacology* 74, 6 (1996): 761–768.

Davidi, Lital, Eyal Shimoni, Inna Khozin-Goldberg, Ada Zamir, and Uri Pick. "Origin of β-carotene-rich plastoglobuli in Dunaliella bardawil." *Plant physiology* 164, 4 (2014): 2139–2156.

de Jesus Raposo, Maria Filomena, Alcina Maria Miranda, Bernardo de Morais, and Rui Manuel Santos Costa de Morais. "Influence of sulphate on the composition and antibacterial and antiviral properties of the exopolysaccharide from *Porphyridium cruentum*." *Life Sciences* 101, 1 (2014): 56–63.

de Jesus Raposo, Maria Filomena, Rui Manuel Santos Costa de Morais, and Alcina Maria Miranda Bernardo de Morais. "Health applications of bioactive compounds from marine microalgae." *Life Sciences* 93, 15 (2013): 479–486.

de Jesus Raposo, Maria Filomena, and Alcina Maria Miranda Bernardo de Morais. "Microalgae for the prevention of cardiovascular disease and stroke." *Life Sciences* 125 (2015): 32–41.

de Jesus, Sérgio S., and Rubens Maciel Filho. "Potential of algal biofuel production in a hybrid photobioreactor." *Chemical Engineering Science* 171 (2017): 282–292.

de Morais, Michele Greque, Christopher Stillings, Roland Dersch, Markus Rudisile, Patrícia Pranke, Jorge Alberto Vieira Costa, and Joachim Wendorff. "Preparation of nanofibers containing the microalga Spirulina (Arthrospira)." *Bioresource Technology* 101, 8 (2010): 2872–2876.

de Morais, Michele Greque, Bruna da Silva Vaz, Etiele Greque de Morais, and Jorge Alberto Vieira Costa. "Biologically active metabolites synthesized by microalgae." *BioMed Research International* 2015 (2015).

Demirbas, Ayhan, and M. Fatih Demirbas. "Importance of algae oil as a source of biodiesel." *Energy Conversion and Management* 52, 1 (2011): 163–170.

Di Lena, Gabriella, Irene Casini, Massimo Lucarini, and Ginevra Lombardi-Boccia. "Carotenoid profiling of five microalgae species from large-scale production." *Food Research International* 120 (2019): 810–818.

Dinamarca, Jorge, Orly Levitan, G. Kenchappa Kumaraswamy, Desmond S. Lun, and Paul G. Falkowski. "Overexpression of a diacylglycerol acyltransferase gene in *Phaeodactylum tricornutum* directs carbon towards lipid biosynthesis." *Journal of phycology* 53, 2 (2017): 405–414.

Droop, M. R. "Heterotrophy of carbon." *Algal Physiology and Biochemistry* (1974): 530–559.

Dugas, Matthew B., Justin Yeager, and Corinne L. Richards-Zawacki. "Carotenoid supplementation enhances reproductive success in captive strawberry poison frogs (Oophaga pumilio)." *Zoo Biology* 32, 6 (2013): 655–658.

Elzenga, J. Theo M., Hidde BA Prins, and Jacqueline Stefels. "The role of extracellular carbonic anhydrase activity in inorganic carbon utilization of *Phaeocystis globosa* (Prymnesiophyceae): A comparison with other marine algae using the isotopic disequilibrium technique." *Limnology and Oceanography* 45, 2 (2000): 372–380.

Fan, Jilian, Chengshi Yan, Carl Andre, John Shanklin, Jörg Schwender, and Changcheng Xu. "Oil accumulation is controlled by carbon precursor supply for fatty acid synthesis in *Chlamydomonas reinhardtii*." *Plant and Cell Physiology* 53, no. 8 (2012): 1380–1390.

Fernández-Sevilla, José M., FG Acién Fernández, and E. Molina Grima. "Biotechnological production of lutein and its applications." *Applied Microbiology and Biotechnology* 86, 1 (2010): 27–40.

Fernández, FG Acién, JM Fernández Sevilla, and E Molina Grima. "Photobioreactors for the production of microalgae." *Reviews in Environmental Science and Bio/Technology* 12, 2 (2013):131–151.

Foroughi, Mehdi, Mohsen Akhavanzanjani, Zahra Maghsoudi, Reza Ghiasvand, Fariborz Khorvash, and Gholamreza Askari. "Stroke and nutrition: A review of studies." *International Journal of Preventive Medicine* 4, Suppl 2 (2013): S165.

Fraeye, Ilse, Charlotte Bruneel, Charlotte Lemahieu, Johan Buyse, Koenraad Muylaert, and Imogen Foubert. "Dietary enrichment of eggs with omega-3 fatty acids: A review." *Food Research International* 48, 2 (2012): 961–969.

Fu, Weiqi, Ólafur Guðmundsson, Giuseppe Paglia, et al. "Enhancement of carotenoid biosynthesis in the green microalga *Dunaliella salina* with light-emitting diodes and adaptive laboratory evolution." *Applied Microbiology and Biotechnology* 97, 6 (2013): 2395–2403.

Garcia-Maraver, Angela, Jose A. Perez-Jimenez, Francisco Serrano-Bernardo, and Montserrat Zamorano. "Determination and comparison of combustion kinetics parameters of agricultural biomass from olive trees." *Renewable Energy* 83, (2015):897-904..

Gao, Yihe, Chapin Gregor, Yuanjie Liang, Dawei Tang, and Caitlin Tweed. "Algae biodiesel-a feasibility report." *Chemistry Central Journal* 6, no. 1 (2012): S1–16.

Global CCS Institute. https://hub.globalccsinstitute.com/publications/novel-co2-capture-taskforce-report/211-algal-bio-sequestration, (accessed July 2017).

Goiris, Koen, Koenraad Muylaert, Ilse Fraeye, Imogen Foubert, Jos De Brabanter, and Luc De Cooman. "Antioxidant potential of microalgae in relation to their phenolic and carotenoid content." *Journal of Applied Phycology* 24, 6 (2012): 1477–1486.

Gonçalves, Ana L., José CM Pires, and Manuel Simões. "Lipid production of *Chlorella vulgaris* and *Pseudokirchneriella subcapitata*." *International Journal of Energy and Environmental Engineering* 4, 1 (2013): 14.

Griffiths, Melinda, Susan TL Harrison, Monique Smit, and Dheepak Maharajh. "Major commercial products from micro-and macroalgae." In *Algae Biotechnology*, 269–300. Springer, Cham, 2016.

Guarnieri, Michael T., Ambarish Nag, Sharon L. Smolinski, Al Darzins, Michael Seibert, and Philip T. Pienkos. "Examination of triacylglycerol biosynthetic pathways via de novo transcriptomic and proteomic analyses in an unsequenced microalga." *PloS one* 6, 10 (2011).

Guedes, Ana Catarina, Helena M. Amaro, and Francisco Xavier Malcata. "Microalgae as sources of carotenoids." *Marine Drugs* 9, 4 (2011): 625–644.

Guihéneuf, Freddy, Asif Khan, and Lam-Son P. Tran. "Genetic engineering: A promising tool to engender physiological, biochemical, and molecular stress resilience in green microalgae." *Frontiers in Plant Science* 7 (2016): 400.

Haneklaus, Silvia, C. W. Kerr, Ewald Schnug, K. Saito, L. J. De Kok, I. Stulen, M. J. Hawkesford, A. Sirko, and H. Rennenberg. "A chronicle of sulfur research in agriculture." *Sulfur Transport and Assimilation in Plants in the Postgenomic Era. Leiden, The Netherlands: Backhuys Publishers* (2005): 249–256.

Hardouin, Kevin, Gilles Bedoux, Anne-Sophie Burlot, Claire Donnay-Moreno, Jean-Pascal Bergé, Pi Nyvall-Collén, and Nathalie Bourgougnon. "Enzyme-assisted extraction (EAE) for the production of antiviral and antioxidant extracts from the green seaweed Ulva armoricana (Ulvales, Ulvophyceae)." *Algal Research* 16 (2016): 233–239.

Ho, Shih-Hsin, Wen-Ming Chen, and Jo-Shu Chang. "*Scenedesmus obliquus* CNW-N as a potential candidate for CO_2 mitigation and biodiesel production." *Bioresource Technology* 101, no. 22 (2010): 8725–8730.

Hoiczyk, Egbert, and Alfred Hansel. "Cyanobacterial cell walls: News from an unusual prokaryotic envelope." *Journal of Bacteriology* 182, 5 (2000): 1191–1199.

Hu, Bing, Min Min, Wenguang Zhou, Yecong Li, Michael Mohr, Yanling Cheng, Hanwu Lei et al. "Influence of exogenous CO_2 on biomass and lipid accumulation of microalgae *Auxenochlorella protothecoides* cultivated in concentrated municipal wastewater." *Applied Biochemistry and Biotechnology* 166, 7 (2012): 1661–1673.

Huertas, I. Emma, Brian Colman, and George S. Espie. "Inorganic carbon acquisition and its energization in eustigmatophyte algae." *Functional Plant Biology* 29, 3 (2002): 271–277.

Huertas, I. Emma, Brian Colman, George S. Espie, and Luis M. Lubian. "Active transport of CO_2 by three species of marine microalgae." *Journal of Phycology* 36, 2 (2000): 314–320.

Huertas, I. Emma, George S. Espie, Brian Colman, and Luis M. Lubian. "Light-dependent bicarbonate uptake and CO_2 efflux in the marine microalga Nannochloropsis gaditana." *Planta* 211, 1 (2000): 43–49.

Hügler, Michael, and Stefan M. Sievert. "Beyond the Calvin cycle: Autotrophic carbon fixation in the ocean." *Annual Review of Marine Science* 3 (2011): 261–289.

Ibáñez-Salazar, Alejandro, Sergio Rosales-Mendoza, Alejandro Rocha-Uribe, Jocelín Itzel et al. "Over-expression of Dof-type transcription factor increases lipid production in *Chlamydomonas reinhardtii*." *Journal of Biotechnology* 184 (2014): 27–38.

Ikawa, Miyoshi. "Algal polyunsaturated fatty acids and effects on plankton ecology and other organisms." 6, 2 (2004): 17–44.

Im, Chung-Soon, Diana Vincent, Rika Regentin, and Anna Coragliotti. "Heterotrophic cultivation of hydrocarbon-producing microalgae." U.S. Patent 8,956,852, issued February 17, 2015.

Jaiswal, Pranita, Radha Prasanna, and Ajai Kumar Kashyap. "Modulation of carbonic anhydrase activity in two nitrogen fixing cyanobacteria, Nostoc calcicola and Anabaena sp." *Journal of Plant Physiology* 162, 10 (2005): 1087–1094.

Jeyasingh, Punidan D., Jared M. Goos, Seth K. Thompson, Casey M. Godwin, and James B. Cotner. "Ecological stoichiometry beyond redfield: An ionomic perspective on elemental homeostasis." *Frontiers in Microbiology* 8 (2017): 722.

Jia, Husen, Riichi Oguchi, Alexander B. Hope, James Barber, and Wah Soon Chow. "Differential effects of severe water stress on linear and cyclic electron fluxes through photosystem I in spinach leaf discs in CO_2-enriched air." *Planta* 228, 5 (2008): 803–812.

Jia, Jing, Danxiang Han, Henri G. Gerken, Yantao Li, Milton Sommerfeld, Qiang Hu, and Jian Xu. "Molecular mechanisms for photosynthetic carbon partitioning into storage neutral lipids in Nannochloropsis oceanica under nitrogen-depletion conditions." *Algal Research* 7 (2015): 66–77.

Jiang, Yuelu, Tomomi Yoshida, and Antonietta Quigg. "Photosynthetic performance, lipid production and biomass composition in response to nitrogen limitation in marine microalgae." *Plant Physiology and Biochemistry* 54 (2012): 70–77.

Jiménez, Carlos, Belén R. Cossío, Diego Labella, and F. Xavier Niell. "The feasibility of industrial production of Spirulina (Arthrospira) in Southern Spain." *Aquaculture* 217, 1–4 (2003): 179–190.

Jung, Hyun Ah, Seong Eun Jin, Bo Ra Ahn, Chan Mi Lee, and Jae Sue Choi. "Anti-inflammatory activity of edible brown alga Eisenia bicyclis and its constituents fucosterol and phlorotannins in LPS-stimulated RAW264. 7 macrophages." *Food and Chemical Toxicology* 59 (2013): 199–206.

Kang, Kyong-Hwa, and Se-Kwon Kim. "Beneficial effect of peptides from microalgae on anticancer." *Current Protein & Peptide Science* 14, 3 (2013): 212–217.

Kang, Nam Kyu, Seungjib Jeon, Sohee Kwon, Hyun Gi Koh, et al. "Effects of overexpression of a bHLH transcription factor on biomass and lipid production in Nannochloropsis salina." *Biotechnology for Biofuels* 8, no. 1 (2015): 200.

Kellmann, Ralf, Anke Stüken, Russell JS Orr, Helene M. Svendsen, and Kjetill S. Jakobsen. "Biosynthesis and molecular genetics of polyketides in marine dinoflagellates." *Marine Drugs* 8, 4 (2010): 1011–1048.

Khan, Shakeel A., Mir Z. Hussain, S. Prasad, and U. C. Banerjee. "Prospects of biodiesel production from microalgae in India." *Renewable and Sustainable Energy Reviews* 13, 9 (2009): 2361–2372.

Khozin-Goldberg, I., Z. Cohen, M. Pimenta-Leibowitz, J. Nechev, and D. Zilberg. "Feeding with arachidonic acid-rich triacylglycerols from the microalga *Parietochloris incisa* improved recovery of guppies from infection with Tetrahymena sp." *Aquaculture* 255, 1–4 (2006): 142–150.

Khozin-Goldberg, Inna, Chiara Bigogno, Pushkar Shrestha, and Zvi Cohen. "Nitrogen starvation induces the accumulation of arachidonic acid in the freshwater green alga *Parietochloris incisa* (Trebuxiophyceae) 1." *Journal of Phycology* 38, 5 (2002): 991–994.

Kim, J., & Kim, S. K. (2013). "Bioactive peptides from marine sources as potential anti-inflammatory therapeutics." *Current Protein and Peptide Science*, 14 (3), 177–182.

Klinthong, Worasaung, Yi-Hung Yang, Chih-Hung Huang, and Chung-Sung Tan. "A review: Microalgae and their applications in CO_2 capture and renewable energy." *Aerosol Air Quality Research* 15, 2 (2015): 712–742.

Koller, M., A. Muhr, and G. Braunegg. "Microalgae as versatile cellular factories for valued products." *Algal Research* 6 (Part A) (2014): 52–63.

Kong, Wei-Bao, Hao Song, Shao-Feng Hua, Hong Yang, Qi Yang, and Chun-Gu Xia. "Enhancement of biomass and hydrocarbon productivities of *Botryococcus braunii* by mixotrophic cultivation and its application in brewery wastewater treatment." *African Journal of Microbiol Research* 6 (2012): 1489–1496.

Korb, Rebecca E., Peter J. Saville, Andrew M. Johnston, and John A. Raven. "Sources of inorganic carbon for photosynthesis by three species of marine diatom." *Journal of Phycology* 33, 3 (1997): 433–440.

Ku, Chai Siah, Yue Yang, Youngki Park, and Jiyoung Lee. "Health benefits of blue-green algae: Prevention of cardiovascular disease and nonalcoholic fatty liver disease." *Journal of Medicinal Food* 16, 2 (2013): 103–111.

Kula, Monika, Magdalena Rys, Katarzyna Możdżeń, and Andrzej Skoczowski. "Metabolic activity, the chemical composition of biomass and photosynthetic activity of *Chlorella vulgaris* under different light spectra in photobioreactors." *Engineering in Life Sciences* 14, 1 (2014): 57–67.

Kumar, Dhananjay, Lalit K. Pandey, and J. P. Gaur. "Metal sorption by algal biomass: From batch to continuous system." *Algal Research* 18 (2016): 95–109.Lam, Man Kee, Keat Teong Lee, and Abdul Rahman Mohamed. "Current status and challenges on microalgae-based carbon capture." *International Journal of Greenhouse Gas Control* 10 (2012): 456–469.

Leblond, Jeffrey D., Hermina Ilea Timofte, Shannon A. Roche, and Nicole M. Porter. "Sterols of glaucocystophytes." *Phycological Research* 59, 2 (2011): 129–134.

Lee, Sanghyun, Yeon Sil Lee, Sang Hoon Jung, Sam Sik Kang, and Kuk Hyun Shin. "Anti-oxidant activities of fucosterol from the marine algae Pelvetia siliquosa." *Archives of Pharmacal Research* 26, 9 (2003): 719–722.

Lenihan-Geels, Georgia, Karen S. Bishop, and Lynnette R. Ferguson. "Alternative sources of omega-3 fats: Can we find a sustainable substitute for fish?" *Nutrients* 5, 4 (2013): 1301–1315.

Li, Yantao, Danxiang Han, Guongrong Hu, David Dauvillee, Milton Sommerfeld, Steven Ball, and Qiang Hu. "Chlamydomonas starchless mutant defective in ADP-glucose pyrophosphorylase hyper-accumulates triacylglycerol." *Metabolic Engineering* 12, 4 (2010): 387–391.

Lin, Jian-Hao, Duu-Jong Lee, and Jo-Shu Chang. "Lutein production from biomass: Marigold flowers versus microalgae." *Bioresource Technology* 184 (2015): 421–428.

Liu, Jin, Zheng Sun, and Feng Chen. "Heterotrophic production of algal oils." In *Biofuels from Algae*, pp. 111–142. London, UK: Elsevier, 2014.
Liu, Ying, Wei Wang, Min Zhang, Peng Xing, and Zhou Yang. "PSII-efficiency, polysaccharide production, and phenotypic plasticity of *Scenedesmus obliquus* in response to changes in metabolic carbon flux." *Biochemical Systematics and Ecology* 38, 3 (2010): 292–299.
Low, K. L., A. Idris, and N. Mohd Yusof. "Novel protocol optimized for microalgae lutein used as food additives." *Food Chemistry* 307 (2020): 125631.
Lowrey, Joshua, Roberto E. Armenta, and Marianne S. Brooks. "Nutrient and media recycling in heterotrophic microalgae cultures." *Applied Microbiology and Biotechnology* 100, 3 (2016): 1061–1075.
Markou, Giorgos, Irini Angelidaki, and Dimitris Georgakakis. "Microalgal carbohydrates: An overview of the factors influencing carbohydrates production, and of main bioconversion technologies for production of biofuels." *Applied Microbiology and Biotechnology* 96, 3 (2012): 631–645.
Marri, Viviana, and Heinz Richner. "Yolk carotenoids increase fledging success in great tit nestlings." *Oecologia* 176, 2 (2014): 371–377.
Martin-Creuzburg, Dominik, Alexander Wacker, and Timo Basena. "Interactions between limiting nutrients: Consequences for somatic and population growth of Daphnia magna." *Limnology and Oceanography* 55, 6 (2010): 2597–2607.
Medina-Jaritz, Nora B., Luis Fernando Carmona-Ugalde, Julio Cesar Lopez-Cedillo, and F. Sandra L. Ruiloba-De Leon "Antibacterial activity of methanolic extracts from *Dunaliella salina* and *Chlorella vulgaris*." *The FASEB Journal* 27, 1 (2013): 1167–1165.
Milledge, John J., Birthe V. Nielsen, and David Bailey. "High-value products from macroalgae: The potential uses of the invasive brown seaweed, *Sargassum muticum*." *Reviews in Environmental Science and Bio/Technology* 15, 1 (2016): 67–88.
Miller, Rachel, Guangxi Wu, Rahul R. Deshpande, Astrid Vieler, Katrin Gärtner, Xiaobo Li, Eric R. Moellering et al. "Changes in transcript abundance in *Chlamydomonas reinhardtii* following nitrogen deprivation predict diversion of metabolism." *Plant Physiology* 154, 4 (2010): 1737–1752.
Min, Min, Bing Hu, Wenguang Zhou, Yecong Li, Paul Chen, and Roger Ruan. "Mutual influence of light and CO_2 on carbon sequestration via cultivating mixotrophic alga *Auxenochlorella protothecoides* UMN280 in an organic carbon-rich wastewater." *Journal of Applied Phycology* 24, 5 (2012): 1099–1105.
Mohan, S. Venkata, M. V. Rohit, P. Chiranjeevi, Rashmi Chandra, and B. Navaneeth. "Heterotrophic microalgae cultivation to synergize biodiesel production with waste remediation: Progress and perspectives." *Bioresource Technology* 184 (2015): 169–178.
Mondal, Madhumanti, Saumyakanti Khanra, O. N. Tiwari, K. Gayen, and G. N. Halder. "Role of carbonic anhydrase on the way to biological carbon capture through microalgae—a mini review." *Environmental Progress & Sustainable Energy* 35, 6 (2016): 1605–1615.
Montechiaro, Federico, Carol J. Hirschmugl, John A. Raven, and Mario Giordano. "Homeostasis of cell composition during prolonged darkness." *Plant, Cell & Environment* 29, 12 (2006): 2198–2204.
Morales-Sánchez, Daniela, Oscar A. Martinez-Rodriguez, John Kyndt, and Alfredo Martinez. "Heterotrophic growth of microalgae: Metabolic aspects." *World Journal of Microbiology and Biotechnology* 31, 1 (2015): 1–9.
Moroney, James V., Nadine Jungnick, Robert J. DiMario, and David J. Longstreth. "Photorespiration and carbon concentrating mechanisms: Two adaptations to high O_2, low CO_2 conditions." *Photosynthesis Research* 117, 1–3 (2013): 121–131.
Moroney, J. V., S. G. Bartlett, and Göran Samuelsson. "Carbonic anhydrases in plants and algae." *Plant, Cell & Environment* 24, 2 (2001): 141–153.

Nidhi, Bhatiwada, Gurunathan Sharavana, Talahalli R. Ramaprasad, and Baskaran Vallikannan. "Lutein derived fragments exhibit higher antioxidant and anti-inflammatory properties than lutein in lipopolysaccharide induced inflammation in rats." *Food & Function* 6, 2 (2015): 450–460.

Okuyama, Yusuke, Kotaro Ozasa, Keiichi Oki, Hoyoku Nishino, Sotaro Fujimoto, and Yoshiyuki Watanabe. "Inverse associations between serum concentrations of zeaxanthin and other carotenoids and colorectal neoplasm in Japanese." *International Journal of Clinical Oncology* 19, 1 (2014): 87–97.

Packer, Aaron, Yantao Li, Tom Andersen, Qiang Hu, Yang Kuang, and Milton Sommerfeld. "Growth and neutral lipid synthesis in green microalgae: A mathematical model." *Bioresource Technology* 102, 1 (2011): 111–117.

Palmqvist, Kristin, Dieter Sültemeyer, Pierre Baldet, T. John Andrews, and Murray R. Badger. "Characterisation of inorganic carbon fluxes, carbonic anhydrase (s) and ribulose-1, 5-biphosphate carboxylase-oxygenase in the green unicellular alga Coccomyxa." *Planta* 197, 2 (1995): 352–361.

Palmqvist, Kristin, Jian-Wei Yu, and Murray R. Badger. "Carbonic anhydrase activity and inorganic carbon fluxes in low-and high-C1 cells of *Chlamydomonas reinhardtii* and *Scenedesmus obliquus*." *Physiologia Plantarum* 90, 3 (1994): 537–547.

Park, Kyoung C., Crystal Whitney, Jesse C. McNichol, Kathryn E. Dickinson, Scott MacQuarrie, Blair P. Skrupski, Jitao Zou, Kenneth E. Wilson, Stephen JB O'Leary, and Patrick J. McGinn. "Mixotrophic and photoautotrophic cultivation of 14 microalgae isolates from Saskatchewan, Canada: Potential applications for wastewater remediation for biofuel production." *Journal of Applied Phycology* 24, 3 (2012): 339–348.

Parmar, Asha, Niraj Kumar Singh, Avani Kaushal, Sagar Sonawala, and Datta Madamwar. "Purification, characterization and comparison of phycoerythrins from three different marine cyanobacterial cultures." *Bioresource Technology* 102, 2 (2011): 1795–1802.

Patrinos, Ari. "Prospects for algae" [PowerPoint slides]. (2017) Retrieved from https://www.acs.org/content/dam/acsorg/policy/acsonthehill/briefings/cellulosicbiofuels/patrinos-presentation.pdf.

Perez-Garcia, Octavio, Froylan ME Escalante, Luz E. de-Bashan, and Yoav Bashan. "Heterotrophic cultures of microalgae: Metabolism and potential products." *Water Research* 45, 1 (2011): 11–36.

Perez-Garcia, Octavio, Yoav Bashan, and Maria Esther Puente. "Organic carbon supplementation of sterilized municipal wastewater is essential for heterotrophic growth and removing ammonium by the microalga Chlorella vulgaris." *Journal of Phycology* 47, 1 (2011): 190–199.

Pignolet, Olivier, Sébastien Jubeau, Carlos Vaca-Garcia, and Philippe Michaud. "Highly valuable microalgae: Biochemical and topological aspects." *Journal of Industrial Microbiology & Biotechnology* 40, 8 (2013): 781–796.

Pleissner, Daniel, Wan Chi Lam, Zheng Sun, and Carol Sze Ki Lin. "Food waste as nutrient source in heterotrophic microalgae cultivation." *Bioresource Technology* 137 (2013): 139–146.

Ponomarenko, L. P., I. V. Stonik, N. A. Aizdaicher, T. Yu Orlova, G. I. Popovskaya, G. V. Pomazkina, and V. A. Stonik. "Sterols of marine microalgae Pyramimonas cf. cordata (Prasinophyta), Attheya ussurensis sp. nov.(Bacillariophyta) and a spring diatom bloom from Lake Baikal." *Comparative Biochemistry and Physiology Part B: Biochemistry and Molecular Biology* 138, 1 (2004): 65–70.

Popper, Zoë A., Gurvan Michel, Cécile Hervé, et al. "Evolution and diversity of plant cell walls: From algae to flowering plants." *Annual Review of Plant Biology* 62 (2011): 567–590.

Price, G. Dean, John R. Coleman, and Murray R. Badger. "Association of carbonic anhydrase activity with carboxysomes isolated from the cyanobacterium Synechococcus PCC7942." *Plant Physiology* 100, no. 2 (1992): 784–793.

Radakovits, Randor, Robert E. Jinkerson, Susan I. Fuerstenberg, Hongseok Tae, Robert E. Settlage, Jeffrey L. Boore, and Matthew C. Posewitz. "Draft genome sequence and genetic transformation of the oleaginous alga Nannochloropsis gaditana." *Nature Communications* 3, 1 (2012): 1–11.

Rawat, Ismail, Sanjay K. Gupta, Amritanshu Shriwastav, Poonam Singh, Sheena Kumari, and Faizal Bux. "Microalgae applications in wastewater treatment." In *Algae Biotechnology*, pp. 249–268. Springer, Cham, 2016.

Razzak, Shaikh A., Mohammad M. Hossain, Rahima A. Lucky, Amarjeet S. Bassi, and Hugo de Lasa. "Integrated CO_2 capture, wastewater treatment and biofuel production by microalgae culturing—a review." *Renewable and Sustainable Energy Reviews* 27 (2013): 622–653.

Reynoso-Camacho, R., Eva Gonzalez-Jasso, R. Ferriz-Martínez, B. Villalón-Corona, G. F. Loarca-Pina, L. M. Salgado, and M. Ramos-Gomez. "Dietary supplementation of lutein reduces colon carcinogenesis in DMH-treated rats by modulating K-ras, PKB, and β-catenin proteins." *Nutrition and Cancer* 63, 1 (2011): 39–45.

Richmond, Amos, ed. *Handbook of microalgal culture: Biotechnology and applied phycology*. John Wiley & Sons, 2008.

Richmond, Amos, and Qiang Hu. *Handbook of microalgal culture: Applied phycology and biotechnology*. Iowa: John Wiley & Sons, 2013.

Rossi, Federico, and Roberto De Philippis. "Exocellular polysaccharides in microalgae and cyanobacteria: Chemical features, role and enzymes and genes involved in their biosynthesis." In *The physiology of microalgae*, pp. 565–590. Springer, Cham, 2016.

Rotatore, C., and B. Colman. "The active uptake of carbon dioxide by the unicellular green algae Chlorella saccharophila and C. ellipsoidea." *Plant, Cell & Environment* 14, 4 (1991): 371–375.

Rotatore, C., B. Colman, and M. Kuzma. "The active uptake of carbon dioxide by the marine diatoms Phaeodactylum ticornutum and Cyclotella sp." *Plant, Cell & Environment* 18, no. 8 (1995): 913–918.

Rotatore, Caterina, and Brian Colman. "Active uptake of CO_2 by the diatom Navicula pelliculosa." *Journal of Experimental Botany* 43, 4 (1992): 571–576.

Ruocco, Nadia, Susan Costantini, Stefano Guariniello, and Maria Costantini. "Polysaccharides from the marine environment with pharmacological, cosmeceutical and nutraceutical potential." *Molecules* 21, 5 (2016): 551.

Ryckebosch, Eline, Sara Paulina Cuéllar Bermúdez, et al. "Influence of extraction solvent system on the extractability of lipid components from the biomass of Nannochloropsis gaditana." *Journal of Applied Phycology* 26, 3 (2014): 1501–1510.

Ryckebosch, Eline, Koenraad Muylaert, and Imogen Foubert. "Optimization of an analytical procedure for extraction of lipids from microalgae." *Journal of the American Oil Chemists' Society* 89, 2 (2012): 189–198.

Sang, Min, Ming Wang, Jianhui Liu, Chengwu Zhang, and Aifen Li. "Effects of temperature, salinity, light intensity, and pH on the eicosapentaenoic acid production of Pinguiococcus pyrenoidosus." *Journal of Ocean University of China* 11, 2 (2012): 181–186.

Sasso, Severin, Georg Pohnert, Martin Lohr, Maria Mittag, and Christian Hertweck. "Microalgae in the postgenomic era: A blooming reservoir for new natural products." *FEMS Microbiology Reviews* 36, 4 (2012): 761–785.

Sayre, Richard. "Microalgae: The potential for carbon capture." *Bioscience* 60, 9 (2010): 722–727.

Scott, Spencer D., Roberto E. Armenta, Kevin T. Berryman, and Andrew W. Norman. "Use of raw glycerol to produce oil rich in polyunsaturated fatty acids by a thraustochytrid." *Enzyme and Microbial Technology* 48, 3 (2011): 267–272.

Sharma, Kalpesh K., Sourabh Garg, Yan Li, Ali Malekizadeh, and Peer M. Schenk. "Critical analysis of current microalgae dewatering techniques." *Biofuels* 4, 4 (2013): 397–407.

Sharwood, Robert E., Oula Ghannoum, and Spencer M. Whitney. "Prospects for improving CO_2 fixation in C3-crops through understanding C4-Rubisco biogenesis and catalytic diversity." *Current Opinion in Plant Biology* 31 (2016): 135–142.

Shrestha, Gajendra, Larry L. St. Clair, and Kim L. O'Neill. "The immunostimulating role of lichen polysaccharides: A review." *Phytotherapy Research* 29, 3 (2015): 317–322.

Sing, Sophie Fon, Andreas Isdepsky, Michael A. Borowitzka, and Navid Reza Moheimani. "Production of biofuels from microalgae." *Mitigation and Adaptation Strategies for Global Change* 18, 1 (2013): 47–72.

Sivaramakrishnan, Ramachandran, and Aran Incharoensakdi. "Enhancement of total lipid yield by nitrogen, carbon, and iron supplementation in isolated microalgae." *Journal of Phycology* 53, no. 4 (2017): 855–868.

So, Anthony K-C., George S. Espie, Eric B. Williams, Jessup M. Shively, Sabine Heinhorst, and Gordon C. Cannon. "A novel evolutionary lineage of carbonic anhydrase (ε class) is a component of the carboxysome shell." *Journal of Bacteriology* 186, 3 (2004): 623–630.

Spalding, Martin H. "Microalgal carbon-dioxide-concentrating mechanisms: Chlamydomonas inorganic carbon transporters." *Journal of Experimental Botany* 59, 7 (2008): 1463–1473.

Spolaore, Pauline, Claire Joannis-Cassan, Elie Duran, and Arsène Isambert. "Commercial applications of microalgae." *Journal of Bioscience and Bioengineering* 101, 2 (2006): 87–96.

Spreitzer, Robert J., and Michael E. Salvucci. "Rubisco: Structure, regulatory interactions, and possibilities for a better enzyme." *Annual Review of Plant Biology* 53 (2002): 449–475.

Steffens, Daniela, Dilmar Leonardi, Paula Rigon da Luz Soster et al. "Development of a new nanofiber scaffold for use with stem cells in a third degree burn animal model." *Burns* 40, 8 (2014): 1650–1660.

Suganya, T., Mahendora Varman, H. H. Masjuki, and S. Renganathan. "Macroalgae and microalgae as a potential source for commercial applications along with biofuels production: A biorefinery approach." *Renewable and Sustainable Energy Reviews* 55 (2016): 909–941.

Sültemeyer, Dieter F., Anthony G. Miller, George S. Espie, Henrich P. Fock, and David T. Canvin. "Active CO_2 transport by the green alga Chlamydomonas reinhardtii." *Plant Physiology* 89, 4 (1989): 1213–1219.

Sun, Liqin, Jinling Chu, Zhongliang Sun, and Lihong Chen. "Physicochemical properties, immunomodulation and antitumor activities of polysaccharide from Pavlova viridis." *Life Sciences* 144 (2016): 156–161.

Talero, Elena, Sofía García-Mauriño, Javier Ávila-Román, Azahara Rodríguez-Luna, Antonio Alcaide, and Virginia Motilva. "Bioactive compounds isolated from microalgae in chronic inflammation and cancer." *Marine Drugs* 13, 10 (2015): 6152–6209.

Tang, Guangwen, and Paolo M. Suter. "Vitamin A, nutrition, and health values of algae: Spirulina, Chlorella, and Dunaliella." *Journal of Pharmacy and Nutrition Sciences* 1, 2 (2011): 111–118.

Tanner, Widmar. "The Chlorella hexose/H+-symporters." *International Review of Cytology* 200, (2000): 101–141.

Thomas, Daniya M., Jerry Mechery, and Sylas V. Paulose. "Carbon dioxide capture strategies from flue gas using microalgae: A review." *Environmental Science and Pollution Research* 23, 17 (2016): 16926–16940.

Tuchman, Nancy C., Marc A. Schollett, Steven T. Rier, and Pamela Geddes. "Differential heterotrophic utilization of organic compounds by diatoms and bacteria under light and dark conditions." In *Advances in algal biology: A commemoration of the work of Rex Lowe*, pp. 167–177. Springer, Dordrecht, 2006.

Van Hunnik, E., A. Livne, V. Pogenberg, E. Spijkerman, H. Van den Ende, E. Garcia Mendoza, D. Sültemeyer, and J. W. De Leeuw. "Identification and localization of a thylakoid-bound carbonic anhydrase from the green algae Tetraedron minimum (Chlorophyta) and Chlamydomonas noctigama (Chlorophyta)." *Planta* 212, 3 (2001): 454–459.

Van Wagenen, Jon, Tyler W. Miller, Sam Hobbs, Paul Hook, Braden Crowe, and Michael Huesemann. "Effects of light and temperature on fatty acid production in Nannochloropsis salina." *Energies* 5, 3 (2012): 731–740.

Varela, Joao C., Hugo Pereira, Marta Vila, and Rosa León. "Production of carotenoids by microalgae: Achievements and challenges." *Photosynthesis Research* 125, 3 (2015): 423–436.

Vílchez, Carlos, Eduardo Forján, María Cuaresma, Francisco Bédmar, Inés Garbayo, and José M. Vega. "Marine carotenoids: Biological functions and commercial applications." *Marine Drugs* 9, 3 (2011): 319–333.

Volkman, John K. "Sterols in microalgae." In *The physiology of microalgae*, pp. 485–505. Springer, Cham, 2016.

Vuppaladadiyam, Arun K., Pepijn Prinsen, Abdul Raheem, Rafael Luque, and Ming Zhao. "Microalgae cultivation and metabolites production: A comprehensive review." *Biofuels, Bioproducts and Biorefining* 12, 2 (2018a): 304–324.

Vuppaladadiyam, Arun K., Joseph G. Yao, Nicholas Florin et al. "Impact of flue gas compounds on microalgae and mechanisms for carbon assimilation and utilization." *ChemSusChem* 11, 2 (2018b): 334–355.

Wang, Jinghan, Haizhen Yang, and Feng Wang. "Mixotrophic cultivation of microalgae for biodiesel production: Status and prospects." *Applied Biochemistry and Biotechnology* 172, 7 (2014): 3307–3329.

Wenzel, Anja, Ann-Kristin Bergström, Mats Jansson, and Tobias Vrede. "Survival, growth and reproduction of Daphnia galeata feeding on single and mixed Pseudomonas and Rhodomonas diets." *Freshwater Biology* 57, 4 (2012): 835–846.

Williams, Peter J. le B., and Lieve ML Laurens. "Microalgae as biodiesel & biomass feedstocks: Review & analysis of the biochemistry, energetics & economics." *Energy & Environmental Science* 3, 5 (2010): 554–590.

Xue, Lu-Lu, and Jian-Guo Jiang. "Cultivation of Dunaliella tertiolecta intervened by triethylamine enhances the lipid content." *Algal Research* 25 (2017): 136–141.

Yago, Takahide, Hisayuki Arakawa, Tsutomu Morinaga, Y. Yoshie-Stark, and M. Yoshioka. "Effect of wavelength of intermittent light on the growth and fatty acid profile of the haptophyte Isochrysis galbana." In *Global Change: Mankind-Marine Environment Interactions*, pp. 43–45. Springer, Dordrecht, 2010.

Yan, Dong, Yue Lu, Yi-Feng Chen, and Qingyu Wu. "Waste molasses alone displaces glucose-based medium for microalgal fermentation towards cost-saving biodiesel production." *Bioresource Technology* 102, 11 (2011): 6487–6493.

Yoo, Sang-Ho, Martin H. Spalding, and Jay-lin Jane. "Characterization of cyanobacterial glycogen isolated from the wild type and from a mutant lacking of branching enzyme." *Carbohydrate Research* 337, 21–23 (2002): 2195–2203.

Zhang, Chun-Dan, Wei Li, Yun-Hai Shi, Yuan-Guang Li, Jian-Ke Huang, and Hong-Xia Li. "A new technology of CO_2 supplementary for microalgae cultivation on large scale–A spraying absorption tower coupled with an outdoor open runway pond." *Bioresource Technology* 209 (2016): 351–359.

Zhong, Yu-Juan, Jun-Chao Huang, Jin Liu, Yin Li, Yue Jiang, Zeng-Fu Xu, Gerhard Sandmann, and Feng Chen. "Functional characterization of various algal carotenoid ketolases reveals that ketolating zeaxanthin efficiently is essential for high production of astaxanthin in transgenic Arabidopsis." *Journal of Experimental Botany* 62, 10 (2011): 3659–3669.

Zhu, Xi, Junfeng Rong, Hui Chen, Chenliu He, Wensheng Hu, and Qiang Wang. "An informatics-based analysis of developments to date and prospects for the application of microalgae in the biological sequestration of industrial flue gas." *Applied Microbiology and Biotechnology* 100, 5 (2016): 2073–2082.

Zhu, Xin-Guang, Stephen P. Long, and Donald R. Ort. "What is the maximum efficiency with which photosynthesis can convert solar energy into biomass?" *Current Opinion in Biotechnology* 19, 2 (2008): 153–159.

13 Soil Microbial Dynamics in Carbon Farming of Agro-Ecosystems

In the Era of Climate Change

Jinus S. Senjam, Kangjam Tilotama, Tracila Meinam, Dhanaraj Singh Thokchom, Yumlembam Rupert Anand, Thoudam Santosh Singh, Koijam Melanglen, Hanglem Sonibala Devi, Khumukcham Nongalleima, S. Gurumurthy, and Thiyam Jefferson Singh

CONTENTS

13.1 INTRODUCTION

The backbone of life on the Earth, the carbon, cannot be ignored, and every living creature on the planet earth is made up of it (Kane 2015). Starting from the industrial period, the carbon cycle of the earth has been heavily troubled with the inputs of carbon dioxide mainly through fossil fuel combustion and with the conversion of natural ecosystems to agricultural lands (Canadell et al. 2007). The balance of carbon is maintained by the major biogeochemical cycle through the biotic and abiotic parts of an ecosystem. All the earth's carbon is also prevented from entering the atmosphere and from being stored in the earth's crust entirely, which make the earth's temperature more stable. This stable thermostat system works over a few hundred thousand years as part of the slow carbon cycle (Riebeek and Robert 2011). The principal components of the global carbon cycle are the dynamics of soil microflora and its biome. To regulate the flow of materials to and from the atmosphere, hydrosphere, and pedosphere, key interactions between the biotic and abiotic components take place (Sharma et al. 2012). Besides the many-fold necessities of carbon, its balance has been entwined with a major serious problem, and we must be aware of it in terms of climate change. This state of imbalance in the global carbon cycle is largely due to the continuous industrial emissions of carbon dioxide and other greenhouse gases (GHGs), burning of fossil fuels, deforestation, and the land use system like the conversion of grassland and forestland to agricultural land, which have resulted in the historic losses of soil carbon (Kane 2015) to the atmosphere (FAO 2019). Therefore, the removal of carbon dioxide from the atmosphere or diversion from the emission source and its storage in terrestrial ecosystems and in oceans and other geological formation are necessary for the carbon sequestration process (Kambale and Tripathi 2010). The storage is a long-term technique in the reservoir pool, and soil, which is the largest terrestrial sink and also a larger potential sink, can store atmospheric carbon dioxide (Zomer et al. 2017). The carbon sequestration takes place in several ways such as reduction of global energy use, development of low or no carbon fuel, and sequestration from point sources or atmosphere by natural or engineering techniques (Schrag 2007). However, restoration of the lost soil carbon will not only benefit the environment, but also be given due importance to the producer's bottom line. Moreover, carbon accumulation in the soil will promote soil particles' aggregation, water retention, microbial activity, biogeochemical cycle, and other various processes of importance, thereby increasing the fertility and productivity (McDowell 2019). However, in the context of climate change, there is less attention given to the concept of carbon sequestration in agricultural ecosystems, which is supposed to be the alternative means in offsetting future emissions effect on the GHG concentrations in the atmosphere. In this chapter, the possible ways and the factors that impact the increased rate of removal of carbon dioxide from the atmosphere to accelerate the gigantic tasks, storing carbon through ecosystems like in plant material, decomposing detritus, and organic soil are overviewed. In this way, the soil of the agricultural lands which are highly productive ecosystems can become biological scrubbers through CO_2 sequestration from the atmosphere (Kaur et al. 2016). Furthermore, if we go down deep inside the soil ingredients, we found that all such chore duty of biological scrubbing for carbon farming mainly depends on microbial communities that fix the atmospheric carbon, promote the growth of plants, and enhance organic material transformation or degradation in the environment (Weiman 2015).

13.2 MICROBIAL COMMUNITIES AND CARBON CYCLE

In the terrestrial biosphere, soil microbes are some of the smallest organisms in soil that have key roles in moving and transforming huge amounts of carbon compounds in their ecosystems. The organic carbons are found to lock in permafrost, which is in high latitude, grassland soil, tropical forests, and the agricultural ecosystems. On the other hand, the microbes play a great role in the determination of longevity and stability of carbon and in determining whether or not the carbon is released in the atmosphere as GHGs, which implies the importance of the processes involved in the carbon cycle (Weiman 2015). They also influence the fertility of the soil and the exchange of CO_2 and other gases within the atmosphere. Hence, they are the primary players within the soil food web and excellent indicators of soil health and functioning (Van Den Hoogen et al. 2019). In these integral components of the complex ecosystem–soil microbial communities, they are the hosting ground of fungal and bacterial dominances, protists, and animals (Bonkowski et al. 2009; Muller et al. 2016), and also mycorrhizal associations and microalgae. The aforementioned microbial organisms have been considered as the contributors of soil carbon sequestration. In the process of carbon sequestration, the soil microbes play an important role in the transformation of plant residues into smaller carbon molecules, and they are more likely to be protected and get sequestered in the soil (Six et al. 2006). Different types of carbon that are of different size and have complex chemical nature are produced at every point of the decomposition pathway. The carbons get associated with silt and clay particles and get incorporated into soil aggregates (Rao and Chhonkar 1998). The nutrient cycling is done in order to sustain the agricultural soil productivity as it is the source and sink for the mineral nutrition and biochemical transformations are being carried out (Jenkinson and Ladd 1981). The decomposition of organic matter produces nutrients that are assimilated by microorganisms and get incorporated into biomass. Also, microbes are immobilized in the soil biomass form or are mineralized (Wani and Lee 1995). Microbes are essential for breaking down and transforming the dead organic material into the forms that other organisms can use. This is the reason why the microbial enzyme systems that involved are known as the key 'engines' which drive the biogeochemical cycles of the earth

(Falkowski et al. 2008). Therefore, sequestering a vast amount of carbon which can improve the soil quality and help in benefitting the environment can happen when the microbial communities and environmental conditions that control the transformation of the organic carbon in the soil are understood.

13.2.1 Mechanism of Carbon Sequestration by Soil Microbes

The amount of carbon present in the Earth's soil is more than the amount present in the atmosphere. It is the largest and most stable carbon reservoir in terrestrial ecosystems. Soil carbon sequestration is the process of removing CO_2 from the atmosphere and storing it in the soil carbon pool (Ontl and Schulte 2012). The carbon is transfer from the atmosphere to the soil by carbon fixing autotrophic organisms, which convert carbon dioxide into organic matter. These are driven mainly by the photosynthesizing plants and photo and chemoautotrophic microbes (Lu and Conrad 2005). Then, the fixed carbon is returned to the atmosphere through different pathways which account for respiration for both the autotrophic and heterotrophic organisms (Trumbore 2006).

The main factors that determine the amount of carbon sequestered in the soil are (i) input of organic matter rate, (ii) the decomposability of organic matter inputs, (iii) the depth at which the organic carbon is placed in the soil, and (iv) intra-aggregate or organo-mineral complexes during physical protection. The biotic activities of plants (which are the main source of carbon through litter and root systems), microorganisms (fungi and bacteria), and 'ecosystem engineers' (termites, ants, earthworms) alter the soil organic carbon (SOC) stocks. In the past, decomposition biotic processes were investigated at the levels such as molecular, organismal, and community (Sinsabaugh et al. 2002a, 2002b; Tate 2002). In the meantime, modification of these stocks is done by abiotic processes that are related to the physical structure, porosity, and mineral fraction of the soil (Marie-France et al. 2017). It is estimated that at least three times the carbon which is stored in the atmosphere is equivalent to the global organic carbon stocks in soil (Gougoulias et al. 2014). Annually, the terrestrial ecosystems and the atmosphere exchange about 8% of the total atmospheric carbon pool through a net primary production and respiration of terrestrial heterotrophic organisms (predominantly microbial) (Gougoulias et al. 2014). In other words, at current rates to exhaust atmospheric carbon stocks for primary production, it would take about 12 years if soil microbial respiration stopped [(if all other components of the carbon cycle are ignored, e.g. oceanic CO_2 exchange) (Sylvia et al. 2005)].

More than 100 years ago, the emphasis of several co-workers was on microbes that are important for soil organic matter (SOM) synthesis (Waksman 1936). The SOM consists of the continuum from fresh to progressively decomposing plant, debris which are of microbes and fauna and exudates, including the microbial biomass, which is responsible for exudate primary decomposition and detrital inputs (Dungait et al. 2012). The role of microbes was thought to be restricted only to the decomposition of the plant and animal matter, but later, the role of microbes in the resynthesis of SOM was recognized (Waksman 1936). The quality and the quantity of organic matter are influenced by the soil microorganisms and again by the soil ecology and properties (Hashem et al. 2019). Increasing the SOC has two benefits:

mitigating climate change and improving soil health and fertility. According to the soil, climate, landscape, and change in the same paddock over time depending on the climate and the farming methods, the amount of SOC present in the soil varies (Dignac et al. 2017). The multistep conversion of the dead plant tissues is mediated by the microbes and exudate organic compounds from plant roots into carbon dioxide or SOM. Reduction of carbon present in the atmosphere takes place if more amount is stored in the soil as SOC. This will also reduce global warming (Dignac et al. 2017).

13.3 CARBON SEQUESTRATION IN AGRICULTURAL SECTOR

Emission of the atmospheric carbon dioxide is reduced by the role agro-ecosystems play in carbon sequestration. In the agricultural sector, carbon sequestration is the capacity of forests and agricultural lands to remove carbon dioxide from the atmosphere. Trees and plants absorb carbon dioxide through photosynthesis, and carbon dioxide is stored as carbon in biomass in tree trunks, branches, foliage, and roots and soils (EPA 2008b). Measurement of carbon sequestration is made in terms of the total organic carbon that is stored in the soil. Soil is the largest terrestrial sink for carbon on the planet earth. Carbon sinks are the forests and grasslands, which store a large amount of carbon in their vegetation and in the root systems for a long period of time. Among the different strategies, one strategy to enhance soil carbon sequestration is microbial modulation (Kaur et al. 2016). There are basically two types of sequestration: abiotic and biotic. The abiotic technique involves the injection of CO_2 into deep oceans, geological strata, old coal mines, and oil wells. On the other hand, the biotic techniques involve managing higher plants and microorganisms in order to remove more carbon dioxide from the atmosphere and then fixing this carbon in stable pools of soil. Further, the biotic sequestration is divided into oceanic and terrestrial sequestration. Oceanic sequestration involves the capture of carbon by the photosynthetic activities of organisms like phytoplankton, which convert the carbon into particulate organic material that deposits on the ocean floor. On the other hand, terrestrial sequestration is the transfer of CO_2 from the atmosphere into the biotic and pedologic C pools. Terrestrial sequestration is accomplished through photosynthesis by the sequestration of CO_2 and is stored in living and dead organic matter. Forests, soils, and wetlands are the major terrestrial carbon sinks (Ahmed et al. 2018).

13.4 MICROBIAL FUNGI AND BACTERIA IN CARBON SEQUESTRATION

In soil carbon sequestration, soil fungi maximize the amount of carbon which is allocated to the soil, and they produce compounds that help in improving the aggregate stability (Six et al. 2006). The clay particles present in the soil and the glue-like substances that are generated by the microbes that produce glomalin (produced by arbuscular mycorrhizal fungi (AMF)) bind the soil aggregates together (Oades 1984; Six et al. 2004b; Wilson et al. 2009). Partially decayed plant residues, i.e. small carbon particles, are captured in the centre of aggregates when the aggregates are formed. The carbon-rich materials are physically protected at the centre of the aggregates from the microbial attack. Penetration of microbes is not allowed at the centre of the stable aggregates. The microbial metabolism is disrupted when oxygen and water are at low concentrations (Six et al. 1998, 2000). Six et al. (2000) noted that the fungal-dominated soils have slow carbon turnover rates. The aggregates protect the soil carbon for a long period of time when they are stable and left undisturbed. Actinomycetes, a soil bacteria, play an important role in the processing of organic matter and in the decomposition of carbon, which are highly recalcitrant form like lignin. The decomposition process and the processing of organic matter are essential to both maximizing biomass production and ensuring that carbon is converted into stable forms that remain protected in the soil (Six et al. 2006; Kindler et al. 2006; Liang et al. 2008; Liang and Balser 2011; Potthoff et al. 2008). On the other hand, mesovores, which include soil-dwelling insects, worms, and nematodes, are responsible for processing larger pieces of plant residues into smaller forms that can be metabolized by smaller organisms such as fungi and bacteria. Bacteria and fungi generally comprise about 90% of the total soil microbial biomass, and they are responsible for the majority of SOM decomposition (Six et al. 2006). They govern most of the transformations and the ensuing long-term storage of organic carbon in soils (Table 13.1). Changes in the ratio of fungi to bacteria and the total biomass may affect the storage and fluxes of carbon (C) and nitrogen (N) in the terrestrial ecosystems (Bailey et al. 2002).

TABLE 13.1
Microbial Communities and Agricultural Management Regimes in Soil Carbon Sequestration Dynamics

Nanozeolite and alfalfa and wheat straw residues	Positive effects on improvements of carbon pools and increased carbon sequestration in soil	Aminiyan et al. (2018)
Bt and non-Bt corn residue-amended soils	Increase carbon (C) mineralization	Fang et al. (2007)
Nanozeolite and alfalfa straw	Increase in carbon pools and improvement of aggregation stability in the soils which are treated and incubated in the laboratory condition	Aminiyan et al. (2015a, 2015b)
Streptomyces	Able to grow on carbon monoxide by first oxidizing CO to CO_2 by their only carbon source, and CO_2 is fixed by using RuBisCo and phosphoribulose kinase	Bell et al. (1987)
Purple sulphide-oxidizing bacteria such as *Chromatium vinosum*, *Sinorhizobium meliloti*, and *Bradyrhizobium japonicum*	Role in microbial autotrophy	Tolli and King (2005), Selesi et al. (2005), Nanba et al. (2004)

(*Continued*)

TABLE 13.1 (*Continued*)
Microbial Communities and Agricultural Management Regimes in Soil Carbon Sequestration Dynamics

Bradyrhizobium japonicum, *Ralstonia eutropha*, *Azospirillum lipoferum*, *Rhodopseudomonas palustris*, *Xanthophyta*, and *Bacillariophyta*	Reported the new perception for the importance of microbial autotrophy in terrestrial C cycling	Yuan et al. (2012)
Thermophilic actinomycete, Streptomyces strain G26	Studied the CO metabolism	Bell et al. (1987)
Chloroflexus aurantiacus	A possible intermediate in the assimilation of CO_2 and acetate. It secretes 3-hydroxypropionate	Holo (1989)
Cyanobacteria, chloroflexi, bacteroidetes/chlorobi, proteobacteria, green sulphur and green non-sulphur bacteria, purple sulphur and purple non-sulphur bacteria, *Azospirillum* sp., *Rhizobium* sp., and *Pseudomonas* sp	Key source of microbial CO_2 assimilation in case of arid soil	Yuan et al. (2012), Atomi (2002), Yousuf et al. (2012), Chowdhury et al. (2007)
Avicennia marina and *Rhizophora stylosa*	Carbon sequestration and microbial communities in mangrove soils	Carlson et al. (2019)
Gram-negative bacteria, gram-positive bacteria, and Actinomycetes	Size and structure of soil microbial community can be altered by planting *Caragana korshinskii* in abandoned cropland, and such changes are closely related to humus formation and carbon sequestration	Xiang et al. (2017)
Gram-negative bacteria, gram-positive bacteria, and Actinomycetes	Soil microbial community composition and their assimilation of plant-derived carbon can be independent of C3 and C4 vegetation changes	Mellado-Vázquez et al. (2019)
G- (Gram-negative) bacteria	Root exudation has high affinity for the plant-derived carbon	Denef et al. (2009); Mellado-Va'zquez et al. (2016)
Saprotrophic fungi	Able to decompose plant litter as well as SOM and root exudates	Treonis et al. (2004), Garcia-Pausas and Paterson (2011); Mellado-Vázquez et al. (2016)

(*Continued*)

TABLE 13.1 (*Continued*)
Microbial Communities and Agricultural Management Regimes in Soil Carbon Sequestration Dynamics

Actinobacteria and Acidobacteria (which are dominated communities)	They promote soil C storage not only by the production of polysaccharides for soil structural stability, but also by the lifestyle (slow growth and lower metabolic activities)	Trivedi et al. (2013)
AMF	Under increased CO_2, it can increase organic carbon decomposition	Cheng, et al. (2012)
AMF	For soil carbon storage, short-term liability but long-term benefits	Verbruggen, et al. (2013)
Associated fungi and roots	Boreal forest in long-term carbon sequestration	Clemmensen, K.E. et al. (2013)
UGmax (biofertilizer)	Improves the SOM status and carbon sequestration	Dębska et al. (2016)
AMF	Enhances carbon sequestration and soil aggregate formation	Oades (1984); Six et al. (2004b); Wilson et al. (2009)
Tillage	Breaks apart aggregates quickly, exposing soil carbon for microbial attack	Grandy and Robertson, (2006, 2007)
Periodic green fallows or cover crops	Balanced carbon towards a net gain rather than a net loss	McDaniel et al. (2014), Tiemann et al. (2015)
Crop rotations and using legume cover crops	Increased the diversity of soil carbon and complexity for more stable	Wickings et al. (2012)
AMF	Glomalin – a very sticky protein – is produced, which helps in binding soil aggregates together to protect soil carbon	Rillig (2004)
AMF	Carbon levels and soil aggregation are positively correlated	Wilson et al. (2009)
Reducing tillage	The capacity of soils to sequester carbon is improved	Six et al. (2006)
Chronically fertilized systems	Soil CO_2 respiration is reduced	Neff et al. (2002)
NT	Especially at the soil surface increased soil carbon rapidly and stabilization of newly added C	West and Post (2002), Fabrizzi et al. (2009), Nicoloso et al. (2018)

(*Continued*)

TABLE 13.1 (*Continued*)
Microbial Communities and Agricultural Management Regimes in Soil Carbon Sequestration Dynamics

Actinomycetes, AMF, fungi, gram-positive bacteria, and gram-negative bacteria	Relationship between the microbial community and both C and N components in the soil is analysed	Liu et al. (2019)
Fertilizers (organic)	Labile organic C fractions and total SOC contents increased significantly	Li et al. (2018)
Tropical forest in the upper elevation	Declines in C groups associated with increased soil C storage and increased fungal phospholipid fatty acid with N additions	Cusack et al. (2011)
Forest in the lower elevation	There were significant losses of labile C chemical groups and increase in phospholipid fatty acid biomass of gram-negative bacteria	Cusack et al. (2011)
Photoautotrophic, chemolithoautotrophic, *Nitrospira*, and gamma-proteobacteria microbial communities	In arid ecosystem, energy for CO_2 fixation is obtained by the oxidation of nitrite, carbon monoxide, iron, or sulphur	Agarwal et al. (2014)
High affinity for AMF, such as perennial grasses and selecting crops with high belowground biomass	can aid in the stabilization of SOC and aggregates	Liang et al. (2012); Jesus et al. (2015); Ontl et al. (2015); Tiemann et al. (2015)
Switchgrass and miscanthus for perennial cropping systems	SOC is increased at the respective rates of 0.8 and 1.3 Mg C $ha^{-1} year^{-1}$ in the 0–15 cm depth	McGowan et al. (2019)
NT for the long term	Higher soil carbon and nitrogen contents, total phospholipid fatty acid, and phosphatase activities at the 0–5 cm depth than those under the CT treatment	Mathew et al. (2012)
Mulching	By adding mulch, increased carbon concentration, and SOM for CS and crop protection against cold stress, crop residues are widely applied in the form of mulch	Windeatt et al. (2014)
Especially N nutrients and crop residues	For sequestering C in soils up to 21.3%–32.5%	Windeatt et al. (2014)

(*Continued*)

TABLE 13.1 (*Continued*)
Microbial Communities and Agricultural Management Regimes in Soil Carbon Sequestration Dynamics

Manure applications	SOC sequestration rates of three long-term (>49) years of manure applications ranged from 10 to 22 kg C $ha^{-1}year^{-1}t^{-1}$ of dry solids, whereas SOC sequestration rates with shorter-term experiments (8–25 years of farmyard manure, cattle slurry, and boiler litter) were from 30 to 200 kg C ha^{-1} $yr.^{-1}t^{-1}$ of dry solids	Powlson et al. (2011)
Crop residues	Soil aggregation and increase in the SOM and C storage	Novelli et al. (2017)
Fast-growing trees or grass from marginal cropland	Increased carbon storage on land	Schahczenski and Hill (2008)
Landscape level (slope position)	SOC also decreases as slope increases	Feyissa et al. (2013), Güner et al. (2012)
For inter-cropped systems Trees grown with legumes or cereals	Mitigate global climate change by enhancing carbon (C) sequestration and reducing GHG emissions from the terrestrial ecosystem	He et al. (2013), Paquette et al. (2006)

13.4.1 Archaea

Using carbon dioxide as a source of carbon and hydrogen as a source of energy, methane is produced by archaea that live in anaerobic conditions. These archaea are termed as CO_2-type hydrogenotrophic methanogens. Some examples are *Methanothermobacter thermautotrophicus*, *Methanosarcina barkeri*, *Methanobrevibacter aboriphilus*, and *Methanothermobacter marburgensis* (Rittmann et al. 2015). Bioremediation and sequestration of carbon dioxide are also done by methanogens. After the process of activated sludge in the treatment plants, the sludge is used as the inoculum in the process of sequestration and the production of methane (Yasin et al. 2013). The gas fermentation product range will be increased by the source of thermostable carbonic anhydrase enzymes, i.e. thermophiles that are suitable for capturing industrial carbon dioxide and also for identifying new archaeal isolates (Smith and Ferry 2000; Migliardini et al. 2014).

13.4.2 Clostridia

Clostridia, the gram-positive bacteria, are the obligate anaerobes. The roles they play are in carbon cycle, human health carbon cycle, production of carbohydrates and organic chemicals, anaerobic degradation, and acidogenesis (Tracy et al. 2012).

The ability to fix CO_2 and CO can be seen in *Clostridium spp.* For example, acetogenic bacteria, *Clostridium autoethanogenum*, fixes CO_2 (in the presence of H_2) and CO into central metabolite acetyl-CoA *via* the Wood–Ljungdahl pathway (Liew et al. 2016). *Clostridium spp.* can be used for managing carbon because of its ability to use carbon source in various ways (including carbohydrates, carbon dioxide, and carbon monoxide). *Clostridium spp.* also produces industrial products such as ethanol, butanol, acetate, acetone, lactate, caproate, valeroate, and caprylate, through different metabolic pathways. Also, it is known to be tolerant to toxic metabolites (Tracy et al. 2012; Ezeji et al. 2007; Qureshi et al. 2007).

13.4.3 Proteobacteria

They are gram-negative bacteria. The capability of sequestering carbon can be seen in many of these bacteria through different metabolic pathways. α-Proteobacteria such as *Xanthobacter flavus*, *Oligotropha carboxidovorans*, *Rhodobacter capsulatus*, and *Rhodobacter sphaeroides* use Calvin cycle for sequestering carbon (Meijer 1994; Ding and Yokota 2004). The ability to fix carbon inside the cytoplasm is seen in these bacteria in the form of polyhydroxyalkanoates (PHAs), which are known as bioplastics. The PHA molecules comprise short chains of poly-3-hydroxybutyrate-co-3-hydroxyvalerate (PHBV) and poly-3-hydroxybutyrate (PHB) (Albuquerque 2011). They are fully biodegradable as they are synthesized biologically by *Ralstonia eutropha*. When compared with the chemically synthesized polymers, they have a lower impact on the environment. Furthermore, proteobacteria can also produce compounds that have medicinal properties through various carbon utilization pathways (Brigham et al. 2010; Pohlmann et al. 2006).

13.4.4 Green Sulphur Bacteria

They are non-motile (except *Chloroherpeton thalassium*, which may glide) and capable of anoxygenic photosynthesis (Bryant). They use sulphide ions as electron donors unlike in plants (Sakurai et al. 2010). In order to fix carbon dioxide, they use the reverse tricarboxylic acid cycle (Tang and Blankenship 2010). Inorganic sulphur compounds or molecular hydrogen is used as an electron donor by the Chlorobium bacteria in reducing carbon dioxide present in the cell material. Generally, it is said that the carbon dioxide fixation mechanism in the autotrophic bacteria is identical to the reductive pentose phosphate cycle, which occurs in green plants (Elsden 1962). It is found that in the cell-free extracts of *Chlorobium thiosulfatophilum*, most of the enzymes of this cycle are present (Smillie et al. 1962). Recently, two new CO_2 fixation reactions that are ferredoxin dependent have been described in the cell-free extracts of *C. thiosulfatophilum* (Evans and Buchanan 1965; Buchanan and Evans 1965):

$$\text{Acetyl} - \text{CoA} + CO_2 + 2\text{ ferredoxin} * e + 2H + \ldots\ldots \text{pyruvate} + \text{CoA}$$
$$+ 2\text{ ferredoxin}$$

$$\text{Succinyl} - \text{CoA} + CO_2 + 2\text{ ferredoxin} * e + 2H+ \ldots\ldots\alpha - \text{oxoglutarate} + \text{CoA} + 2\text{ ferredoxin}$$

The enzymes that catalyse the reactions are pyruvate synthase and α-oxoglutarate synthase, respectively. On the basis of the above two reactions, Evans et al. (1966) recommended a cyclic pathway for CO_2 assimilation. They named it reductive carboxylic acid cycle. The principle of this cycle is a reversal of the tricarboxylic acid cycle. Van Niel (1949) envisaged the operation of the reverse tricarboxylic acid cycle as a mechanism of CO_2 fixation. Evans et al. (1966) proposed that the reductive carboxylic acid cycle may be mainly related to the synthesis of precursors of amino acids, lipids, and porphyrins, and the reductive pentose phosphate cycle to the synthesis of carbohydrates.

13.4.5 Purple Sulphur Bacteria

They are part of a group of Proteobacteria that are capable of photosynthesis and are collectively cited as purple bacteria. They are anaerobic or microaerophilic. Often, they are found in stratified water environment comprising hot springs, stagnant water bodies, etc (Daldal et al. 2008). They play an important role in primary production, which suggests that these organisms can affect the carbon cycle through the process of carbon fixation (Storelli et al. 2013). In 1931, Van Niel represented the overall reaction of photosynthesis. He outlined his investigation as the following formulas:

$$\text{In purple bacteria: } CO_2 +_{1/2} (HS)^- + H_2O \ldots\ldots\ldots\ldots\text{light}\ldots\ldots (CH_2O) + (HSO_4)^-$$

$$\text{In green bacteria: light } CO_2 + H_2S - \text{light} \rightarrow (CH_2O) + H_2O + 2S$$

Roelefson (1934) and Gaffron (1934) discovered that purple bacteria uses molecular hydrogen as an external reductant, and also thiosulphate and malate, in the light. Van Niel, and Wesseles and French calculated the molecular ratios of hydrogen to carbon dioxide, which were absorbed together by the illuminated purple bacteria. Most values of photosynthetic quotients, H_2/CO_2, obtained with purple bacteria by many investigators, are somewhat larger than 2, which indicates that a possible production of organic matter is reduced beyond the carbohydrate level.

13.4.6 Chemolithotrophs

A chemolithotroph can use inorganic reduced compounds as an energy source (Chang 2016). A majority of chemolithotrophs are proficient in fixing the carbon dioxide (CO_2) through the Calvin cycle (Kuenen 2009). Chemolithotrophs include sulphur oxidizers, nitrifying bacteria, iron oxidizers, and hydrogen oxidizers.

13.4.7 Photoautotrophs and Algae

Several studies have emphasized the need to infer the potential of microalgal cultivation systems in order to reduce CO_2 emissions (Herzog and Drake 1996; Stewart and

Hessami 2005). It has been calculated that approximately half of the atmospheric oxygen is produced by microalgae, which simultaneously use CO_2 to grow photoautotrophically (Tabatabaei et al. 2011). Green algae, cyanobacteria, diatoms, red algae, golden algae, yellow-green algae, brown algae, and euglenoids constitute microalgae (Chen et al. 2009). Microalgae is subdivided mainly based on their habitat and morphology (Cheah 2015). Macroalgae and autotrophic microorganisms are known to contribute significantly to the assimilation of carbon dioxide (Chen et al. 2009) in aquatic systems (Savage et al. 2010). However, generally, they don't have a key role in carbon dioxide fixation. Working with the soil and compost isolates, it was shown that certain actinomycetes, especially *Streptomyces*, can grow on carbon monoxide (Bell et al.1987). Other studies have shown that microbial autotrophy is apparent in many taxa, such as aerobic, CO-oxidizing bacteria (such as *Bradyrhizobium japonicum*, *Sinorhizobium meliloti*) and purple sulphide-oxidizing bacteria (like *Chromatium vinosum*) (Nanba et al. 2004; Selesi et al. 2005; Tolli and King 2005).

The algal species fixes CO_2 through the Calvin–Benson cycle by the enzyme RuBisCo, which converts CO_2 into the complex organic compounds. It was found that microalgal species have a great potential for CO_2 sequestration (Cheah 2015). Microalgal species such as *Anabaena sp.* and *C. vulgaris* can fix CO_2 at the rates of 1.45 and 6.24 g/L/d, respectively (Ghorbani et al. 2014). A growing commendation is there that for generating biomass and capturing carbon, microalgae are among the most productive biological system. Further efficiencies are gained by harvesting 100% of the biomass, much more than being possible in terrestrial biomass production systems. Microalgae are well suited to capture carbon by transporting bicarbonate into the cells. In open ponds, as high as 90% carbon dioxide- or bicarbonate-capturing efficiencies have been documented (Sayre 2010). As an alternative, photosynthesis has long been comprehended as a means to sequester CO_2 (Jeong et al. 2003). The natural sinks where carbon sequestration can be taken place are as follows: (i) forestation (terrestrial), (ii) oceanic fertilization, and (iii) microalgal cultures. Various co-workers have demonstrated the role of photosynthetic organisms in CO_2 biomitigation (Berberoglu et al. 2009; Benemann 1997; Hase et al. 2000).

Cyanobacteria and eukaryotic algae transport and use bicarbonate as a source of carbon dioxide (Spalding 2008). Algae have active bicarbonate pumps and can concentrate bicarbonate in the cell. The bicarbonate is subsequently dehydrated, either spontaneously or by carbonic anhydrase, and the resulting CO_2 is captured through Calvin-cycle activity, ultimately in the form of algal biomass. Between 1.6 and 2 g of CO_2 is captured for every gram of algal biomass produced (Herzog and Golomb 2004). The efficiency of CO_2 capture by algae can vary according to the state of the algal physiology, pond chemistry, and temperature. Carbon dioxide-capture efficiencies as high as 80% to 99% are achievable under optimal conditions and with gas residence times as short as 2 s (Keffer and Kleinheinz 2002). Photorespiration reduces the photosynthetic carbon fixation efficiency by 20% to 30% (Zhu et al. 2008). To reduce the competitive inhibition of oxygen on carbon fixation by RuBisCo, algae actively pump sufficient bicarbonate into the cells to increase internal CO_2 concentrations to the levels above those achievable by equilibrium with air, and competitively inhibit the photorespiration (Badger 1994).

Microalgae have much higher growth rates and CO_2 fixation rate compared to the conventional forestry, agriculture, and aquatic plants (Li et al. 2008), thus making them suitable candidates for removing CO_2. Selected media will be required to have high CO_2 fixation rates and a rapid growth rate, while being easily cultivated on a large scale in order to generate a large biomass yield and produce valuable by-products to offset the costs of carbon mitigation. A novel approach to offsetting emissions is through direct biological carbon mitigation where CO_2 from the flue gases of point sources is used to cultivate the photosynthetic autotrophic organisms. The produced biomass can subsequently be converted into biofuels, biochemicals, food, or animal feed. These useful by-products provide revenue to finance the carbon mitigation process (Farrelly et al. 2013).

13.5 FACTORS AFFECTING CARBON SEQUESTRATION

Soils have a large number of essential functions, some of them for the environment (protection function) and others for human or animal nutrition (production function). Most soil functions are significantly influenced by the quantity and quality of SOM. This factor is essential for soil organisms and their diversity, plant nutrition, water-holding capacity, aggregate stability, and erosion control. More recently, the role of SOC within the global C cycle has received increasing interest (Percival et al. 2000; Bernoux et al. 2002). Carbon farming is a technique that extracts carbon dioxide from the air and stores it in the soil (Berger 2019). The ability of agricultural lands to store or sequester carbon depends on several factors, including climate; soil type; and type of crop or vegetation, cover, and management practices (EPA 2008). The carbon farming method consists of composting, planting cover crops, reducing tillage, leaving crop residues on the land, and manage to graze (Berger 2018). On the other hand, temperature, rainfall, land management, soil nutrition, soil type, and agroclimatic regions all influence SOC levels. Although the importance of soil microorganisms in global C cycling is well known, only a few researchers have attempted to combine the chemical and microbiological views of the C cycling (Kandeler et al. 2001). Quantification of possible measures to increase C stocks in soils; no or minimum tillage (Sainju et al. 2002); conversion of arable land into pasture or forest; and changes in crop rotations (Blair and Crocker 2000; Gregorich et al. 2001) is currently being intensively pursued. Changes in organic C dynamics in soils are intimately connected with or even driven by changes in microbial activities and sustainable agricultural practices (Sinsabaugh et al. 2002).

13.5.1 Land Uses and Management

In an ecosystem, the carbon sequestration, the soil fertility, and the biodiversity are interlinked. The biodiversity of the aboveground and also of the underground changes as the land use pattern changes with time. Poor land management, deforestation, and land degradation contribute a lot to carbon dioxide emissions. This can be checked by reducing deforestation, by improving the practices of agriculture and forestry, and by adopting conservation practices of the land (FAO 1999). Alteration of carbon balance occurs when the natural systemic system is converted into a managed system (Yihenew and Getachew 2013). Also, the negative impact of agriculture on stock can be noticed (Berry 2011; Girshowing et al. 2008; Lemenih et al. 2005; Melero et al. 2011; Yeshanew et al. 2007). Nonetheless, agriculture is among the land use practices that emit and sequester CO_2 (Berry 2011).

The soil has a carbon-carrying capacity and an equilibrium carbon content, which depend on the nature of vegetation, precipitation, and temperature of the land (Jobbágy and Jackson 2000). The carbon inflow–outflow equilibrium in the soil is unsettled by the land use change until a new equilibrium is ultimately attained in the new ecosystem (Guo and Gifford 2002). During the process, the soil carbon stocks may change: either as a source of carbon or as a carbon sink (Van der Werf et al. 2009). However, the stock of carbon in a given soil depends on the number of carbon inputs per year (biomass added to the soil each year) together with the carbon input rates at which they are transformed into SOM (composition rate) and the amount of SOM which gets decomposed each year (decomposition rate). The composition and decomposition rates mainly depend on the biophysical factors such as soil texture and climate, whereas the kinds of biomass and their amounts that are added to the soil each year largely depend on land use (types of crops or vegetation), management of weeds and pests, and irrigation and fertilizer use (Gonçalves and Carlyle 1994; Verheijen et al. 2005). Farming practices that include minimal disturbance of soil and encourage the process of carbon sequestration will slow down or reverse the carbon loss from the fields (EPA 2008).

13.5.2 No-Tillage

There is an increased stock of carbon in agricultural soils when conservation agricultural practices (CAPs) are used in the field (West and Post 2002; Lal 2004; Smith 2004; Luo and Zhou 2010). Among all the CAP options, the conventional tillage (CT) and no-tillage (NT) are deemed to be efficient strategies (Smith et al. 1998; Paustian et al. 2000; Six et al. 2004a). As an alternative to CT, NT practice can cause organic carbon accumulation in the soil. Thus, this will mitigate climate change through carbon sequestration (Powlson et al. 2011. NT depends on any chemical herbicides, specialized planting equipment, and genetically modified seeds in order to reduce or eliminate the need for tillage equipment. Soil aggregates remain intact, physically protecting carbon as the soil in these systems stays undisturbed (West and Post 2002). Even though there are other valid reasons to use CT, information that it promotes carbon sequestration (Baker et al. 2007). Soil quality is largely regulated by SOM content, which is a dynamic pool. Also, it responds effectively to changes such as primary tillage, soil management, and carbon inputs resulting from the biomass production. The amount of carbon released from the soils is directly dependent on the volume of the soil, which is disturbed during the tillage operations. Therefore, for better carbon conservation, minimum tillage is recommended (Sharma et al. 2012).

13.5.3 Land Cover Management

It is a relatively inexpensive and practical means that can promote both above- and belowground carbon sequestration processes in terrestrial ecosystems at the local and regional scales (Metting et al. 2001). The balance towards a net gain of carbon can be tipped by the use of cover crops or periodic green allows (McDaniel et al. 2014; Tiemann et al. 2015). The shoots and roots of the cover crops are used to feed the fungi, earthworms, bacteria, and other soil organisms, which increases soil carbon levels with time. Cover crops keep the soil in place. They bolster soil health, improve the water quality, and reduce pollution from the agricultural activities (www.sare.org).

13.5.4 Crop Rotation

Greater SOC and nitrogen contents are achieved when crop management practices like rotations of crops are done with plants that produce high residue and surface residue cover, which are maintained with reduced tillage. This process will also improve the soil productivity. The increase in the complexity and diversity of soil carbon occurs when plant diversity is introduced to crop rotations and legume cover crops are used (Wickings et al. 2012). Crop rotations, especially those that include cover crops, sustain soil quality and productivity by enhancing soil C, N, and microbial biomass, making them a cornerstone for sustainable agro-ecosystems (Mc Daniel et al. 2014).

13.5.5 Regenerative Agriculture and Conservation Farming

Regenerative agriculture is defined as a combination of diverse cover crops, NT, multiple crop rotations, and the in-farm fertility, which will increase the soil carbon and sustain the soil health. Soil carbon is also increased through the conservation farming practices that conserve moisture, improve the yield potential, and also reduce erosion and fuel cost. Direct seeding, perennial forage crops, rotational grazing, field windbreaks, reduced summer fallow, and proper straw management are some examples of practices that can reduce carbon dioxide emissions and at the same time increase soil carbon (AARD 2000). The soil carbon is also increased by using higher-yielding crops, which also maximizes the yield potential.

13.5.6 Reforestation

The global carbon (C) cycle is significantly impacted by the terrestrial ecosystems, comprising vegetation and the uplands and wetlands soil. Terrestrial ecosystems are under natural conditions and are a sink of atmospheric carbon dioxide (CO_2). Nonetheless, the conversion of natural ecosystems to managed ecosystems weakens the ecosystem carbon stocks and aggravates the gaseous emissions. Now, it is time to implement restorative land use and soil management systems in order to strengthen the recarbonization of the terrestrial biosphere (Lal et al. 2018). Restoration and conversion of wastelands into vegetative lands can solve the global warming (Juwarkar et al. 2010) since the photosynthetic turnover exceeds 20 times the annual increase in atmospheric carbon dioxide. The CO_2 concentration in the atmosphere can be reduced by planting fast-growing plants on a huge scale on marginal land. Conversion of swamp plants into hummus or peat also reduces carbon concentration in the atmosphere (Dyson 1977). Conversion of marginal cropland to trees or grass maximizes the carbon storage on the land (Schahczenski and Hill 2008). Young and temperate forests are found to be more effective carbon sinks than the old rainforests since the newly deforested areas are mostly open and sunny. Also, they are easily recolonized by the fast-growing species. Such plants can extract carbon from the air and incorporate it into the biomass far quicker than the mature trees (Morgan 2019). Carbon footprint can be reduced by planting trees and conserving forests. Forests absorb carbon dioxide more than they release, and they are known as carbon sinks (Friedel 2017).

The main goals of reforestation are (i) for increasing the forests' landmass, (ii) for increasing the carbon density of the existing forests, and (iii) for promoting the use of more forest products and the use of biochar. Reforestation can also reduce the carbon deficit, which is caused by years of agricultural production (Ontl and Schulte 2012). Reforestation may also affect many soil properties primarily through fundamental changes in organic matter and nutrient cycling (Sauer et al. 2012). The excellent option for mitigating CO_2 emissions is carbon sequestration of forests having rapid growth rates like in tropical plantations and natural succession (Montagnini and Porras 1998). Rising atmospheric carbon dioxide (CO_2) levels associated with climate change are caused by the rainforests which hold a significant amount of the carbon that is stored in terrestrial ecosystems (Ontl and Schulte 2012). The additional benefits such as biodiversity habitat, air filtration, water filtration, flood control, and enhanced soil fertility are from reforestation. The balance between the accumulation of biomass and litter, losses from respiration and decomposition of litter, and soil carbon occurs when there is carbon sequestration following reforestation. Generally, the soil provides a more stable carbon store than the plant biomass. It continues to accumulate carbon even after forest maturity. Reforestation helps in good forest management, which includes the modification of rotation length, eradicating disease, avoiding losses from pests, managing the soil carbon pool, preventing fire and extreme weather, and maintaining biodiversity (O'Driscoll 2018).

13.5.7 Organic Amendments

Many soil characteristics, including colour, nutrient-holding capacity (cation- and anion-exchange capacity), nutrient turnover, and stability, are influenced by the total organic carbon, which in turn affects water relations, aeration, and workability. Carbon sequestration in croplands and rangelands needs SOM (a certain amount of organic matter present in the soil). Organic amendments like manures (Spokas and Reicosky 2009), different types of compost, and good agricultural practices such as cover crops, nutrient management, and mulching are effectively used for this purpose (Farooqi et al. 2018). The application of such practices and other organic by-products from a feedlot or modified organic materials (such as biochar, composts, and manure pellets) are widely practised in sole cropping systems (Spokas and Reicosky 2009). Organic amendments enhance the physical and chemical properties of soils (Hansen et al. 2016), also improve nutrient cycling, increase C sequestration, and reduce GHG emissions. The application of biochar obtained from various feedstocks has been shown to increase SON content, reduce CO_2 and N_2O emissions, and increase CH_4

uptake, as compared to no application of biochar. The effect of biochar on GHG emissions within amended soils relies on both biochar and soil properties (Spokas and Reicosky 2009). Also, composting, despite emitting GHGs during storage, and dried pelleting are two methods of conserving nutrients in manure and facilitating their slow release into the soil (Ball et al. 2004). As compared to the application of manure, the application of compost emits less CO_2 and N_2O, which are widely available by-products from livestock production systems. Manure management and its application to soil play a critical role in GHG emissions, including CH_4 and N_2O (Zhongqi et al. 2016), due to its high nutrient and C contents. While the application of pelleted manure leads to more N_2O emission compared to the application of raw manure, the inclusion of compost into the soil provides better nutrient input compared to raw manure from the perspective of C sequestration (Eghball and Ginting 2003). In short, enrichment planting and the application of organic soil amendments will be better options than the application of raw manure for enhancing C sequestration and reducing GHG emissions (Bharat et al. 2018). Also, SOC content can be enhanced by applying organic fertilizers, thus resulting in increased biomass production. This biomass production restores carbon into the soil (FAO 1999). The SOC sequestration rates clearly decreased in the treatment of chemical fertilizers after about 20 years. The overall mean SOC sequestration efficiency of the wide range of C inputs via roots, stubble, straw, and manure was equally high, with an overall mean efficiency of 16% over the 29-year period. The C sequestration efficiencies of wheat straw and cattle manure are higher than that of pig manure. The balance between organic C inputs into the soil (via crop residues, organic amendments in compost, animal manure, etc.) and organic C decomposition by soil microbes is the net SOC sequestration. Its efficiency is commonly expressed by the relationship between annual C input and SOC accumulation rate, which is an indicator of soil C sequestration ability (McLauchlan 2006). As a result, information about the C sequestration efficiency is useful for seeking high-efficiency management strategies for improving the SOC stock and soil fertility. Climate, the quantity and the quality of added organic materials, and soil-inherent properties (Hua et al. 2014) regulate C sequestration efficiency. Organic systems of production also increase the SOM levels and also eliminate the emissions from the production and transportation of synthetic fertilizers. Components of organic agriculture can be implemented with other sustainable farming systems, like CT, to further increase climate change mitigation potential (Jeff and Holly 2008).

13.5.8 Landscape-Level (Slope Position) Effects on Carbon Inputs

Slope position impacts soil moisture and nutrient levels, with subsequent impacts on the root growth of plants that may have consequences for soil carbon (Ehrenfeld et al. 1992). The variation in the carbon sequestration capacity across slope positions occurs due to the combined effects of changes in carbon inputs and the losses from land use, landscape-level effects on carbon input, and loss rates during land management (Ontl and Schulte 2012). Getu (2012) observed that the aboveground biomass carbon and the belowground biomass carbon tend to increase with slope. SOC also reduces as slope increases (Feyissa et al. 2013; Güner et al. 2012). The potential of a material to move down increases as the steep slope increases, which causes erosion (Barthès and Roose 2002; Castillo-Santiago et al. 2003) Steep slope soils are more vulnerable to erosion (Hancock et al. 2010). It was noted that vegetative biomass also contributed to SOC concentrations. Hence, Yohannes et al. (2015) reported that high SOC in lower slope might be attributable to the diversity of tree species and canopy, less erosion rate, and litterfall accumulation. In deep soil, the responses of SOC concentration to slope in various vegetation covers are different (Han et al. 2010). When organic carbon sources sequestered in soils are compared at different slope positions, it was found that soil aggregates have been playing a role in carbon sequestration process, which depends upon landscape positions and the soil profile depth. In soil structure, soil aggregate is one important characteristic and it is nearly linked to soil erodibility, soil biota, soil nutrient availability, soil water retention, and buffering capacity. The aforementioned characteristics influence the accumulation of soil carbon as they provide physical protection to the SOC (Wu et al. 1990; Beare et al. 1994; Fox and Le Bissonnais 1998; Wang et al. 2001; Six et al. 2000, 2004; Eynard et al. 2006). It was found that the soil aggregation is enhanced in 0–20 cm layer and the aggregates can absorb C into deep layers in the depositional environment (toe of the slope under the protection from human disturbances. The impacts of deposition, soil aggregates, erosion, and vegetation restoration on SOC accumulation and redistribution on soil erosion and deposition are huge. SOC sources that are eroded from upland and get redeposited in low-lying areas can be physically conserved against decomposition through the soil aggregation (Six et al. 2004; Yadav and Malanson 2007; Berhe et al. 2007).

13.5.9 Topographic Heterogeneity

The soil carbon sequestration capacity may get influenced by local controls on the processes of ecosystems at the scale of a watershed or crop field. The local scales may vary due to rainfall, infiltration, soil erosion, deposition of sediments, and soil temperature and landscape heterogeneity (Thompson and Kolka 2005). The SOC content decreases with the depth at all topographic positions (Parras-Alcántara et al. 2015). It was found that the mountain ecosystems show a high degree of topographic heterogeneity. This leads to heterogeneity in the snowpack, and soil properties, microclimates, and plant responses to environmental change may not be consistent across the landscape (Ackerly et al. 2010; Ford et al. 2013; Geiger et al. 2003). There is a shift in the distribution of herbaceous plants, shrubs, and trees as a result of climate change. Topography, soil properties, and snowpack are the fine-scale variables that can play an important role in modulating the vegetation responses to warming, primarily in the mountainous regions (Bourgeron et al. 2015).

13.5.10 Land Use Pattern

Agricultural activities are used as both GHG sources and GHG sinks. Agricultural GHG sinks are the reservoirs of carbon that have been removed from the atmosphere using the biological carbon sequestration process (IPCC 2007). CT, improved cropping change, land use change, organic production, land restoration, irrigation, and water management are some of the innovative farming practices from which the farmers can address climate change (Schahczenski and Hill 2008). 'Land use pattern' can be defined as the layout of the uses of the land. It can be used for agriculture, forest, pasture, agroforestry, etc (Lochan 2018). Continuous practice of monoculture, deforestation, and poor farming practices, often over-reliant on chemical inputs, have shredded the land of all its goodness. However, agriculture can be one of the major solutions to mitigate the losses. Other ways of keeping carbon in the soil are NT and lack of soil disturbances. Furthermore, cover cropping, tree plantation, and crop rotation help in preventing soil erosion. These kinds of practices help in keeping carbon in the soil, underground, and also out of the atmosphere (Hillsdon 2019). Perennial plants keep live roots in the soil year-round, thereby sequestering more carbon. The more carbon, the more water will be stored in the soil. Also, more perennial plants can grow and store carbon. In addition, decomposition of perennial plants in the soil also helps in adding organic material to the earth. Perennials that are deeper-rooted can reach a moisture inaccessible to annual grasses, thereby keeping the land green for much time and making macronutrients available. Their powerful roots also help in decompacting the soil (Berger, 2018). The carbon sinks such as forests and stable grasslands can store large amounts of carbon in their vegetation and root systems for long periods of time (EPA 2008b). The tropical highland rainforest and the soil dry valleys illustrate lower humification index, when compared with other agro-ecologies. It was estimated that the diversified production systems of crops with livestock are more stable for carbon stocks. They may also be essential in helping the farmers to adapt the effects of climate change (Segnini et al. 2011). The tropical forest vegetation is a large and a significant carbon storage as it contains more carbon per unit area than any other land covers (Yohannes et al. 2015). Research on grazing practices and the production of meat animals, particularly cattle, has gained considerable interest in carbon sequestration potential. When governed correctly, herds of grazing animals can actually maximize the annual pasture biomass production and redistribute the carbon throughout pastures, which is in the more processed form of manure. This leads to rapid increases in soil carbon. Techniques such as repeatedly moving cattle to new pastures, high stocking densities, and prevention of overgrazing make the pasture plants have high biomass. In addition, generally, this style of production does not require tillage. This means that soil aggregates are not disrupted and that their carbon stays physically protected from disturbance (Kane 2015). The soils under the pasture have greater nitrogen availability, nitrogen stocks, greater carbon, and lower C: N ratios than soils under the transitional vegetation and forests (Garten et al. 2002). The intentional integration of trees with crops and/or livestock in an agricultural production system, which is called as agroforestry, can enhance carbon sequestration and reduce GHG emissions from the terrestrial ecosystems. This will also mitigate global climate change. Agroforestry systems in subtropics store more C in vegetation and soils when compared to systems with only trees or trees grown with legumes or cereals as inter-cropped systems (He et al. 2013; Paquette et al. 2006). Combining woody vegetation with cropping and livestock production through agroforestry systems increases the total production and enhances food and nutrition security. This also mitigates the climate change effects (Nair 2007, 2011).

In agroforestry systems, carbon sequestration and GHG emissions are complex and depend on various biophysical factors such as climatic conditions, vegetation characteristics, soil properties, and water regime. An agroforestry system depends on the region-specific biophysical condition to enhance the carbon sequestration and to reduce GHG emissions. When compared to other climatic regions, agroforestry systems in temperate regions have higher carbon pools (Mutuo et al. 2005; Oelbermann et al. 2004).

13.5.11 Increased Temperature Effect on Carbon Sequestration of Soil

The carbon balance may be affected by the increased temperature because of the reduction in the rate of photosynthesis and limited water availability. But when water is available, an increase in temperature may increase the productivity of the plant, which in turn affects the carbon balance (Maracchi et al. 2005). Also, the activities of soil microorganisms are affected by climate change directly (changes in precipitation, extreme climatic condition) and indirectly (changes in soil physiochemical condition affecting plant diversity and productivity, carbon supply to the soil, activity and structure of the microbial community that participated in decomposition and release of carbon from the soil). A rise in temperature may lead to increased rates of decomposition by SOM which produces more CO_2, thus resulting in a positive effect on the changes in climate (Pataki et al. 2003).

13.5.12 Plant Respiration Impact on Carbon Sequestration

The terrestrial agricultural ecosystem's carbon cycle is influenced by the balance between respiration and photosynthesis (Prentice et al. 2004). We are unable to grasp the carbon climate system in a world where CO_2 is ever-increasing, and how the balance between sinks of terrestrial ecosystem like photosynthesis and its sources like respiration, including microbial respiration, affects it (Jin 2007). By photosynthesis, CO_2 is removed from the atmosphere and converted to organic carbon. This organic carbon is then converted to CO_2 by respiration (Sharma et al. 2012). By photosynthesis, a large amount of carbon is fixed which regulates the net productivity of ecosystem strongly. It also regulates the carbon cycles at global and regional levels, which in turn regulates the concentration of CO_2 and the dynamic of climate in the earth system. Soil respiration may not be a direct mechanism of storage of carbon on land, but it helps us to understand CO_2 sequestration and global carbon trading market (SD 2017).

13.5.13 Arid and Semi-Arid Climate

The formation rate of inorganic carbon is low in arid and semi-arid climates as CO_2 from the air in the soil is sequestered by changing it first into secondary carbonates (Lal et al. 2008). Due to the high rate of decomposition and primary productivity because of adequate rainfall and warm temperature, the tropics have intermediate SOC level. However, due to less primary productivity, arid regions have low SOC level (Ontl and Schulte 2012).

13.5.14 Temperate Ecosystems

In the temperate ecosystems, when moisture and temperature levels are high during summer, they can have high primary productivity, and when they have cool temperature the rest of year after summer, decomposition rates slow down and overtime organic matter builds up slowly(Ontl and Schulte 2012). In temperate grassland, stocks of sequestered carbon are mainly found below the ground in soils and roots (biomass of roots is in top 30 cm of soil). But, because of different rates of death, decomposition rates, growth of root, which occurs simultaneously, the determination of the accurate Carbon, which is transferred from different sources to the soil, is quite difficult (Reeder and Schuman 2002). Besides death and decomposition, plant roots contribute to soil carbon by rhizodeposition from sloughing, mucilage production, and exudation of living roots (Van veen et al. 1991; Reeder and Schuman 2002). The amount of atmospheric CO_2 is lowest when the Earth is cold. The photosynthetic rate exceeds the decomposition rate, which results in an increased level of SOC in the wet and cold climate of northern latitudes (Ontl and Schulte 2012).

13.5.15 Wetland Agriculture Land

Wetlands are considered a potential carbon sink, but when not properly managed, they become GHG source. In the wetland, the resulting carbon storage from the balance between the input of carbon (like organic matter) and the output of carbon (such as methanogenesis, decomposition) depends upon geography and topography, hydrological regime, the presence of plant types, soil moisture, morphology, pH, and temperature (Adhikari et al. 2009). A large amount of carbon is sequestered in wetlands and ponds as there is a reduction in decomposition in waterlogged soils due to the lack of oxygen. The amount of carbon stored in the soils can be increased drastically by crop rotation, nutrient management, cover cropping, and CT (Sharma et al. 2012). Disturbance and degradation of wetlands result in higher rate of carbon stored in them to be decomposed and released to the atmosphere as GHGs. So, to retain the present carbon reserve, protecting wetlands is a way to avoid the emission of CO_2 and GHGs.

13.6 OUTLOOK AND CONCLUSION

Therefore, taking everything into account, we conclude that the role of different types of microbes in the carbon sequestration in different agro-ecosystems is magnificent. From the carbon cycle to nutrient cycling and the synthesis of organic matter, the role of microbes is undoubtedly important. Microbial modulation has been considered as one of the strategies to enhance soil C sequestration. For carbon farming, methods such as composting, planting cover crops, reducing tillage, leaving crop residues on the land, and managing grazing are used in the agro-ecosystems. NT, land cover management, crop rotation, regenerative agriculture and conservation farming, reforestation, and use of organic amendments are known to conserve SOC. It is found out that minimum disturbance of soil will result in the maximum conservation of SOC. In the era of climate change, the effects of increased temperature on carbon sequestration are also well deciphered. It is found that the rate of decomposition of SOM and the production of GHGs are increased, thereby contributing to global warming. From the study of sequestration of carbon in different ecosystems such as arid and semi-arid regions, temperate regions, and wetland ecosystem, it is found that sequestration of carbon is low in arid regions and high levels of SOC are found in the temperate and wetland ecosystems. As highlighted by the above studies, knowing the approaches of carbon capture and storage is essential in the context of mitigating climate change. This will help the present and the generations to come reverse the climate change effects. The effective methods of restoring carbon in the soil such as methods of generating clean energy and cleaner methods of farming, which existed before humans tampered with the environment, are vital. After all the destruction, ignorance is not bliss in the era of climate change. Hence, we should try to fill in the knowledge gaps and help in the mitigation process.

REFERENCES

AARD (Alberta Agriculture and Rural Development). 2000. Greenhouse Gas Emissions and Alberta's Cropping Industry-Things You Need to Know. https://www1.agric.gov.ab.ca/$Department/ deptdocs.nsf/all/cl9706/$FILE/Cropping.pdf (accessed December 12, 2019)

Ackerly, D.D., Loarie, S.R., Cornwell, W.K., et al. 2010. The geography of climate change: Implications for conservation biogeography. *Diversity and Distributions* 16: 476–487. doi: 10.1111/j.1472–4642.2010.00654.x.

Adhikari, S., Bajracharaya, R., & Sitaula, B. 2009. A review of carbon dynamics and sequestration in wetlands. *Journal of Wetlands Ecology* 2(1): 42–46. doi: 10.3126/jowe.v2i1.1855.

Agarwal L., Qureshi A., Kalia V.C., et al. 2014. Arid ecosystem: Future option for carbon sinks using microbial community intelligence. *Current Science* 106: 1357–1363.

Ahmed A.A.Q., Odelade K.A., & Babalola O.O. 2019. Microbial inoculants for improving carbon sequestration in agroecosystems to mitigate climate change. In: W. Leal Filho (eds) *Handbook of Climate Change Resilience*. Springer, Cham. pp. 1–21. doi: 10.1007/978-3-319-71025-9_119-1.

Albuquerque, M.G.E., Martino, V., Pollet, E., Avérous, L., & Reis, M.A.M. 2011. Mixed culture polyhydroxyalkanoate (PHA) production from volatile fatty acid (VFA)-rich streams: Effect of substrate composition and feeding regime on PHA productivity, composition and properties. *Journal of Biotechnology* 151: 66–76.

Aminiyan, M.M., Hosseini, H., & Heydariyan, A. 2018. Microbial communities and their characteristics in a soil amended by nanozeolite and some plant residues: Short time in-situ incubation. *Eurasian Journal of Soil Science* 7 (1):9–19.

Aminiyan, M.M., Sinegani, A.A.S., & Sheklabadi, M., 2015a. Aggregation stability and organic carbon fraction in a soil amended with some plant residues, nanozeolite, and natural zeolite. *International Journal of Recycling of Organic Waste in Agriculture* 4(1): 11–22.

Aminiyan, M.M., Sinegani, A.A.S., & Sheklabadi, M., 2015b. Assessment of changes in different fractions of the organic carbon in a soil amended by nanozeolite and some plant residues: Incubation study. International *Journal of Recycling of Organic Waste in Agriculture*. 4(4): 239–247.

Atomi, H. 2002. Microbial enzymes involved in carbon dioxide fixation. *Journal of Bioscience and Bioengineering* 94(6):497–505.

Badger, M.R., & Price, G.D. 1994. The role of carbonic anhydrase in photosynthesis. *Annual Review of Plant Physiology and Plant Molecular Biology* 45: 369–392.

Bailey, V.L., Smith, J.L., & Bolton, H., Jr. 2002. Fungal-to-bacterial ratios in soils investigated for enhanced C sequestration. Soil Biology and Biochemistry 34: 997–1007 doi: 10.1016/S0038–0.

Baker, J.M. Ochsner, T.E., Venterea, R.T., & Griffis, T.J. 2007. Tillage and soil carbon sequestration—What do we really know? *Agriculture, Ecosystems & Environment* 118(1–4): 1–5. https://doi.org/10.1016/j.agee.2006.05.014.

Ball, B., McTaggart, I., & Scott, A. 2004. Mitigation of greenhouse gas emissions from soil under silage production by use of organic manures or slow-release fertilizer. *Soil Use and Management.* 20: 287–295. doi:10.1111/j.1475–2743.2004.tb00371.x

Barthès, B., & Roose, E., 2002. Aggregate stability as an indicator of soil susceptibility to runoff and erosion; validation at several levels. *Catena* 47 (2): 133–149.

Beare, M.H., Hendrix, P.F., Cabrera, M.L. & Coleman, D.C. 1994. Aggregate-protected and unprotected organic matter pools in conventional- and no-tillage soils. *Soil Science Society of America Journal* 58: 787–795. doi:10.2136/sssaj1994.03615995005800030021x.

Bell, J.M., Falconer, C., Colby, J., & Williams, E. 1987. CO metabolism by a thermophilic actinomycete, Streptomyces strain G26. *The Journal of General Microbiology* 133: 3445–3456.

Benemann, J. 1997. CO_2 mitigation with microalgae systems. *Energy Conversion and Management* 38: 475–479

Berberoglu, H., Gomez, P.S., & Pilon, L. 2009. Radiation characteristics of Botryococcus braunii, Chlorococcum littorale, and Chlorella sp. used for CO_2 fixation and biofuel production. *Journal of Quantitative Spectroscopy* 110: 1879–1893.

Berger J.J. 2018. Could our farms become the world's great untapped carbon sink? https://healthyforests.files.wordpress.com/2018/10/sustain-europe-aut-win-2019-johnjberger-carbon-farming.pdf.

Berger J.J. 2019. Can soil microbes slow climate change? https://www.scientificamerican.com/article/can-soil-microbes-slow-climate-change/.

Berhe, A.A., Harte, J., Harden, J.W., & Torn, M.S. 2007. The significance of the erosion induced terrestrial carbon sink. *BioScience* 57: 337–346.

Bernoux, M., da Conceição Santana Carvalho, M., Volkoff, B., & Cerri, C.C. 2002. Brazil's carbon stocks. *Soil Science Society of America Journal* 66: 888–896.

Berry, N. 2011. Whole farm Carbon accounting by smallholder, lesson from plan Vivo project presentation at the small scale mitigation whole farm and landscape accounting workshop. FAO, Rome, 27–28th October, 2011.www.fao.org/climatechnage/micca/72531/en/.

Bharat, M., Shrestha, S.X. et al. 2018. Enrichment Planting and Soil Amendments Enhance Carbon Sequestration and Reduce Greenhouse Gas Emissions in Agroforestry Systems: A Review. https://www.mdpi.com/1999-4907/9/6/369/pdf

Blair, N., & Crocker, G.L. 2000. Crop rotation effects on soil carbon and physical fertility of two Australian soils. *Australian Journal of Soil Research* 38: 71–84

Bonkowski, M., Villenave, C., & Griffiths, B. 2009. Rhizosphere fauna: The functional and structural diversity of intimate interactions of soil fauna with plant roots. *Plant Soil.* 321: 213–233. doi:10.1007/s11104-009-0013-2.

Bourgeron, P.S., Humphries, H.C., & Liptzin, D. 2015. The forest alpine ecotone: A multi-scale approach to spatial and temporal dynamics of treeline change at Niwot Ridge. *Plant Ecology and Diversity* 8 (5–6): 763–79. doi:10.1080/17550874.2015.1126368.

Brigham, C.J., Budde, C.F., Holder, J.W., et al. 2010. Elucidation of β-oxidation pathways in Ralstonia eutropha H16 by examination of global gene expression. *Journal of Bacteriology* 192 (20): 5454–5464. doi: 10.1128/JB.00493–10.

Canadell, J.G., Le Quere, C., Raupach, M.R., et al. 2007. Contributions to accelerating atmospheric CO_2 growth from economic activity, carbon intensity, and efficiency of natural sinks. *Proceedings of the National Academy of Sciences* 104: 18866–18870.

Carlson D. 2018. Carbon Sequestration and Microbial. *Communities in Mangrove Ecosystems*. Stanford University. http://surj.stanford.edu/wp-content/uploads/2018/09/DaryllCarlson.pdf

Castillo-Santiago, M.A., Hellier, A., Tipper, R., & De Jong, B.H.J. 2003. Carbon emissions from land use change: An analysis of causal factors in Chiapas, Mexico. *Mitigation and Adaptation Strategies for Global Change* 12: 1–30.

Chang, Kenneth, 2016. "Visions of life on mars in earth's depths". *New York Times*. Archived from the original on September 12, 2016. (Retrieved November 12, 2019).

Cheah, W.Y., Show, P.L., Chang, J-S., Ling, T.C., & Juan, J.C. 2015. Biosequestration of atmospheric CO_2 and flue gas-containing CO_2 by microalgae. *Bioresource Technology* 184: 190–201. doi: 10.1016/j.biortech.2014.11.026.

Chen, P., Min Min, Yifeng Chen, et al. 2009. Review of the biological and engineering aspects of algae to fuels approach. *International Journal of Agricultural and Biological Engineering* 2(4): 1–30. doi: 10.3965/j.issn.1934–6344.2009.04.001–030.

Cheng, L, Booker, FL, & Tu, C, et al. 2012. Arbuscular mycorrhizal fungi increase organic carbon decomposition under elevated CO_2. *Science*. 337 (6098): 1084–1087. doi: 10.1126/science.1224304.

Chowdhury, S.P., Schmid, M., Hartmann, A., & Tripathi, A.K., 2007. Identification of diazotrophs in the culturable bacterial community associated with roots of Lasiurus sindicus, a perennial grass of Thar Desert, India. *Microbial Ecology* 54(1): 82–90.

Clemmensen, K.E., Bahr, A., & Ovaskainen, O. et al. 2013. Roots and associated fungi drive long-term carbon sequestration in boreal forest. *Science* 339(6127): 1615–1618. doi: 10.1126/science.1231923.

Cusack, D.F., Silver, W.L., Torn, M.S., & McDowell W.H. 2011. Effects of nitrogen additions on above- and belowground carbon dynamics in two tropical forests. *Biogeochemistry* 104: 203. doi: 10.1007/s10533-010-9496-4.

Daldal, F., Thurnauer M.C., & Hunter C.N. 2008. Advances in photosynthesis and respiration, In: C.N. Hunter, F. Daldal, M.C. Thurnauer & J.T. Beatty (eds) *The Purple Phototrophic Bacteria* vol. 28, pp. 1–1013. doi: 10.1007/978-1-4020-8815-5.

Dębska, B., Długosz, J., Piotrowska-Długosz, A. et al. 2016. The impact of a bio-fertilizer on the soil organic matter status and carbon sequestration—results from a field-scale study. *Journal of Soils Sediments*. 16: 2335–2343 doi:10.1007/s11368-016-1430-5

Denef, K., Roobroeck, D., Manimel Wadu, M.C.W., Lootens P., & Boeckx P. 2009. Microbial community composition and rhizodeposit-carbon assimilation in differently managed temperate grassland soils. *Soil Biology and Biochemistry* 41(1): 144–153

Dignac, M., Derrien, D., Barré, P. et al. 2017. Increasing soil carbon storage: Mechanisms, effects of agricultural practices and proxies. A review. *Agronomy for Sustainable Development* 37:14 doi: 10.1007/s13593-017-0421-2

Ding, L., & Yokota, A. 2004. Proposals of *Curvibacter gracilis* gen. nov., sp. nov. and *Herbaspirillum putei* sp. nov. for bacterial strains isolated from well water and reclassification of [*Pseudomonas*] *huttiensis*, [*Pseudomonas*] *lanceolata*, [*Aquaspirillum*] *delicatum* and [*Aquaspirillum*] *autotrophicum* as *Herbaspirillum huttiense* comb. nov., *Curvibacter lanceolatuscomb.* nov., *Curvibacter delicatus* comb. nov. and *Herbaspirillum autotrophicum* comb, Nov. *International Journal of Systematic Evolutionary Microbiology* 54: 2223–2230.

Dungait, J.A.J., Hopkins, D.W., Gregory, A.S., & Whitmore, A.P. 2012. Soil organic matter turnover is governed by accessibility not recalcitrance. *Global Change Biology* 18: 1781–1796. doi: 10.1111/j.1365–2486.2012.02665.x

Dyson, F.J. 1977. Can we control the carbon dioxide in the atmosphere. *Energy* 2(3): 287–291.

Eghball, B., & Ginting, D. 2003. *Carbon Sequestration Following Beef Cattle Feedlot Manure, Compost, and Fertilizer Applications.* Nebraska Beef Cattle Reports, University of Nebraska: Lincoln, NE.

Ehrenfeld, J.G., Kaldor, E., & Parmelee, R.W. 1992. Vertical distribution of roots along a soil topo sequence in the New Jersey pinelands. *Canadian Journal of Forest Research.* 22: 1929–1936.

Elsden, S.R. 1962. In: I.C. Gunsalus, & R.Y. Stanier, (ed.) *The Bacteria*, vol. 3. Academic Press, New York and London, p. 1.

EPA. 2008a. Carbon Sequestration in Agriculture and Forestry. www.epa.gov/sequestration/index.html

EPA. 2008b. Carbon Sequestration in Agriculture and Forestry. www.epa.gov/sequestration/index.html.

Evans, M.C.W., Buchanan, B.B., & Arnon, D.I. 1966. A new ferredoxin-dependent carbon reduction cycle in a photosynthetic bacterium. *Proceedings of the National Academy of Sciences of the United States of America* 55: 928.

Eynard, A., Schumacher, T.E., Lindstrom, M.J., Malo, D.D., & Kohl, R.A. 2006. Effects of aggregate structure and organic C on wettability of Ustolls. *Soil & Tillage Research* 88: 205–216.

Ezeji, T., Qureshi, N., & Blaschek, H.P. 2007. Butanol production from agricultural residues: Impact of degradation products on *Clostridium beijerinckii* growth and butanol fermentation. *Biotechnology Bioengineering* 97: 1460–1469.

Fabrizzi, K.P., Rice, C.W., & Amado, T.J.C. 2009. Protection of soil organic C and N in temperate and tropical soils: Effect of native and agroecosystems. *Biogeochemistry.* 92: 129–143. doi:10.1007/s10533-008-9261-0.

Falkowski, P.G., Fenchel, T., & Delong, E.F. 2008. The microbial engines that drive Earth's biogeochemical cycles. *Science.* 320: 1034–1039.

Fang, M., Motavalli, P.P., Kremer, R.J., & Nelson, K.A. 2007. Assessing changes in soil microbial communities and carbon mineralization in Bt and non-Bt corn residue-amended soils. *Applied Soil Ecology.* 37(1–2): 150–160.

FAO (Food and Agriculture Organization of the United Nations. 1999. Prevention of land degradation, enhancement of carbon sequestration and conservation of biodiversity through land use change and sustainable land management with a focus on Latin America and the Caribbean. *Proceedings of an IFAD/FAO Expert Consultation IFAD*, Rome, Italy, 15 April 1999. http://www.fao.org/3/a-bc909e.pdf

Farrelly D.J., Everard C.D., Fagan C.C., & McDonnell K.P. 2013. Carbon sequestration and the role of biological carbon mitigation: A review. *Renewable and Sustainable Energy Reviews.* 21: 712–727. doi: 10.1016/j.rser.2012.12.038.

Farooqi Z.U.R., Sabir M., Zeeshan N., & Naveed K. et al. 2018. Enhancing carbon sequestration using organic amendments and agricultural practices. *Carbon Capture, Utilization and Sequestration*, Ramesh K. Agarwal, IntechOpen. doi: 10.5772/intechopen.79336.

Feyissa, A., Soromessa, T., & Argaw, M. 2013. Forest carbon stocks and variations along altitudinal gradients in Egdu forest: Implications of managing forests for climate change mitigation. *Science, Technology and Arts Research Journal* 2(4): 40–46.

Ford, K.R., Ettinger, A.K., Lundquist, J.D., Raleigh, M.S., & Hille Ris L.J. 2013. Spatial heterogeneity in ecologically important climate variables at coarse and fine scales in a high-snow mountain landscape. *PLoS ONE.* 8: e65008.

Fox, D.M., & Le Bissonnais, Y. 1998. Process-based analysis of aggregate stability effects on sealing, infiltration, and inter-rill erosion. *Soil Science Society of America Journal.* 62: 717–724.

Friedel, M. 2017. Forests as Carbon Sinks. [online] American Forests. Available at: https://www.americanforests.org/blog/forests-carbon-sinks/ [Accessed 1 Nov. 2019].

Gaffron, H., & Wohl, K. 1936. Zur Theorie der Assimilation. *Naturwissenschaften* 24: 81–90 doi:10.1007/BF01473561.

Garcia-Pausas, J., & Paterson, E. 2011. Microbial community abundance and structure are determinants of soil organic matter mineralisation in the presence of labile carbon. *Soil Biology and Biochemistry* 43(8): 1705–1713.

Garten, C.T., & Ashwood, T.L. 2002. Landscape level differences in soil carbon and nitrogen: Implications for soil carbon sequestration. *Global Biogeochemical Cycles* 16(4): 1114, doi: 10.1029/2002GB001918.

Geiger, R., Aron, R.H., & Todhunter, P. 2003. *The Climate Near the Ground*. Lanham, MD: Rowman & Littlefield Publishers, Inc. p. 589. doi: 10.1007/978-3-322-86582-3.

Getu, S. 2012. *Carbon Stocks in Different Pools in Natural and Plantation Forests of Chilimo, Central Highland of Ethiopia*. Unpublished M.Sc thesis, Addis Ababa University, Addis Ababa.

Ghorbani, A., Rahimpour, H.R., Ghasemi, Y., Zoughi, S., & Rahimpour, M.R. 2014. A review of carbon capture and sequestration in Iran: Microalgal biofixation potential in Iran. *Renewable and Sustainable Energy Reviews,* 35: 73–100.

Gonçalves, J.L.M., & Carlyle, J.C. 1994. Modelling the influence of moisture and temperature on net nitrogen mineralization in a forested sandy soil. *Soil Biology and Biochemistry*. 26: 1557–1564.

Gougoulias, C., Clark, J.M., & Shaw, L.J. 2014. The role of soil microbes in the global carbon cycle: Tracking the below-ground microbial processing of plant-derived carbon for manipulating carbon dynamics in agricultural systems. *Journal of the Science of Food and Agriculture* 94(12): 2362–2371. doi:10.1002/jsfa.6577.

Grandy, A.S., & Robertson, G.P. 2006. Aggregation and organic matter protection following tillage of a previously uncultivated soil. *Soil Science Society of America Journal* 70(4): 1398–1406.

Grandy, A.S., & Robertson, G.P. 2007. Land-use intensity effects on soil organic carbon accumulation rates and mechanisms. *Ecosystems* 10: 59–74.

Gregorich, E.G., Drury, C.F., & Baldock, J.A. 2001. Changes in soil carbon under long-term maize in monoculture and legume-based rotation. *Canadian Journal of Soil Science*. 81: 21–31.

Güner, S.T., Çömez, A., & Özkan, K. 2012. Predicting soil and forest floor carbon stocks in Western Anatolian Scots pine stands, Turkey. *African Journal of Agricultural Research*. 7: 4075–4083.

Guo, L.B., & Gifford, R.M. (2002), Soil carbon stocks and land use change: A meta analysis. *Global Change Biology*, 8: 345–360. doi:10.1046/j.1354-1013.2002.00486.x.

Han, X., Tsunekawa, A., Tsubo, M., & Li, S. 2010. Effects of land-cover type and topography on soil organic carbon storage on Northern Loess Plateau, China. *Acta Agriculturae Scandinavica, Section B Soil & Plant Science*. 60.4: 326–334.

Hancock, G.R., Murphy, D., & Evans, K.G. 2010. Hillslope and catchment scale soil organic carbon concentration: An assessment of the role of geomorphology and soil erosion in an undisturbed environment. *Geoderma*. 155: 36–45. doi: 10.1016/j.geoderma.2009.11.021.

Hansen, V., Müller-Stöver, D., Munkholm, L.J. et al. 2016. The effect of straw and wood gasification biochar on carbon sequestration, selected soil fertility indicators and functional groups in soil: An incubation study. *Geoderma*. 269: 99–107.

Hase, R., Oikawa, H., Sasso, C., Morito, M., and Watabe, Y. 2000. Photosynthetic production of microalgal biomass in a race way system under greenhouse conditions in Sendai City. *Journal of Bioscience and Bioengineering* 89: 157–163.

Hashem, A., Tabassum, B., & Fathi Abd A.E. 2019. *Bacillus subtilis:* A plant-growth promoting Rhizobacterium that also impacts biotic stress. *Saudi Journal of Biological Sciences*. 26(6): 1291–1297. doi:10.1016/j.sjbs.2019.05.004

He, Y., Qin, L., Li, Z. et al. 2013. Carbon storage capacity of monoculture and mixed-species plantations in subtropical China. Forest Ecology and Management 295: 193–198.

Herzog, H., & Golomb, D. 2004. Carbon capture and storage from fossil fuel use. *Encyclopedia of Energy* 1: 1–11.

Herzog, H.J., & Drake, E.M. 1996. Carbon dioxide recovery and disposal from large energy systems. *Annual Review of Environment and Resources* 21: 145–166.

Hillsdon, M. 2019. Turning Agriculture from Climate Culprit to Carbon Sink. 27 May 2019, http://www.ethicalcorp.com/turning-agriculture-climate-culprit-carbon-sink.

Holo, H. 1989. *Chloroflexus aurantiacus* secretes 3-hydroxypropionate, a possible intermediate in the assimilation of CO_2 and acetate. *Archives of Microbiology*, 151: 252. doi: 10.1007/BF00413138.

Hua, K., Wang, D., Guo, X., & Guo, Z. 2014. Carbon sequestration efficiency of organic amendments in a long-term experiment on a vertisol in Huang-Huai-Hai Plain, China. *PloS one*. 9(9): e108594. doi: 10.1371/journal.pone.0108594.

IPCC. 2007. Climate change 2007: Agriculture. In: B. Metz, O.R. Davidson, P.R. Bosch, R. Dave, & L.A. Meyer (eds) *Contribution of Working Group III to the Fourth Assessment Report of the Intergovernmental Panel on Climate Change*, Cambridge, UK and New York: Cambridge University Press, pp. 497–540

Jenkinson, D.S., & Ladd, J.N. 1981. Microbial biomass in soil: Measurement and turnover. In: E.A. Paul & J.N. Ladd (eds) *Soil Biochemistry*, vol. 5. Marcel Dekker, New York, pp. 415–471.

Jeong, M.J., Gillis, J.M., & Hwang, J.Y. 2003. Carbon dioxide mitigation by microalgal photosynthesis. *Bulletin of the Korean Chemical Society* 24(12): 1763–1766.

Jesus, E.C., Liang C., & Quensen J.F., et al. 2015. Influence of corn, switchgrass, and prairie cropping systems on soil microbial communities in the Upper Midwest of the United States. *GCB Bioenergy*. 8: 481–494. doi: 10.1111/gcbb.12289.

Jin, V.L., & Evans, R.D. 2007. Elevated CO_2 increases microbial carbon substrate use and nitrogen cycling in Mojave Desert soils. *Global Change Biology* 13: 452–465.

Jobbágy, E.G., & Jackson, R.B. 2000. The vertical distribution of soil organic carbon and its relation to climate and vegetation. *Ecological Applications* 10: 423–436.

Juwarkar, A.A., Mehrotraa, K.L., Nair, R. et al. 2010. Carbon sequestration in reclaimed manganese mine land at Gumgaon, India. *Environmental Monitoring and Assessment* 160: 457–464. doi: 10.1007/s10661-008-0710-y.

Kambale J.B., & Tripathi V.K. 2010. Biotic and abiotic processes as a carbon sequestration strategy. *Journal of Environmental Research and Development* 5(1): 240–251.

Kandeler, E., Tscherko, D., Stemmer, M., Schwarz, S., & Gerzabek, M.H. 2001. Organic matter and soil microorganisms - investigations from the micro-to the macro-scale. *Bodenkultur*. 52: 117–131.

Kane D. 2015. Carbon Sequestration Potential on Agricultural Lands: A Review of Current Science and Available Practices. http://sustainableagriculture.net/wpcontent/uploads/2015/12/Soil_C_review_Kane_Dec_4-final-v4.pdf.

Kaur L., Khajuria R., & Kaushik A. 2016. Microbial Carbon Sequestration. In: J. Singh & P. Gehlot (eds) *Microbes: In Action* Chapter: 14. Agrobios, Jodhpur, pp. 227–240.

Keffer, J.E., & Kleinheinz, G.T. 2002. Use of *Chlorella vulgaris* for CO_2 mitigation in a photobioreactor. *Journal of Industrial Microbiology and Biotechnology* 29: 275–280.

Kindler, R., Miltner, A., Richnow, H.-H., & Kastner, M. 2006. Fate of gram-negative bacterial biomass in soil–mineralization and contribution to SOM. *Soil Biology and Biochemistry* 38: 2860–2870. doi: 10.1016/j.soilbio.2006.04.047.

Kuenen, G. 2009. Oxidation of inorganic compounds by chemolithotrophs. In: J. Lengeler, G. Drews, & H. Schlegel. (eds) *Biology of the Prokaryotes*. Stuttgart: John Wiley & Sons, p. 242.

Lal, R. 2004. Soil carbon sequestration impacts on global climate change and food security. *Science* 304: 1623–1627. doi: 10.1126/science.1097396.

Lal, R., Smith, P., Jungkunst, H.F. et al. 2018. The carbon sequestration potential of terrestrial ecosystems. *Journal of Soil and Water Conservation*. 73(6): 145A–152A. doi: 10.2489/jswc.73.6.145A.

Lemenih, M., Karltun, E., & Olsson, M. 2005. Soil organic matter dynamics after deforestation along a farm field chronosequence in southern highlands of Ethiopia. *Agriculture, Ecosystems & Environment* 109: 9–19.

Li, J., Wu, X., Gebremikael, M.T. et al. 2018. Response of soil organic carbon fractions, microbial community composition and carbon mineralization to high-input fertilizer practices under an intensive agricultural system. *PLoS One*. 13(4): e0195144. doi:10.1371/journal.pone.0195144.

Li, Y., Horsman, M., Wu, N., Lan, C.Q, & Dubois-Calero, N. 2008. Biofuels from Microalgae. *Biotechnology Progress*. 24: 815–820. doi:10.1021/bp070371k.

Liang, C., & Balser, T.C. 2011. Microbial production of recalcitrant organic matter in global soils: Implications for productivity and climate policy. *Nature Reviews Microbiology*. 9: 75. doi: 10.1038/nrmicro2386-c1.

Liang, C., Jesus, E.C., & Duncan, D.S. et al. 2012. Soil microbial communities under model biofuel cropping systems in southern Wisconsin, USA: Impact of crop species and soil properties. *Applied Soil Ecology* 54: 24–31. doi:10.1016/j.apsoil.2011.11.015.

Liang, C., Fujinuma, R., & Balser, T.C. 2008. Comparing PLFA and amino sugars for microbial analysis in an Upper Michigan old growth forest. *Soil Biology and Biochemistry* 40: 2063–2065. doi: 10.1016/j.soilbio.2008.01.022.

Liew, F., Henstra, A.M., Winzer, K. et al. 2016. Insights into CO_2 fixation pathway of *Clostridium autoethanogenum* by targeted mutagenesis. *mBio* 7 (3): e00427–e00416. doi: 10.1128/mBio.00427-16.

Liu, M., Sui, X., Hu, Y., & Feng, F. 2019. Microbial community structure and the relationship with soil carbon and nitrogen in an original Korean pine forest of Changbai Mountain, China. *BMC Microbiology* 19: 218. doi: 10.1186/s12866-019-1584-6.

Lochan S. 2018. What is Land Use Pattern? Types of Land Use in India. https://www.sansarlochan.in/en/land-use-pattern-types/.

Lu, Y., & Conrad, R. 2005. *In situ* stable isotope probing of methanogenic archaea in the rice rhizosphere. *Science*. 309(5737): 1088–1090. doi: 10.1126/science.1113435.

Luo, Y., & Zhou, X. 2010. *Soil Respiration and the Environment*. Academic Press, Burlington.

Maracchi, G., Sirotenko, O., & Bindi, M. 2005. Impacts of present and future climate variability on agriculture and forestry in the temperate regions: Europe. *Climate Change* 70: 117–135.

Marie-France, D., Delphine, D., & Barré P., et al. 2017. Increasing soil carbon storage: Mechanisms, effects of agricultural practices and proxies. A review. *Agronomy for Sustainable Development* 37 (2):14. doi: 10.1007/s13593-017-0421-2.

Mathew, R.P., Feng, Y., Githinji, L., Ankumah, R., & K.S. Balkcom 2012. Impact of No-Tillage and Conventional Tillage Systems on Soil Microbial Communities. *Applied and Environmental Soil Science*. 548620:10. doi: 10.1155/2012/548620.

McDaniel, M.D., Tiemann, L.K. and Grandy, A.S. 2014. Does agricultural crop diversity enhance soil microbial biomass and organic matter dynamics? A meta-analysis. *Ecological Applications*. 24: 560–570. doi: 10.1890/13-0616.1.

McDowell J. 2019. Cover crops and carbon sequestration: Benefits to the producer and the planet. (2019, March 11). Retrieved November 1, 2019, from https://cropwatch.unl.edu/2019/cover-crops-and-carbon-sequestration-benefits-producer-and-planet.

McGowan, A.R., Nicoloso, R.S., Diop, H.E., Roozeboom, K.L., & Rice, C.W. 2019. Soil organic carbon, aggregation, and microbial community structure in annual and perennial biofuel crops. *Agronomy Journal* 111: 128–142 doi:10.2134/agronj2018.04.0284.

McLauchlan, K.K. 2006. Effects of soil texture on soil carbon and nitrogen dynamics after cessation of agriculture, *Geoderma* 136: 289–299. doi: 10.1016/j.geoderma.2006.03.053.

Meijer W.G. 1994. The Calvin cycle enzyme phosphoglycerate kinase of *Xanthobacter flavus* required for autotrophic CO_2 fixation is not encoded by the cbb operon. *Journal of Bacteriology* 176(19): 6120–6126. doi: 10.1128/jb.176.19.6120-6126.1994.

Melero, S., López, B.R., López, B.L., Muñoz, R.V., Moreno, F., & Murillo, J.M.. 2011. Long-term effect of tillage, rotation and nitrogen fertilizer on soil quality in a mediterranean vertisol. *Soil & Tillage Research* 114: 97–107. doi: 10.1016/j.still.2011.04.007.

Mellado-Vázquez, P.G., Lange, M., & Gleixner, G. 2019. Soil microbial communities and their carbon assimilation are affected by soil properties and season but not by plants differing in their photosynthetic pathways (C3 vs. C4). *Biogeochemistry* 142: 175–187. doi:10.1007/s10533-018-0528-9.

Mellado-Vázquez, P.G., Lange, M., Bachmann, D. et al. 2016. Plant diversity generates enhanced soil microbial access to recently photosynthesized carbon in the rhizosphere. *Soil Biology and Biochemistry* 94:122–132.

Metting, F.B., Smith, J.L., Amthor, J.S., & Izaurralde, R.C. 2001. Science needs and new technology for increasing soil carbon sequestration. *Climatic Change*. 51: 11–34. doi:10.1023/A:1017509224801.

Migliardini, F., De Luca, V., Carginale, V. et al. 2014. Biomimetic CO_2 capture using a highly thermostable bacterial α-carbonic anhydrase immobilized on a polyurethane foam. *Journal of Enzyme Inhibition and Medicinal Chemistry*. 29(1): 146–150. doi: 10.3109/14756366.2012.761608.

Montagnini, F., & Porras, C. 1998. Evaluating the role of plantations as carbon sinks: An example of an integrative approach from the humid tropics. *Environmental Management*. 22: 459–470 doi:10.1007/s002679900119.

Morgan, E.D. 2019. Why new forest are better at sequestering carbon than old ones. https://psmag.com/environment/young-trees-suck-up-more-carbon-than-old-ones.

Muller, D.B., Vogel, C., Bai, Y., & Vorholt, J.A. 2016. The plant microbiota: Systems-level insights and perspectives In: N.M. Bonini (ed) *Annual Review of Genetics,* vol. 50. Palo Alto, CA, Annual Reviews, pp. 211–234. doi: 10.1146/annurev-genet-120215-034952.

Mutuo, P.K., Cadisch, G., Albrecht, A. et al. 2005. Potential of agroforestry for carbon sequestration and mitigation of greenhouse gas emissions from soils in the tropics. *Nutrient Cycling in Agroecosystems* 71: 43–54 doi:10.1007/s10705-004-5285-6.

Nair, P.K.R. 2011. Agroforestry systems and environmental quality. *Journal of Environmental Quality* 40: 784–790. doi: 10.2134/jeq2011.0076.

Nair, P.R. 2007. The coming of age of agroforestry. *Journal of the Science of Food and Agriculture* 87: 1613–1619. doi: 10.1002/jsfa.2897.

Nanba, K., King, G.M., & Dunfield, K. 2004. Analysis of facultative lithotroph distribution and diversity on volcanic deposits by use of the large subunit of ribulose 1,5-bisphosphate carboxylase/oxygenase. *Applied and Environmental Microbiology* 70(4): 2245–2253. doi:10.1128/aem.70.4.2245-2253.2004.

Nicoloso, R.S., C.W. Rice, T.J.C. Amado, C.N. Costa, & E.K. Akley. 2018. Carbon saturation and translocation in a no-till soil under organic amendments. *Agriculture, Ecosystems & Environment* 264: 73–84. doi:10.1016/j.agee.2018.05.016.

Novelli, L.E., Caviglia, O.P., Pineiro, G. 2017. Increased cropping intensity improves crop residue inputs to the soil and aggregate-associated soil organic carbon stocks. *Soil & Tillage Research* 165: 128–136. doi: 10.1016/j.still.2016.08.008.

O'Driscoll, D. 2018. Carbon Abatement Potential of Reforestation. K4D Helpdesk, Brighton, UK: Institute of Development Studies. https://assets.publishing.service.gov.uk/media/5c6bd659e5274a72bac384ea/Carbon.pdf.

Oades, J.M. 1984. Soil organic matter and structural stability: Mechanisms and implications for management. *Plant Soil* 76(1–3): 319–337.

Oelbermann, M., Voroney, R.P., and Gordon, A.M. 2004. Carbon sequestration in tropical and temperate agroforestry systems: A review with examples from Costa Rica and Southern Canada. *Agriculture, Ecosystems & Environment* 104: 359–377.

Ontl, T.A. and Schulte, L.A. 2012. Soil Carbon Storage. *Nature Education Knowledge*. 3(10): 35.

Paquette, A., Bouchard, A., & Cogliastro, A. 2006. A survival and growth of under-planted trees: A meta-analysis across four biomes. *Ecological Applications* 16: 1575–1589.

Parras-Alcántara, L., Lozano-García, B., and Galán-Espejo, A. 2015. Soil organic carbon along an altitudinal gradient in the Despeñaperros Natural Park, southern Spain, *Solid Earth* 6: 125–134. doi:10.5194/se-6-125-2015.

Pataki, D.E., et al. 2003. Tracing changes in ecosystem function under elevated carbon dioxide conditions. *BioScience* 53: 805–818.

Paustian, K., Six, J., Elliott, E.T., & Hunt, H.W. 2000. Management options for reducing CO_2 emissions from agricultural soils. *Biogeochemistry*. 48: 147–163.Percival, H.J., Pargitt, R.L., & Scott, N.A. 2000. Factors controlling soil carbon levels in New Zealand grasslands: Is clay content important? *Soil Science Society of America Journal* 64: 1623–1630.

Pohlmann, A., Fricke, W.F., Reinecke, F. et al. 2006. Genome sequence of the bioplastic-producing "Knallgas" bacterium *Ralstonia eutropha* H16, *Nature Biotechnology*. 24: 1257–1262.

Potthoff, M., Dyckmans, J., Flessa, H., & Joergensen, R. 2008. Decomposition of maize residues after manipulation of colonization and its contribution to the soil microbial biomass. Biology and Fertility of Soils 44: 891–895 doi: 10.1007/s00374-007-0266-y.

Powlson, D.S., Glendining, M.J., Coleman, K., & Whitmore A.P. 2011. Implications for soil properties of removing cereal straw: Results from long-term studies. *Agronomy Journal*. 103: 279–287 doi: 10.2134/agronj2010.0146s.

Prentice, I.C., Farquhar, G.D., Fasham, M.J.R., et al. 2001. The carbon cycle and atmospheric carbon dioxide. In: J.T. Houghton (ed) *Climate Change. The Scientific Basis. Contribution of Working Group I to the Third Assessment Report of the Intergovernmental Panel on Climate Change*. Cambridge, Cambridge University Press, pp. 183–238.

Qureshi, N., Saha, B.C., & Cotta, M.A. 2007. Butanol production from wheat straw hydrolysate using *Clostridium beijerinckii*. *Bioprocess Biosystem Engineering*. 30: 419–427 DOI: 10.1007/s00449-007-0137-9.

Rao, D.L.N., & Chhonkar, P.K. 1998. Organic matter in relation to soil biological quality and sustainability. *Bulletin of the Indian Society of Soil Science*. 19: 80–89.

Reeder, J.D., & Schuman, G.E., 2002. Influence of livestock grazing on C sequestration in semi-arid mixed-grass and short-grass rangelands. *Environmental Pollution* 116: 457–463.

Riebeek H., & Robert S. 2011. Introduction the slow carbon cycle the fast carbon cycle changes in the carbon cycle effects of changing the carbon cycle studying the carbon cycle. https://earthobservatory.nasa.gov/features/CarbonCycle.

Rillig, M.C. 2004. *Arbuscular mycorrhizae*, glomalin, and soil aggregation. *Canadian Journal of Soil Science* 84(4): 355–363.

Rittmann, S., Seifert, A., & Herwig, C. 2015. Essential prerequisites for successful bioprocess development of biological CH_4 production from CO_2 and H_2. *Critical Reviews in Biotechnology* 3(2): 141–151. doi: 10.3109/07388551.2013.820685.

Roelefson, P.A. 1934. Metabolism of purple sulphur bacteria. Proc. Acad. Soc. (Amsterdam) 37, 660–669; Chem. Abstr. 29, 2994.

Sainju, U.M., Singh, B.P., & Whitehead, W.F. 2002. Long-term effects of tillage, cover crops, and nitrogen fertilization on organic carbon and nitrogen concentrations in sandy loam soils in Georgia, USA. *Soil & Tillage Research* 63: 167–179.

Sakurai H., Ogawa T., Shiga M., & Inoue K. 2010. Inorganic sulfur oxidizing system in green sulfur bacteria. *Photosynthesis Research* 104(2–3): 163–176. doi: 10.1007/s11120-010-9531-2.

Sauer, T.J., James, D., Cambardella, E., Cynthia, A., & Hernandez-Ramirez, G. 2012. Soil properties following reforestation or afforestation of marginal cropland. Publications from USDAARS / UNL Faculty. 1295.

Savage, D.F., Afonso, B., Chen, A.H., & Silver, P.A. 2010. Spatially ordered dynamics of the bacterial carbon fixation machinery. *Science*. 5;327(5970):1258–1261. doi: 10.1126/science.1186090.

Sayre, R. 2010. Microalgae: The potential for carbon capture. *BioScience*. 60(9): 722–727. doi: 10.1525/bio.2010.60.9.9.

Schahczenski, J., & Hill, H. 2008. Agriculture, climate change and carbon sequestration. https://www.nrcs.usda.gov/Internet/FSE_DOCUMENTS/nrcs141p2_002437.pdf.

Schrag, D.P. 2007. Preparing to capture carbon. *Science*. 315: 812–813.

Segnini, A., Posadas, A., Quiroz, R., et al. 2011. Soil carbon stocks and stability across an altitudinal gradient in southern Peru. *Journal of Soil and Water Conservation* 66(4): 213–220.

Selesi, D., Schmid, M., Hartmann, A. 2005. Diversity of green-like and red-like ribulose-1,5-bisphosphate carboxylase/oxygenase large-subunit genes (cbbL) in differently managed agricultural soils. *Applied and Environmental Microbiology* 71: 175–184.

Sharma P., Abrol, V., Abrol, S., & Kumar, R. 2012. Climate change and carbon sequestration in dryland soils, resource management for sustainable agriculture, Vikas Abrol and Peeyush Sharma, IntechOpen, doi: 10.5772/52103. Available from: https://www.intechopen.com/books/resource-management-for-sustainable-agriculture/climate-change-and-carbon-sequestration-in-dryland-soils.

Sinsabaugh, R.L., Carreiro, M.M., & Alvarez, S. 2002a. Enzyme and microbial dynamics of litter decomposition. In: R.G. Burns, & R.P., Dick (eds) *Enzymes in the Environment - Activity, Ecology, and Applications*. Marcel Dekker, New York, pp. 249–265.

Sinsabaugh, R.L., Carreiro, M.M., Repert, D.A., 2002b. Allocation of extracellular enzymatic activity in relation to litter composition, N deposition, and mass loss. *Biogeochemistry* 60: 1–24. doi: 10.1016/j.pedobi.2005.06.003.

Six, J., Bossuyt, H., Degryze, S., & Denef, K. 2004b. A history of research on the link between (micro) aggregates, soil biota, and soil organic matter dynamics. *Soil & Tillage Research* 79: 7–31.

Six, J., Elliott, E.T., & Paustian, K. 2000.Soil macroaggregate turnover and microaggregate formation: A mechanism for C sequestration under no –tillage agriculture. *Soil Biology and Biochemistry* 32(14): 2099–2103.

Six, J., Elliott, E.T., Paustian, K., & Doran, J.W. 1998. Aggregation and soil organic matter accumulation in cultivated and native grassland soils. *Soil Science Society of America Journal* 62(5): 1367.

Six, J., Frey, S.D., Thiet, R.K., & Batten, K.M. 2006. Bacterial and fungal contribution to carbon sequestration in agroforestry. *Soil Science Society of America Journal* 70(2): 555

Six, J., Ogle, S.M., Breidt, F.J., Conant, R.T., Mosier, A.R., & Paustian, K. 2004a. The potential to mitigate global warming with no-tillage management is only realized when practiced in the long term. *Global Change Biology* 10: 155–160.

Smillie, R.M., Rigopoulos, N., & Kelly, H. 1962. Enzymes of the reductive pentose phosphate cycle in the purple and in the green photosynthetic sulphur bacteria. *Biochimica et Biophysica Acta* 56: 612.

Smith, P. 2004. Carbon sequestration in croplands: The potential in Europe and the global context. *European Journal of Agronomy* 20: 229–236.

Smith, K.S., & Ferry, J.G. 2000. Prokaryotic carbonic anhydrases. *FEMS Microbiology Reviews* 24: 335–366.

Smith, P., Powlson, D.S., Glendining, M.J., & Smith, J.U. 1998. Preliminary estimates of the potential for carbon mitigation in European soils through no-till farming. *Global Change Biology* 4: 679–685.

Spalding M.H., 2008. Microalgal carbon-dioxide-concentrating mechanisms: Chlamydomonas inorganic carbon transporters. *Journal of Experimental Botany*, 59(7): 1463–1473. doi: 10.1093/jxb/erm128.

Spokas, K.A., & Reicosky, D.C. 2009. Impacts of sixteen different biochars on soil greenhouse gas production. *Annals of Environmental Science* 3: 4.

Stewart, C., & Hessami, M.A. 2005. A study of methods of carbon dioxide capture and sequestration-the sustainability of a photosynthetic bioreactor approach. *Energy Convers Manage*. 46:403–420.

Storelli, N., Peduzzi, S., Saad, M., et al. 2013. CO_2 assimilation in the chemocline of Lake Cadagno is dominated by a few types of phototrophic purple sulfur bacteria. *FEMS Microbiology Ecology* 84 (2): 421–432. doi: 10.1111/1574–6941.12074.

Sylvia, D.M., Hartel, P.G. Fuhrmann, J.J., and Zuberer, D.A. 2005. *Principles and Applications of Soil Microbiology* (2nd ed.). Sylva D.M. (ed) Pearson Prentice Hall, Upper Saddle River, NJ.

Tabatabaei, M., Tohidfar, M., Jouzani, G.S., Safarnejad, M., & Pazouki, M. 2011. Biodiesel production from genetically engineered microalgae: Future of bioenergy in Iran. *Renewable & Sustainable Energy Reviews* 15: 1918–1927.

Tang, K.H., & Blankenship, R.E. 2010. Both forward and reverse TCA cycles operate in green sulfur bacteria. *Journal of Biological Chemistry* 285(46): 35848–35854, doi:10.1074/jbc.M110.157834.

Tate, R.L. III. 2002. Microbiology and enzymology of carbon and nitrogen cycling. In: R.G. Burns, & R.P. Dick (eds) *Enzymes in the Environment- Activity, Ecology, and Applications*. Marcel Dekker, New York, pp. 227–135.

Thompson, J.A., and Kolka, R.K. 2005. Soil carbon storage estimation in a central hardwood forest watershed using quantitative soil-landscape modeling. *Soil Science Society of America Journal* 69: 1086–1093.

Tiemann, L.K., Grandy, A.S., Atkinson, E.E. Marin-Spiotta, E., & McDaniel, M.D. 2015. Crop rotational diversity enhances below ground communities and functions in an agroecosystem. *Ecology Letters* 18(8): 761–771.

Tracy, B.P., Jones, S.W., Fast, A.G., Indurthi, D.C., & Papoutsakis, E.T. 2012. Clostridia: The importance of their exceptional substrate and metabolite diversity for biofuel and biorefinery applications. *Current Opinion in Biotechnology* 23: 364–381.

Treonis, A.M., Ostle, N.J., Stott, A.W., Primrose, R, Grayston, S.J., & Ineson, P. 2004. Identification of groups of metabolically-active rhizosphere microorganisms by stable isotope probing of PLFAs. *Soil Biology and Biochemistry* 36(3): 533–537.

Trivedi, P., Anderson, I.C., & Singh, B.K. 2013. Microbial modulators of soil carbon storage: Integrating genomic and metabolic knowledge for global prediction. *Trends in Microbiology* 21(12): 641–651. doi: 10.1016/j.tim.2013.09.005.

Trumbore, S. 2006. Carbon respired by terrestrial ecosystems – recent progress and challenges. *Global Change Biology* 12: 141–153. doi:10.1111/j.1365–2486.2006.01067.x.

van den Hoogen, J., Geisen, S., Routh, D. et al. 2019. Soil nematode abundance and functional group composition at a global scale. *Nature*. 572: 194–198. doi:10.1038/s41586-019-1418-6.

Van der Werf, G.R., Morton, D.C., DeFries, R.S., et al. 2009. CO_2 emissions from forest loss. *Nature Geoscience* 2:737–738. doi.org/10.1038/ngeo671

van Niel, C.B. 1949. In: J. Franck, & W.E. Loomis (ed) *Photosynthesis in Plants*. The Iowa State College Press, Ames, IA, p. 437.

van Veen, J.A., Liljeroth, E., Lekkerkerk, L.J.A., and van de Geijn, S.C., 1991. Carbon fluxes in plant–soil systems at elevated atmospheric CO_2 levels. *Ecological Applications* 1: 175–181.

Verbruggen, E., Veresoglou, S.D., Anderson, I.C., et al. 2013, *Arbuscular mycorrhizal* fungi – short-term liability but long-term benefits for soil carbon storage? *New Phytologist* 197: 366–368. doi:10.1111/nph.12079.

Verheijen, F., Bellamy, P., Kibblewhite, M.G. and Gaunt, J. 2005. Organic carbon ranges in arable soils of England and Wales. *Soil Use and Management* 21: 2–9.

Waksman, S.A. 1936. Humus: Origin, Chemical Composition and Importance in Nature. Williams and Wilkins, Baltimore, MD.

Wang, X., Yost, R.S., & Linquist, B.A. 2001. Soil aggregate size affects phosphorus desorption from highly weathered soil and plant growth. *Soil Science Society of America Journal* 65: 139–146.

Wani, S.P., & Lee, K.K. 1995. In: P.K. Thampan, (ed.) *Microorganisms as Biological Inputs for Sustainable Agriculture in Organic Agriculture*. Peekay Tree Crops Development Foundation, Cochin, pp. 39–76.

Weiman, S. 2015. Microbes help to drive global carbon cycling and climate change. *Microbe*. 10(6): 233–238 doi: 10.1128/microbe.10.233.1.

West, T.O., & Post, W.M. 2002. Soil organic carbon sequestration rates by tillage and crop rotation: A global data analysis. *Soil Science Society of America Journal* 66: 1930–1946.

Wickings, K., Grandy, A.S., Reed, S.C., & Cleveland, C.C. 2012. The origin of litter chemical complexity during decomposition (N Johnson, Ed.). *Ecology Letters* 15(10): 1180–1188.

Wilson, G.W.T., Rice, C.W., Rillig, M.C., Springer, A., & Hartnett, D.C. 2009. Soil aggregation and carbon sequestration are tightly correlated with the abundance of *Arbuscular mycorrhizal* fungi: Results from long-term field experiments. *Ecology Letters* 12: 452–461. doi: 10.1111/j.1461-0248.2009.01303.x.

Windeatt, J.H., A.B. Ross, P.T. Williams, et al. 2014. Characteristics of biochars from crop residues: Potential for carbon sequestration and soil amendment. *Journal of Environmental Economics and Management* 146: 189–197. doi:10.1016/j.jenvman.2014.08.003.

Wu L, Vomocil, J.A., & Childs, S.W. 1990. Pore size, particle size, aggregate size, and water retention. *Soil Science Society of America Journal* 54: 952–956.

www.sare.org. Cover crops for sustainable crop rotations. pp1-4.

Xiang, Y., Cheng, M., Huang, Y., An, S., & Darboux, F. 2017. Changes in soil microbial community and its effect on carbon sequestration following afforestation on the loess plateau, China. *International Journal of Environmental Research and Public Health*. 14(8): 948. doi:10.3390/ijerph14080948.

Yadav, V., & G. Malanson. 2007. Progress in soil organic matter research: Litter decomposition, modelling, monitoring and sequestration. *Progress in Physical Geography* 31: 131–154.

Yasin, N.H.M., Fukuzaki, M., Maeda, T. et al. 2013. Biohydrogen production from oil palm frond juice and sewage sludge by a metabolically engineered Escherichia coli strain. *International Journal of Hydrogen Energy* 38(25): 10277–10283. doi: 10.1016/j.ijhydene.2013.06.065.

Yeshanew, A., Zech, W., Guggenberger, G., Tekalign, M. 2007. Soil aggregation, and total and particulate organic matter following conversion of native forests to continuous cultivation in Ethiopia. *Soil Tillage Res*. 94: 101–108.

Yihenew, G.S., and Getachew, A. 2013. Effects of different land use systems on selected physicochemical properties of soils in North Western Ethiopia. *The Journal of Agricultural Science* 5(4): 112–120.

Yohannes, H., Soromessa, T., & M. Argaw. 2015. Carbon stock analysis along slope and slope aspect gradient in Gedo Forest: Implications for climate change mitigation. *Journal of Earth Science and Climatic Change* 6: 305.

Yousuf, B., Sanadhya, P., Keshri, J., & Jha, B. 2012. Comparative molecular analysis of chemolithotrophic bacterial diversity and community structure from coastal saline soils, Gujarat, India. *BMC Microbiology* 12: 150.

Yuan, H., Ge, T., Chen, C., O'Donnell, A.G., Wu, J. 2012. Significant role for microbial autotrophy in the sequestration of soil carbon. *Applied and Environmental Microbiology* 78 (7): 2328–2336. doi: 10.1128/AEM.06881-11.

Zhongqi, H.E., Pagliari, P.H., & Waldrip, H.M. 2016. Applied and environmental chemistry of animal manure: A review. *Pedosphere*. 26: 779–816.

Zhu, X.G., Long, S.P., Ort, D.R. 2008. What is the maximum efficiency with which photosynthesis can convert solar energy into biomass? *Current Opinion in Biotechnology* 1: 153–159.

Zomer, R.J., Bossio, D.A., & Sommer, R. et al. 2017. Global sequestration potential of increased organic carbon in cropland soils. *Scientific Reports* 7: 15554 doi:10.1038/s41598-017-15794-8.

SUPPORTING REFERENCES

Alexandra Bot FAO Consultant, & Benites J. 2005. FAO Land and Plant Nutrition Management Service FOOD AND AGRICULTURE ORGANIZATION OF THE UNITED NATIONS Rome, https://www.nature.com/scitable/knowledge/library/soil-carbon-storage-84223790/. (accessed March 12, 2019).

Ayma-Romay, A.I., Bown, H.E. 2019. Biomass and dominance of conservative species drive above-ground biomass productivity in a mediterranean-type forest of Chile. *Forest Ecosystem* 6: 47. doi:10.1186/s40663-019-0205-z

Bardgett, R., Freeman, C. & Ostle, N. 2008. Microbial contributions to climate change through carbon cycle feedbacks. *The ISME Journal*. 2: 805–814 doi:10.1038/ismej.2008.58.

CF (CO_2 Foundation). 2019. Reforestation and Afforestation (Carbon Sequestration 101). https://co2foundation.org/reforestation-and-afforestation/. (accessed March 12, 2019).

Conrad, R. 2009. The global methane cycle: Recent advances in understanding the microbial processes involved. *Environmental Microbiology Reports* 1: 285–292. doi:10.1111/j.1758–2229.2009.00038.x.

Gelaw, A.M., Singh, B.R., & Lal, R. 2014. Soil organic carbon and total nitrogen stocks under different land uses in a semi-arid watershed in Tigray, Northern Ethiopia. *Agriculture Ecosystems & Environment* ELSEVIER 188: 256–263.

Hawken, P. 2017. Drawdown: The Most Comprehensive Plan Ever Proposed to Reverse Global Warming, New York. https://www.drawdown.org/solutions/food/regenerative agriculture. (accessed March 12, 2019).

Jansson, C., & Northen, T. 2010. Calcifying cyanobacteria-the potential of biomineralization for carbon capture and storage. *Current Opinion in Biotechnology* 21: 1–7.

Jones, M.B., & Donnelly, A. 2004. Carbon sequestration in temperate grassland ecosystems and the influence of management, climate and elevated CO_2. *New Phytologist,* 164: 423–439. doi:10.1111/j.1469-8137.2004.01201.x.

NRCS. 2016. Agriculture, Climate Change and Carbon Sequestration By Jeff Schahczenski and Holly Hill NCAT Program Specialists © 2008 NCAT. https://www.nrcs.usda.gov/Internet/FSE_DOCUMENTS/ nrcs141p2_002437.pdf. (accessed March 12, 2019).

Pluske, W., Murphy, D., & Sheppard, J. 2019. Soil Quality. Total Organic Carbon. http://www.soilquality.org.au/factsheets/organic-carbon. (accessed March 12, 2019).

Powlson, D., Stirling, C., Jat, M. et al. 2014. Limited potential of no-till agriculture for climate change mitigation. *Nature Climate Change* 4: 678–683. doi:10.1038/nclimate2292.

Revealing the Role of Microbial Communities in Carbon Cycling, Joseph Graber, Genomic Science Program, Biological Systems Science Division, Office of Biological and Environmental Research at the US Department of Energy Office of Science, July 2011, available at http://genomicscience.energy.gov/carboncycle/ BSSDCarbonCycleFlyer_sm.pdf. Accessed December 2019. (accessed March 12, 2019).

Sustainable Agriculture Research & Education. Cover Crops and Carbon Sequestration. https://www.sare.org/Learning-Center/Topic-Rooms/Cover-Crops/Ecosystem-Services-from-Cover-Crops/Cover-Crops-and-Carbon-Sequestration. (accessed March 12, 2019).

Vernimmena, R.R.E., Verhoefa, H.A., & Verstratenb, J.M. et al. 2007. *Nitrogen mineralization, nitrification and denitrification potential in contrasting lowland rain forest types in Central Kalimantan*, Indonesia.

Young Carbon Farmers. The importance of carbon in the soil. https://www.futurefarmers.com.au/young-carbon-farmers/carbon-farming/importance-of-carbon-in-the-soil. (accessed March 12, 2019).